Radioisotopes in Biology

Second Edition

The Practical Approach Series

Related **Practical Approach** Series Titles

Non-isotopic Methods in Molecular Biology

Please see the **Practical Approach** series website at
http://www.oup.com/pas
for full contents lists of all Practical Approach titles.

No. 252

Radioisotopes in Biology
Second Edition
A Practical Approach

Edited by

R. J. Slater
Department of Biosciences,
University of Hertfordshire,
College Lane, Hatfield AL10 9AB

OXFORD
UNIVERSITY PRESS

OXFORD

UNIVERSITY PRESS

Great Clarendon Street, Oxford OX2 6DP

Oxford University Press is a department of the University of Oxford.
It furthers the University's objective of excellence in research, scholarship,
and education by publishing worldwide in

Oxford New York

Athens Auckland Bangkok Bogotá Buenos Aires Calcutta Cape Town
Chennai Dar es Salaam Delhi Florence Hong Kong Istanbul Karachi
Kuala Lumpur Madrid Melbourne Mexico City Mumbai Nairobi Paris
São Paulo Singapore Taipei Tokyo Toronto Warsaw

with associated companies in Berlin Ibadan

Oxford is a registered trade mark of Oxford University Press in the UK and
in certain other countries

Published in the United States by Oxford University Press Inc., New York

British Library Cataloging in Publication Data
Data available

Library of Congress Cataloging in Publication Data

Radioisotopes / edited by R. J. Slater.—2nd ed.
     p. cm.—(Practical approach series; 252)
   Includes bibliographical references and index.
   1. Radioactive tracers in biology—Laboratory manuals. I. Slater,
   Robert J. II. Series.
   QH324.3. R33 2001   570'.28–dc21   2001036672

1 3 5 7 9 10 8 6 4 2

ISBN 0-19-963827-6 (Hbk.)
ISBN 0-19-963826-8 (Pbk.)

Typeset in Swift by Footnote Graphics, Warminster, Wilts
Printed in Great Britain on acid-free paper
by Bath Press Ltd (Avon)

Preface

Radioisotopes play a major role in the biosciences. Their continuing, and in some specific fields increasing, use is one reason for publication of this second edition. In addition, there are constantly improving technologies for labelling and detection, and changes in legislation that make this publication timely. The general aim of the volume is essentially as the first edition. That is, to present information that is relevant to most, if not all, radiotracer experiments, including: the decision to choose then use a radioisotope, the establishment of a suitable working environment conducive to safe laboratory practice, compliance with relevant legislation, and detection. The book then goes on to describe many laboratory protocols from areas of research that commonly use radioactivity.

The general principles behind experimental design and choice of radionuclide are relatively unchanged and are discussed in the Introduction. Some aspects of the properties of radiation and the essential aspects of laboratory design and safe handling are discussed in Chapter 2, with a summary of supplementary legislative information supplied in Chapters 10 and 11. Since the first edition, there has been an increased awareness of the value of, and legal requirement, for risk assessment. The issues concerned and some practical advice on how to carry out risk assessment are covered in Chapter 3, new to this edition. The principles and practice of detection by autoradiography with X-ray film and scintillation counting are covered in Chapters 4 and 5 respectively. Additional information on autoradiography with emulsions is provided in part of Chapter 7. The remainder of the book is concerned with laboratory protocols specific to particular branches of the biosciences that rely heavily on radioisotopes: nucleic acid and protein labelling, subcellular localisation of biological molecules, immunoassay techniques and pharmacological experiments. The last item is new to this edition and covers *in vitro* techniques, particularly those associated with receptor biology. Finally the volume includes a short Appendix of reference material.

As with all books in this series, the emphasis is on the practical considerations rather than a detailed theoretical discussion. The book should be valuable to a wide range of radioisotope users from undergraduate project and research students to laboratory supervisors and radiation protection staff.

Robert Slater
August 2001

Contents

CONTENTS

CONTENTS

6 *In vitro* labelling of nucleic acids and proteins   *131*

Martin W. Cunningham, A. Patel, A. C. Simmonds and D. Williams

Protocol list

Abbreviations

ADC	analogue–digital converter
ALI	annual limit on intake
AP	alkaline phosphatase
BSA	bovine serum albumin
CHO	Chinese hamster ovary
CIAP	calf intestinal alkaline phosphatase
c.p.m.	counts per minute
c.p.s.	counts per second
DEPC	diethylpyrocarbonate
DMEM	Dulbecco's modified Eagle's medium
DMSO	dimethylsulfoxide
ddNTP	dideoxynucleoside triphosphates
dNTP	deoxynucleoside triphosphate
DOT	Department of Transportation
d.p.m.	disintegrations per minute
d.p.s.	disintegrations per second
DTT	dithiothreitol
ECL	enhanced chemiluminescence
ELISA	enzyme-linked immunosorbent assay
EPA	Environmental Protection Agency
ERDA	Energy Research and Development Administration
ESCR	external standards channels ratio
FCS	fetal calf serum
FDA	Food and Drugs Administration
FRC	Federal Radiation Council
FRET	fluorescence resonance energy transfer
FWHM	full width of band at half its maximum height
GRF	growth hormone releasing factor
HPLC	high performance liquid chromatography
HPRI	human placental ribonuclease inhibitor
HRPO	horseradish peroxidase
HSE	Health and Safety Executive

HTRF	homogeneous time-resolved fluorescence
IAEA	International Atomic Energy Authority
IBMX	isobutylmethylxanthine
IC#	isotope's centre number
ICC	immunocytochemistry
ICRP	International Commission on Radiological Protection
IGF-1	insulin-like growth factor 1
IRMA	immunoradiometric assay
ISH	*in situ* hybridization
LLW	low-level waste
LSC	liquid scintillation counter
mAb	monoclonal antibody
MCA	multichannel analyser
NOS	nitric oxide synthase
NRC	Nuclear Regulatory Commission
NSB	non-specific binding
NSP	*N*-succinimidyl propionate
PAGE	polyacrylamide gel electrophoresis
PBS	phosphate-buffered saline
PEG	polyethylene glycol
PMSF	phenylmethylsulfonyl fluoride
PMT	photomultiplier tube
PNK	polynucleotide kinase
PPO	2,5 diphenyloxazole
PVDF	polyvinylidene difluoride
PVP	polyvinylpyrrolidone
rDNA	recombinant DNA
RIA	radioimmunoassay
RPA	Radiation Protection Advisor
RPS	Radiation Protection Supervisor
SCR	sample channels ratio
SDS	sodium dodecyl sulfate
SEITU	*S*-ethylisothiourea
SIS	spectral index of the sample
tSIE	transformed spectral index of the external standard
SQP(E)	standard quench parameter
SPA	scintillation proximity assay
TCA	trichloroacetic acid
TdT	terminal deoxynucleotidyl transferase
UV	ultraviolet
VLLW	very-low-level waste

Chapter 1
Introduction

ROBERT J. SLATER

Department of Biosciences, University of Hertfordshire, College Lane,
Hatfield AL10 9AB, UK

1 The decision to use a radioisotope

On the grounds of safety, whenever an experiment is being considered the need
for a radioisotope must be questioned: is a radioisotope absolutely essential to
the experiment planned? Or put another way, are non-radioactive methods
available that will achieve the same end?

If the answer to these questions suggests that there is a non-radioactive
technique that is equally effective, then that is the one that must be used. This is
a legal requirement. Only if a radioisotope is absolutely essential, or if its use results in a significant improvement in benefit, is the use of a radioisotope justified.
This complies with current thinking and official recommendations regarding
radiation hazard, that is the ALARA principle: as low as is reasonably achievable.
Although legislatively applied dose limits for radiation workers apply (see Chapter 2), the basic principle of safe laboratory practice is that as little radioisotope
should be used as possible.

2 Laboratory facilities

Having decided that a radioisotope is required, it is necessary to satisfy the legal
requirements for its use and to establish that the necessary facilities are available. The most commonly used isotopes, ^3H and ^{14}C, are of low energy, represent
little or no external hazard and are unlikely to require a controlled laboratory.
Iodine-125 and ^{32}P, on the other hand, can represent a significant hazard and a
controlled facility may be required. For a detailed discussion see Chapters 2 and
3 and Appendix 1.

3 The choice of labelled compound

Looking at a radioisotope catalogue for the first time can be daunting. A particular compound may appear as several entries with alternative radio-nuclides (e.g.
^3H or ^{14}C), labelling position, specific activity (i.e. Ci or Bq per mol), pack size,
formulation (e.g. ethanol or aqueous solutions) and cost. In making a choice
the researcher often follows the example of a previously reported experiment.
Although this can be of great help, it is not necessarily so. It assumes that the

previous worker made the correct choice, it may not be appropriate to the current safety standards, design of laboratory available, or take into account newly available products. Factors that influence choice vary considerably from one experiment to the next and on the facilities available, but a summary of some general features is given below.

3.1 The choice of radionuclide

The vast majority of radioisotope experiments in the biological sciences involve one of the following: ^{3}H, ^{14}C, ^{35}S, ^{33}P, ^{32}P, ^{131}I, and ^{125}I. These are presented in order of increasing toxicity (see Chapter 2 and Appendix 1) so, all other factors being equal, use a ^{3}H compound where possible. In some cases the choice will be constrained by the licensing of your laboratory and the advice of your departmental Radiation Protection Supervisor must be sought.

All the radioisotopes listed above are β emitters except for the iodine isotopes which emit electromagnetic radiation (^{131}I emits both β and γ radiation); avoid these if possible because of difficulties in shielding (see Chapter 2). Half-lives are given in Appendix 1; the shorter the half-life the greater the potential specific activity; sometimes, therefore, ^{14}C compounds are inappropriate as they cannot be produced at high specific activity (see Section 3.3).

The energy of the β-radiation emitted increases in the order shown above, ^{3}H being the weakest. This affects the efficiency and method of detection and is discussed in more detail in later chapters. A very important, and sometimes overlooked, consequence is that ^{3}H cannot be detected by an end-window ionization counter (e.g. Geiger counter) often used as a contamination monitor (see Appendix 2). Other consequences of the low β energy of ^{3}H are its relatively low efficiency in scintillation counting and general unsuitability for autoradiography other than for specialist purposes, such as *in situ* autoradiography or as an exercise in patience.

High energy radiation, such as from ^{32}P, is more suitable for autoradiography but because the radiation spreads further there is a corresponding reduction in resolution (see Chapter 4). A summary of some of the factors involved in choosing a radionuclide is given in *Table 1*. For some applications (e.g. DNA sequencing) the use of ^{35}S derivatives is a useful compromise providing high specific activity and suitable energy for autoradiography.

When you have chosen your radionuclide you must carry out a risk assessment (Chapter 3).

3.2 Labelling position

Usually there is a choice between different specifically-labelled and uniformly-labelled compounds, the latter generally being cheaper. Few general points can be made about this, as the choice will depend on the precise nature of the experiment; for example, it would be unsuitable to use a compound labelled in say a carboxyl group (^{14}COOH) for a tracer study if the first step was a decarboxylation. It is worth pointing out that a C atom need not necessarily be

Table 1. The relative merits of commonly used β-emitters

	Advantages	Disadvantages
^{3}H	Safety High specific activity possible Wide choice of labelling positions in organic compounds Very high resolution in autoradiography	Low efficiency of detection Isotope exchange with environment Isotope effect
^{14}C	Safety Wide choice of labelling positions in organic compounds Good resolution in autoradiography	Low specific activity
^{35}S	High specific activity Good resolution in autoradiography	Short half-life Relatively long biological half-life
^{33}P	High specific activity Good resolution in autoradiography Less hazardous than ^{32}P	Lower specific activity than ^{32}P Less sensitive than ^{32}P Cost
^{32}P	Ease of detection High specific activity Short half-life simplifies disposal Čerenkov counting	Short half-life affects cost and experimental design External radiation hazard Poor resolution in autoradiography

radioactive to be traceable, it could be tagged with a covalently attached radio-nuclide such as ^{3}H.

3.3 Specific activity

The specific activity of a radioisotope defines its radioactivity related to the amount of material (e.g. Bq/mol, Ci/mmol or d.p.m./μmol). Suppliers of radio-isotopes often offer a range of specific activities for a particular compound, the highest often being the most expensive. The advantages of using a very high specific activity compound are as follows:

- products of a reaction using the labelled precursor can be produced at high specific activity (e.g. in the synthesis of high sensitivity nucleic acid probes, described in Chapter 6);

- small chemical quantities of radiolabelled compound can be added such that the equilibrium of metabolic concentrations is not unduly perturbed; this can be important in *in vivo* labelling experiments where the purpose is to observe a physiological process; addition of radiolabelled compound is a balance between adding sufficient label to guarantee statistically meaningful results but insufficient to unbalance endogenous pool sizes;

- calculating the amount of substance required to make up radioactive solutions of known specific activity is simplified as the contribution to concentration made by the stock radiolabelled solution is often negligible.

Sometimes, however, it is not necessary to purchase the highest specific activity available. For example, enzyme assays *in vitro* often require a relatively high substrate concentration thereby necessitating reduction in specific activity by addition of cold carrier (see Chapter 2, Section 8.1).

3.4 The isotope effect

In choosing a radionuclide and compound, consideration must be given to the isotope effect. That is, a radioactive atom may not behave in a manner directly analogous to its non-radioactive isotope. This is particularly true for tritium, as it is much larger than hydrogen, and can show up, for example, in the rate of reaction involved in forming or breaking a tritium bond. The radioisotope must be positioned on a part of the molecule not involved in the reaction mechanism.

4 Storage and purity

All commercially available radioactive products are accompanied by a specification sheet. This will give details of the formulation and purity of the compound, the quality control data and a recommended means of storage.

Do not store a radioisotope under conditions (e.g. temperature) that differ from that stated. There are several ways in which radiochemicals can decompose during storage. First, there is the entirely unavoidable decomposition caused by radioactive decay

$$\text{e.g. X--}^{14}\text{CH}_3 \quad \text{X--}^{14}\text{NH}_2 \quad \text{X--NH}_2$$

this is referred to as primary (internal) decomposition. Second, there is primary (external) decomposition where emitted radiation is absorbed by other radioactive molecules creating impurities. The extent to which this occurs is dependent on a number of factors: temperature (low temperatures can sometimes increase decomposition by stabilizing transition states), specific activity and energy of the radiation; the lower the energy the more it can be absorbed and the greater the decomposition. High specific activity ^{3}H compounds are therefore highly susceptible to external decomposition. Third, there is secondary decomposition, where labelled molecules interact with excited species such as free radicals produced by the radiation. Fourth, chemical decomposition can occur: radiochemicals are frequently stored or used in minute quantities relative to non-radioactive materials; interaction with, say, glass surfaces which would normally be regarded as of no consequence can result in significant losses. Finally, there is the problem of microbiological spoilage. Every attempt must be made to maintain the radioisotope in a sterile environment (e.g. by retaining the Teflon seal on a storage vial) to prevent uptake or decomposition by micro-organisms.

The following list provides a general guide to storing radioisotopes:

- follow the recommended storage conditions provided by the supplier;
- store radiochemicals at as low a specific activity as possible or appropriate;
- maintain sterility;

- dilute compounds in an appropriate solvent such as that formulated with the isotope and recommended by the supplier;
- include free radical scavengers such as alcohol, sodium formate, glycerol or ascorbic acid, if possible.

5 Experimental design

Before carrying out an experiment with a radioisotope for the first time go through a theoretical exercise to estimate the quantity of radionuclide required and anticipate experimental difficulties. Then do a 'dry run' without radiolabel to identify practical difficulties. Time spent at this stage will considerably increase the chance of success, will reduce unnecessary exposure to radioactivity and will minimize the risk of accident, spillage, and contamination.

For the theoretical exercise consider the following:

- the minimum and maximum quantities of radioactive compound that can be added to the system to allow observation but not disturbance of the process investigated (e.g. substrate concentration for an enzyme assay, endogenous pool sizes for *in vivo* labelling);
- estimate the uptake rate (e.g. of labelled compound by cultured cells);
- estimate the efficiency of the extraction process to recover radioactivity;
- consider the counting efficiency;
- determine the accuracy of counting required and time available for detection (Chapter 5);
- the number of replicates required.

For a dry run, a dye can be used in place of a radiochemical to identify where difficulties may arise in preventing contamination.

Do a risk assessment (Chapter 3) and always remember the safety precautions:

- maximize the **distance** between yourself and the source;
- minimize the **time** of exposure, and **maintain shielding** at all times.

Chapter 2
Radioisotope use

DAVID PRIME and BARRY FRITH

Radiological Protection Service, University of Manchester, Oxford Road, Manchester M13 9PL, UK

1 Introduction

The subject of this chapter is the safe use of radioisotopes in biological tracer work. The emphasis of the treatment will be on the practical aspects but a certain amount of a more fundamental nature will be included where this is necessary to clarify explanations and terminology. Since biological tracers are used in a wide variety of ways, it is only intended to discuss this topic in general terms, paying particular attention to the most commonly used radioisotopes.

2 Radioactivity

2.1 Radioactive decay

The nuclei of atoms consist primarily of combinations of protons and neutrons, collectively called nucleons. The atomic number is the number of protons in a nucleus and the mass number is the sum of the numbers of protons and neutrons. These combinations are only stable in certain ratios. Radioactivity is a method of adjusting this ratio in a nucleus to achieve a stable configuration. This process can take place in one step or in a whole series of steps (for example when uranium decays through a series of radionuclides to a stable isotope of lead). There are a number of different ways in which this readjustment process can occur, several of which are important in biological tracer work.

The radiations that emanate as a result of the decays discussed below all have characteristic energies or energy spectra. Such energies are expressed in units of electron volts (eV) particularly in MeV or keV.

2.1.1 Alpha decay

An alpha particle is the nucleus of a helium-4 atom and therefore consists of two neutrons and two protons. This heavy, slow moving particle is densely ionizing and is particularly effective at causing biological damage. For example, radium-226 decays as follows:

$$^{226}_{88}\text{Ra} \quad \rightarrow \quad ^{222}_{86}\text{Ra} \quad + \quad ^{4}_{2}\text{He}$$
$$\text{radium-226} \quad \text{radon-222} \quad \text{alpha}$$

The extreme toxicity of alpha emitting nuclides, due to this effectiveness at causing biological damage, means that they are unsuitable for tracer work, but are important pollutants of the environment as a result of nuclear power and the most important component of natural background radiation.

2.1.2 Beta decay

Beta decay is a general term correctly used to describe decay in which either a proton or a neutron is lost by one of the following processes:

$$n^0 \quad \rightarrow \quad p^+ \quad + \quad e^-$$
$$\text{neutron} \quad \text{proton} \quad \text{electron}$$

$$p^+ \quad \rightarrow \quad n^0 \quad + \quad e^+$$
$$\text{proton} \quad \text{neutron} \quad \text{positron}$$

$$p^+ \quad \rightarrow \quad e^- \quad + \quad n^0$$
$$\text{proton} \quad \text{electron} \quad \text{neutron}$$

In the first case the net result is that a neutron turns into a proton whilst in the latter two a proton turns into a neutron. The first process is negatron emission although it is often simply called beta emission. The second process is positron emission and the third is electron capture.

Negatron emission

This decay mode is particularly important, since fission products usually decay by this mode and most tracers used in industry, medicine and research are of this type. Tritium, beryllium-7, carbon-14, strontium-90, ruthenium-106 and caesium-137 decay by this mechanism. For example, the decay of tritium is as follows:

$$^{3}_{1}\text{H} \quad \rightarrow \quad ^{3}_{2}\text{He} \quad + \quad e^-$$
$$\text{tritium} \quad \text{helium-3} \quad \text{beta}$$

Another type of radiation known as a neutrino accompanies all forms of beta decay. The only importance of neutrinos for radiation protection purposes is that they carry away some of the energy available. This distribution leads to a spectrum of beta particle energies.

Frequently the nuclide formed as a result of the decay is in an excited state, i.e. it has some excess energy that it gets rid of by emitting another type of radiation called a gamma ray. Gamma rays are electromagnetic radiation of short wavelength and hence high frequency and can accompany most forms of nuclear decay.

Positron emission

The choice of one of the two decay modes that are open to change a proton into a neutron depends on the energy available. If this is greater than 1.02 MeV (twice the rest mass of an electron), then positron emission may occur. This process results in the emission of a positive electron (or positron) from the nucleus of the atom. This particle rapidly meets an electron and the two particles destroy each other with the creation of two gamma rays each of 0.51 MeV, typical of positron emission.

This decay mode is of limited interest in biological sciences since few tracers decay by this means. However, 1.5% of the decays of the activation product zinc-65 are by this mode.

$$^{65}_{30}Zn \rightarrow {}^{65}_{29}Cu + e^+$$

Sodium-22 is sometimes used in biological research and this decays as follows:

$$^{22}_{11}Na \rightarrow {}^{22}_{10}Ne + e^+ + \gamma$$

Electron capture

If there is less than 1.02 MeV available, electron capture is the only decay mode possible whereby a proton is transformed into a neutron. Most (98.5%) of zinc-65 decays are of this type.

$$^{65}_{30}Zn + e^- \rightarrow {}^{65}_{29}Cu + \gamma,X$$

The zinc nucleus captures an electron that combines with a proton to form a neutron. Some of the copper atoms that are formed have excess energy that is emitted as a gamma ray. The capture of an electron is normally from an orbit close to the nucleus; this capture therefore creates a hole that is usually filled by an electron further away from the nucleus. This in turn causes a whole series of further electron transitions as the hole moves away from the nucleus. The emission of an X-ray accompanies each of these electron movements. X-rays are electromagnetic radiations that differ from gamma rays primarily in their mode of formation. With electron capture nuclides the initial electron transitions are of significance, since it is only these which lead to X-rays of sufficient energy to be readily detectable. The products of all this atomic rearrangement are a series of radiations that may well include X-rays, gamma rays and electrons.

Some other important nuclides used in medical research and diagnosis decay by this mode, e.g. iodine-125, chromium-51, cobalt-57. Iodine-125 decays as follows:

$$^{125}_{53}I + e^- \rightarrow {}^{125}_{52}Te + \gamma,X$$

2.1.3 Internal transition

In this type of decay a nucleus in an excited state emits a gamma ray. This is precisely the same as the gamma emission that can accompany other forms of radioactive decay; however, such gamma rays are emitted instantaneously whilst with internal transition the excited state has a significant half-life. This type of excited state is known as a metastable state and its presence in a nuclide is indicated by the symbol m. Caesium-137 decays to barium-137m and it is the radiation from the barium that is of particular importance in the detection and effects of caesium-137:

$$^{137}_{55}\text{Cs} \rightarrow \ ^{137m}_{56}\text{Ba} + e^-$$

$$^{137m}_{56}\text{Ba} \rightarrow \ ^{137}_{56}\text{Ba} + \gamma$$

One isotope of technetium, which is used extensively in nuclear medicine, also decays by this mode.

$$^{99m}_{43}\text{Tc} \rightarrow \ ^{99}_{43}\text{Tc} + \gamma$$

2.2 Units of radioactivity

The modern unit of radioactivity is the becquerel, which is used in legislation and scientific publications. Unfortunately, an obsolete unit, the curie, is still used by the suppliers of radioactive materials.

1 becquerel (Bq) = 1 disintegration per second (d.p.s.)
1 curie (Ci) = 3.7×10^{10} d.p.s. = 37 GBq

2.3 Properties of radiations

All forms of radiation follow an inverse square law, that is the intensity of radiation decreases with the square of the distance or:

$$I \propto \frac{1}{d^2}$$

where I is the intensity of the radiation and d is the distance from the source.

The above formula strictly only applies to radiation emitted from a point source (i.e. equally in all directions); if shielding restricts the radiation into a narrow beam, a process known as collimation, then the distances referred to in *Table 1* apply.

2.3.1 Beta radiation

The nature of beta interactions with matter means that it is possible to absorb this kind of radiation completely. *Table 1* does not include radiation with energy

Table 1 The properties of different types of ionizing radiation

Radiation	Energy (MeV)	Range (cm)	Shielding
Alpha	4–8	2.5–8 (air)	paper
Beta	0.1–3	15–1600 (air)	low atomic number materials
Gamma and X-rays	0.03–3	HVL (air) 1.3–13 m	lead or high density materials
		HVL (Pb) 0.02 mm–1.5 cm	

Half value layer (HVL) is the thickness of material required to reduce the original radiation intensity to one half.

as low as tritium since this does not constitute an external hazard. Even the most energetic of tritium beta particles have insufficient energy to penetrate the outer dead layer of skin.

The absorption of higher energy beta particles has another important feature: the production of secondary X-rays. It is these X-rays, also called *Bremsstrahlung* radiation, that can cause a radiation monitor to show an enhanced background. All interactions of beta particles with matter lead to this type of radiation but the production is greater the higher the energy of the beta particles and the higher the atomic number of the absorbing matter. It is therefore desirable to use a material with a low atomic number, such as plastic or Perspex, for shielding beta emitters.

2.3.2 X-rays and gamma rays

The interaction of this kind of radiation with matter is such that total absorption does not occur, hence the use of 'half value layer' as described in *Table 1*. The absorption is dependent on the mass of material through which the radiation passes, so the denser the material the more efficient it is as an absorber. For this reason, high-density materials such as lead, tungsten and even uranium are used for shielding gamma- and X-ray-emitting nuclides. Uranium is itself radioactive, but it still finds shielding applications for high activity sources.

2.4 Radiation dose and equivalent dose

The physical unit of absorbed radiation dose is the gray (Gy), which is the absorption of 1 joule of energy per kilogram of mass. This unit is used for expressing physical measurements of radiation dose and should also be used to describe acute biological effects of radiation. However, different radiations have different efficiencies of causing long-term biological effects for the same physical radiation dose. This has led to the introduction of the concept of 'equivalent dose'. The same equivalent dose gives the same risk of biological effect irrespective of the type of radiation. The unit of equivalent dose is the sievert (Sv) and the equivalent dose in sieverts is obtained by multiplying the absorbed dose by a weighting factor (w_R) which reflects the different biological effectiveness of radiations: the dose in sieverts is the dose in grays multiplied by w_R. w_R is 1 for beta particles, gamma and X-rays; 20 for alpha particles and 5–20 for neutrons.

2.5 External radiation doses from beta emitters

Apart from tritium and a few other rare exceptions, all negative beta emitters pose some degree of external hazard that becomes greater as the energy of the radiation increases. Various formulae may be used to estimate the dose rate from a source, an example of which is:

$$D_\beta = 760A$$

where D_β is the unshielded dose rate in μGy/h at a distance of 10 cm from a point source, and A is the source activity in MBq.

This formula is approximate in that it applies strictly only to point sources and ignores the absorption of beta particles. These approximations will be minimized when the formula is used for high-energy beta emitters at short distances.

Example: A person handling a solution of phosphorus-32 spills 10 MBq on to a glove. What will the dose rate be 1 mm from this glove?

Dose rate at 10 cm = 760 $\times$ 10 $\times$Gy/h

Since the radiation weighting factor for beta radiation is 1, this will be the same as 7.6 mSv/h. The inverse square rule can then be applied in order to calculate the dose at 1 mm:

$$\text{Dose rate at 1 mm} = 7.6 \times 10^2/0.1^2$$
$$= 7.6 \times 100 \times 100 \text{ mGy/h}$$
$$= 76 \text{ Gy/h}$$

The dose rate from a beta source spilt on to skin will be correspondingly larger.

This simple calculation shows that radiation doses from beta sources can be very high indeed and pose a considerable local radiation hazard.

2.6 External radiation doses from X-rays and gamma rays

The usual method of calculating dose rates from emitters of electromagnetic radiation is to use the specific dose rate constant, Γ. The dose rate for electromagnetic radiation (Dem) from a point source of gamma rays or X-rays is

$$D_{em} = \Gamma \frac{A}{d^2}$$

where d is the distance in m, A is the activity of the source in MBq and Γ is the specific dose rate constant, in μGy·m^2/MBq·h

Example: What is the dose rate from an unshielded source of 40 MBq caesium-137 at a distance of 25 cm?

From *Table 2*, the specific gamma ray constant for ^{137}Cs is 8.92×10^{-2} μGy·m^2/MBq·h. Therefore, the dose rate is $8.92 \times 10^{-2} \times 40/(25/100)^2$, i.e. 57 μGy/h.

What thickness of lead shielding is required to reduce the dose rate to 7.5 μGy/h?

From *Table 2*, the lead half value layer for ^{137}Cs/^{137m}Ba radiation is 0.8 cm. If one half value layer reduces the radiation dose to one half, two will cause a reduction to one quarter, three to one eighth, etc. In general, n half value layers will reduce the radiation dose by a factor of $1/2^n$. In this example the attenuation required is 7.5/57, i.e. 0.13, so $1/2^n = 0.13$ where n is the number of half value layers required. From this it follows that $n = 2.9$ and the thickness of lead required is $2.9 \times 0.8 = 2.3$ cm.

This type of calculation can be used to estimate the shielding required.

Table 2 Minimum ALIs, gamma ray doses and beta and gamma shielding data for some commonly encountered radioisotopes

Nuclide	Gamma dose (10^{-2} μGy·m²/ MBq·h)	Thickness of lead (mm)		Thickness (mm) required for total absorption of beta and electrons		ALI (MBq)
		half value	tenth value	glass	plastic	
Carbon-14				0.2	0.3	34
Caesium-137	8.9	8	24	2.1	3.8	1.5
Chromium-51	0.5	2	7	<0.1	<0.1	530
Cobalt-57	2.5	<1	1	<0.1	<0.1	21
Cobalt-60	35.7	16	46	0.4	0.7	5.9
Iodine-125	1.9	<1	<1	<0.1	<0.1	1.3
Iodine-131	5.7	3	11	0.9	1.6	0.9
Phosphorus-32				3.4	6.3	6.3
Sulfur-35				0.2	0.3	15
Technetium-99m	1.6	<1	1	0.2	0.3	690
Tritium				<0.1	<0.1	480

2.7 Internal radiation doses from radionuclides

The calculation of radiation doses from the ingestion of radionuclides can be very complicated. The International Commission on Radiological Protection (ICRP) has published a large amount of information on the subject, including metabolic models for every common radionuclide. For most purposes it is only necessary to understand the concept of annual limit on intake (ALI) to use these data.

The ingestion of one ALI will result in a person receiving not more than a committed effective dose, equivalent to the dose limit (20 mSv). For details see paragraph 174 ICRP 60 (3). A comparison of ALIs will therefore give both the doses received from ingestion or inhalation and an indication of the relative toxicities of radionuclides.

The above table includes data from the ICRP (1) and from the extremely useful publication the *Radionuclide and Radiation Protection Data Handbook 1998* (2).

> *Example*: An incident results in a person ingesting 5 MBq of iodine-125. What radiation dose has been received?
> From *Table 2*, the ALI for ^{125}I is 1.3 MBq and this is an effective dose limit. The ingestion will therefore result in an effective dose of 5/1.3 × 20 mSv or 77 mSv to the thyroid.

If an incident occurs in which the ingestion of a radioactive material takes place or is suspected, biological monitoring (e.g. of urine) should be carried out to check any dose calculations.

3 Biological effects of radiation and the basis of legislation

3.1 Effective dose

Although the use of equivalent dose will ensure that the risk of irradiation by, for example, 1 mSv to a specific part of the body remains the same whatever the type of radiation, different parts of the body do have differing sensitivities to biological effects of radiation. To take this into consideration, the dose to each organ and tissue group of the body is calculated and then multiplied by a weighting factor (see *Table 3*) (3) that reflects the risk associated with that particular part of the body. The effective dose is the sum of these weighted doses. Effective dose is the unit usually used when talking of radiation doses to individuals, although it is often called simply 'dose' or 'radiation dose'.

The units of radiation dose are large for most practical purposes; for example, radiation background is usually expressed in microsieverts.

3.2 Tissue level effects

3.2.1 Whole body effects

This sort of effect is important only if there is a serious accident or in nuclear warfare. Below 1 Gy there are no clinical symptoms of whole body radiation, although it is possible to detect chromosome aberrations at lower doses (minimum 10 mGy).

3.2.2 Local effects

The local effect of greatest importance is that of radiation burns. The threshold for burns is 3 Gy. At this dose level, erythema (reddening) can occur which may

Table 3 Organ weighting factors

Tissue or organ	Weighting factor
Gonads	0.2
Red bone marrow	0.12
Colon	0.12
Lung	0.12
Stomach	0.12
Bladder	0.05
Breast	0.05
Liver	0.05
Oesophagus	0.05
Thyroid	0.05
Skin	0.01
Bone surface	0.01
Remainder	0.05
Total	1

persist for a few days before disappearing. At higher doses, above 10 Gy, the burn becomes more serious, with the skin breaking down, producing ulceration or blistering. Such burns are slow to heal and in serious cases skin grafting may be necessary. Beta emitters have been shown in an earlier section to be able to give a high local dose.

3.2.3 Effects *in utero*

The relative importance of the adverse effects of radiation *in utero* at different stages of gestation are shown in *Table 4*. The main incidence of mental retardation occurs between 8 and 15 weeks.

Table 4 Effects on development *in utero*

Stage of pregnancy	Time	Effect	Threshold
Pre-implantation	⩽6 days	embryonic death	<0.1 Gy
Organogenesis	6–60 days	malformations	0.05 Gy
	8–15 weeks	mental retardation	0.1 Sv
Full course	~9 months	cancer induction	none

Some *in utero* effects are not thought to have a threshold. There cannot be an absolutely safe level of radiation at which such effects do not occur and the legislation seeks to keep the occurrence of these at an 'acceptable level' (3).

3.3 Carcinogenic effects

There is a great deal of evidence for carcinogenic effects of ionizing radiation. This includes the survivors of the atomic weapons from Hiroshima and Nagasaki; early research workers; the recipients of several types of medical treatments; and the effects of diagnostic X-rays. From this type of data, estimates can be made of the likely incidence of radiation-induced cancers. These estimates are being continually refined with resulting adjustments in the risk estimates. The current estimate from the ICRP (3) is an average risk of fatal cancer of 4.0×10^{-2} Sv^{-1} for adult workers and 5.0×10^{-2} Sv^{-1} for the whole population. The following factors are important when considering the possibility of radiation-induced cancer:

- colon and skin are the most likely cancers to be induced, but the latter is of low fatality;
- lung, stomach and colon cancer cause most deaths;
- sensitivity to cancer induction varies with age, the young being at highest risk;
- since women are more prone to breast cancer and certain other radiation-induced cancers than men, they are more likely to suffer from fatal cancer (by a factor of approximately 1.3).

3.4 Genetic effects

No genetic effects of radiation have been demonstrated conclusively in human populations. Extrapolation from animal experiments indicates that such effects do occur but that their incidence is less than that of cancer. The current estimate (5) of risk of significant genetic effects is 0.4×10^{-2} Sv^{-1} cases in the first two generations after irradiation. The corresponding figure for all generations is about 1.0×10^{-2} Sv^{-1}.

Several observed effects could have been caused by ionizing radiations. For example, some researchers have suggested that the high incidence of Down syndrome in Kerala (India) is due to the high natural background radiation. Also childhood leukaemia has been suggested as a possible genetic effect of radiation, although this is now considered to be unlikely (4).

4 Biological basis of dose limits

Legislation can easily deal with tissue level effects such as described above by ensuring that any limit is below the threshold for that particular effect. Such effects with thresholds and where the severity increases with increasing radiation dose are called deterministic or non-stochastic effects.

However, it is much more difficult to apply legislation to effects occurring without a threshold where the probability of the effect increases with increasing dose (stochastic effects). Once the radiation doses that persons or populations have received have been estimated, the risk of harm can be calculated. To carry out this calculation, it is necessary to know quantitatively the biological effects of radiation. The calculated risk can then be weighed against the benefits accrued. This risk can be justified or reduced to an acceptable level. This philosophy also requires the definition of acceptable levels of risk that will differ for different groups. Such acceptable levels of risk can be calculated from accident statistics. The ICRP examined these types of statistics in 1977 and suggested a set of acceptable risk figures. ICRP recommended a risk of 50 deaths per million per year for occupationally exposed workers and 1–10 deaths per million per year for the public (5).

4.1 Risk rates for biological effects

The following figures (Tables 5, 6, 7 and 8) have been derived from many studies of the induction of cancer and other effects by ionizing radiation (3). The recently revised study of data from survivors of the Hiroshima and Nagasaki atomic bombs is particularly significant. The type of model used to calculate a risk rate from the data is important.

Table 5 Effects on development *in utero*

Effect	Danger period	Incidence
Mental retardation	8–15 weeks	30 IQ points Sv^{-1}
Cancer induction	whole pregnancy	2×10^{-2} Sv^{-1} to age 10

Table 6 Genetically significant effects

Incidence to equilibrium	10×10^{-3} Sv^{-1}

Table 7 Incidence of fatal cancers

Tissue group	Fatal cancer rate (Sv^{-1})
Leukaemia	4.5×10^{-3}
Bone	0.5×10^{-3}
Lung	9.0×10^{-3}
Thyroid	0.75×10^{-3}
Breast	2.5×10^{-3}
Colon	9.5×10^{-3}
Oesophagus	3.5×10^{-3}
Stomach	11×10^{-3}
Liver	2.0×10^{-3}
Bladder	2.0×10^{-3}
Skin	0.2×10^{-3}
All other tissues	4.55×10^{-3}
Total	50×10^{-3}

The absolute risk model contains the assumption that the risk of cancer remains constant with time after a latency period, during which the risk is assumed to be zero. This is a step function: it is assumed that no cancers occur within the latency period and a given dose produces a set number of cancers irrespective of who is being irradiated. A second, more likely model, assumes that the relative risk of cancer remains constant with time again after a latency period and the number of cancers is related to dose and other risks (such as cigarette smoking). The first model, therefore, adds a number of cancers as a result of radiation exposure, whilst the second model multiplies the cancer rate by a factor as a result of exposure. These effects are also assumed to occur without a threshold.

The radiation limits are effective doses corresponding to a level of risk above which the consequences for the individual would be widely regarded as unacceptable. Exposures that are not unacceptable are then subdivided into 'tolerable' (i.e. tolerated but not welcome) and 'acceptable'.

A dose limit represents the boundary between 'unacceptable' and 'tolerable'. To provide a quantitative basis for the choice of a dose limit, the ICRP took account of a range of quantifiable factors in its approach to detriment. The factors considered were:

- the lifetime attributable probability of death;
- the time lost if the attributable death occurs;
- the reduction of life expectancy;
- the annual distribution of the attributable probability of death;
- the increase in the age-specific mortality rate, i.e. in the probability of dying in a year at any age, conditional on reaching that age.

Table 8 Probability of attributable death and loss of life expectancy at age 18 years resulting from subsequent occupational exposure (Loss of life = Time lost × Probability of death)

Annual dose (mSv)	Approximate life dose (Sv)	Probability of death (%)	Time lost by death (years)	Loss of life (years)
10	0.5	1.81	13	0.23
20	1	3.57	13	0.46
30	1.5	5.28	13	0.68
50	2.5	8.56	13	1.11
50 (1977 data)	2.5	3.13	20	0.63

The new limit suggested in 1991 by the ICRP of a lifetime dose of 1 Sv or 20 mSv/year implies a lower risk than the old limit of 50 mSv/year based on 1977 data (3).

5 Working with unsealed sources

5.1 Getting started

Researchers who propose to undertake work with radioisotopes and labelled compounds must first consider their responsibilities under the law. In Britain, two pieces of legislation are of major importance: The Radioactive Substances Act 1993 (6), often referred to as the Principle Act, and the Ionising Radiations Regulations 1999 (7). The former deals with the acquiring and disposal of radioactive material, the latter with its use and manipulation. Similar legislation exists in other European countries and in the USA.

There are, however, materials which, although radioactive, are of such a low radiotoxicity or low radioactivity that exemptions are granted with regard to parts or all of the legal requirements, for example in Britain, in general, there is no requirement to comply with the Ionising Radiations Regulations 1999 when the activity concentration of the radioactive material being used may be 'disregarded for the purposes of radiation protection' (usually taken as <100 Bq/g unless the material is a significant radiological hazard). The majority of tracer work involves the use of material of very much higher activity concentration, and the use of isotopes in the production of labelled compounds requires a level of activity concentration which is higher still.

If, in Britain, the researcher decides that the proposed work falls subject to legal controls (if in doubt advice should be sought from a Radiation Protection Adviser or the National Radiological Protection Board (NRPB)), he or she is required to notify the Health and Safety Executive (HSE) of their intention to start work, giving at least 28 days' notice.

The organization of working procedures, and methods, in the isotope laboratory is based on the fundamental, and legal, requirements to restrict the risks associated with working with isotopes by reducing dose equivalents to 'as low as reasonably achievable', the ALARA principle, while at the same time ensuring that dose limits are not exceeded. Without prejudice to the ALARA principle, the practical implementation of this basic concept centres around two, for the want of a better phrase, trigger points, which correspond to one tenth, and three tenths of the dose limit.

In essence, the risks to workers who might receive no more than one tenth the dose limit from exposure to external radiation or from the ingestion of radioactive material are restricted and controlled by the strict control of the working environment, and those workers who might receive three tenths of the dose limit have their exposure controlled by additional personal dose assessment.

The machinery for achieving control at these two levels of risk is both administrative and technical. UK regulations require that a chain of communication be established between the HSE and the controlling authority or employer, through a Radiological Protection Adviser, appointed by the controlling authority, and Radiological Protection Supervisor to the worker. The maintenance of this chain is very important, particularly between the worker and the Radiological Protection Supervisor, who must know what radioactive work is going on in his or her area of responsibility.

Local Rules, which describe how the work will be undertaken in compliance with the regulations, must be written, and must include a hazard assessment and a contingency plan. They form the hinge-pin of the administrative control, but more of this later.

5.2 Designation of working areas

The initial designation of working areas is the first step in the implementation of control procedures. Working areas must be designated as 'supervised' or 'controlled' and a record kept of their location. Whether a working area is designated supervised or controlled depends on the likelihood that a worker who works in the area continually would receive three tenths of the dose equivalent limit. Initially the Radiological Protection Adviser may use his judgement in designating areas, but often revisions are required once the area is in use since only then can the measurements of dose rate and contamination be made, which are crucial to the decision-making process.

Bearing in mind that the hazard in the open source laboratory comes from both external and internal radiation, there are limiting conditions of dose rate and surface and air contamination above which an area cannot be designated supervised.

5.2.1 Controlled area designation

An area should be designated as controlled:

(1) If the external dose rate exceeds 7.5 $\times$Sv/h averaged over the working day, or

- if only the hands can enter the area 75 $\times$Sv/h averaged over 8 h; or
- there is a significant risk of the spread of contamination outside the area; or
- access is controlled whilst work is under way; or
- workers are liable to receive an effective dose of >6 mSv per year.

(2) If the radiation dose exceeds 7.5 $\times$ Sv h^{-1} averaged over a minute and

- the work is site radiography; or
- employees untrained in radiation protection are likely to enter that area.

(none of this applies to radioactive substances dispersed in a human body).

Other criteria, which are best evaluated by the Radiological Protection Adviser, may also be used to determine the designation of an area.

5.2.2 Supervised area designation

An employer shall designate as a 'supervised area' any area under his control, not being an area designated as a controlled area when:

- it is necessary to keep the conditions of the area under review to determine whether the area should be designated as a controlled area; or
- any person is likely to receive an effective dose greater than 1 mSv a year or an equivalent dose greater than one tenth of any relevant dose limit due to radioactivity being taken into the body.

Just what these area designation criteria mean at a practical level will be discussed later, but the concept of supervised and controlled areas is introduced here because it leads logically on to the classification of workers.

6 Designation of 'classified workers'

Some workers who work with radioactive materials are designated as classified radiation workers. The designation depends upon the likelihood that the worker will receive three tenths the dose equivalent limit. It would seem to follow, therefore, that non-classified workers are those who spend their time at work in supervised areas, and classified workers are those who work in controlled areas. This is generally what happens, but there may be times when a non-classified worker works in a controlled area for short periods to, say, dispense an isotope from a stock solution, or to feed experimental animals. This is acceptable provided the worker is operating under an 'Approved System of Work' which has been designed to restrict the person's dose to less than three tenths the dose equivalent limit. In these circumstances, personal monitoring of the non-classified worker is required in order to demonstrate compliance with this requirement.

The classification of workers is an administrative mechanism used to indicate the type, degree and extent to which monitoring is carried out to assure the required level of risk.

7 Controlling the risks to radiation workers

7.1 The non-classified worker

The reader is reminded that the aim here is to assure the wellbeing of the non-classified worker by control of the working environment. Regular recorded monitoring of external dose, and surface and air contamination is required to ensure that conditions in the working area comply with those for a supervised area. Dose rates are measured with an instrument which is suitable for the purpose and has been calibrated by an 'Approved Laboratory'. This usually means an ion chamber, proportional counter, scintillation counter or Geiger–Müller (G–M) device, of which there are many available, but in the case of tritium contamination monitoring estimates are made by taking swabs (wipe tests) of the area. The measurement of the activity on the swab by liquid scintillation counting and the use of an approximate efficiency factor of ten is used to determine the activity in the area tested. A routine for monitoring must be laid down in the Local Rules, and any results that are around one tenth of the derived limit or above must be recorded.

7.2 The classified worker

In addition to monitoring the working environment in a controlled area, personal monitoring of the worker is required so as to assess the classified worker's individual dose equivalent. Film badges, or thermoluminescent dosimeters, the monitoring of air in the individual's breathing zone, and the assessment of personal contamination by external measurement for skin contamination, and bioassay for internal contamination, are undertaken routinely by an 'Approved Laboratory'. The findings must be recorded and the records kept for 50 years after the last entry.

What has been said up to now is intended to make the isotope user aware of the general requirements of the Ionising Radiations Regulations 1999, and to show something of the framework of controls which enable isotope work to be under taken within acceptable risk limits. There is, however, the requirement to reduce the risks to 'as low as reasonably achievable'. We must use the experimental and manipulative skills that we have to reduce the already low risk still further.

8 The tracer study

Most isotope experiments follow a systematic procedure; the design of the experiment, the selection of the isotope tracer, the purchase of the isotope, the receiving and opening, the dispensing from the stock solution and dilution to working solution strength, the introduction of the isotope into the experiment followed by sampling and measurement, and finally the decontamination of equipment and monitoring of the working area and personnel. If we work through an imaginary experiment step by step and examine the ways in which we can reduce the dose and thence the risk at each stage, we should then have a useful working guide.

8.1 Experimental design

Tracer studies yield information about the movement of stable materials in a system. Reactions and reaction rates are studied by determining how much and how quickly radioactive tracer, and therefore stable material, moves from place to place. Two parameters of the isotope solution or labelled compound solution are very important in achieving good experimental results. These are activity concentration and specific activity. Activity concentration is the activity per unit mass or volume of the medium containing the activity, while specific activity is the activity per unit mass of the compound being studied, be it compound, ion or element. Specific activity is a measure of the ratio of active, labelled species to the non-active or stable species which we often refer to as carrier.

The ability to measure the radioactivity in samples taken from the experimental system with a sensitivity and precision, and in some rare cases accuracy, is another important requirement for a successful experiment, but it is also fundamental in considering the appropriate activity concentration and specific activity of the isotope or labelled compound to be used. The statistics of counting is dealt with elsewhere (see Chapter 3), but in the broadest terms it is necessary to aim for radioactive sources for counting, that is sources made from aliquots taken from the experiment, which count at a rate significantly greater than the counter background, and generate 10^4–10^5 counts in a reasonably short counting period in order to achieve a counting error (sometimes called a random error since it is due to the inherent randomness of radioactivity) that is small by comparison with other systematic experimental errors. The required specific activity is linked to the radioactivity represented by the above number of counts in the counting period used divided by the mass of the compound of interest in the aliquot taken. The choice of specific activity will depend on the concentration of compound and sensitivity required (the higher the specific activity, the greater the sensitivity).

So far we have made the, not unreasonable, assumption that the tracer behaves exactly as the non-active material. In almost all cases the errors attendant on this assumption are negligible, but we should be aware of a mass effect, which is a problem when reaction rates are being studied where the mass of the tracer is very different from that of the stable species. Problems of significant magnitude are encountered only when elemental hydrogen is being traced using tritium, and as most tracer work involves large molecules the problem seldom worries the biochemist.

There are obvious advantages to be gained, in terms of safety, from using isotope and labelled compound solutions of the lowest activity concentration and specific activity, and adjustments to these parameters should be made as early as practicable in the experiment. The adjustment of activity concentration is straightforward enough, and the reduction of specific activity is achieved by adding carrier in amounts calculated using the formulae

$$W = Ma(1/A' - 1/A)$$

where W is the mass of carrier in milligrams to be added to an ampoule of M MBq of a compound of molecular weight a, to change the original specific activity, A (in MBq/mmol) to the required specific activity, A' (in MBq/mmol).

8.2 The purchase of isotopes and labelled compounds

Buying isotopes and labelled compounds of the appropriate activity concentration and specific activity has been considered. The choice of isotope must be considered next. Often there are several which might be used. Do have regard for the relative radiotoxicity of the possible candidates before placing your order. How much activity should be purchased? The answer is not more than is needed to achieve the experimental goal. The cost reduces dramatically with increasing pack size so that ten units of activity cost little more than twice the cost of one unit, but do not be tempted to buy more than you need. It might make good sense at the purchasing stage of the experiment, but it is contrary to the ALARA principle, and the additional cost of, say, operating a controlled area as opposed to a supervised area, and the additional cost of waste disposal, by no means insignificant, not to mention the increased potential hazard and the dubious storage properties of the labelled compound in a resealed container, and other factors will probably outweigh any initial advantages.

8.3 Receiving the isotope

The receipt of radioactivity must be recorded and its temporary storage must be in a suitably secure place which is chosen having regard to the need to control dose rates and contain the activity under likely conditions of use and, to some extent, abuse. The store should be used for active material only and should contain no inflammable or explosive substances. Because of the concentration of radioactive material, the store is likely to be a high dose rate location, and consideration must be give to the possible need to designate it as a controlled area. If the store is an enclosure other than one that can be entered, for example a refrigerator into which only the hands may enter, then due regard should be given to this in making your decision.

8.4 Isotope manipulations

The manipulation of isotope stock solutions, when dose rates, activity concentrations and specific activities are at the highest levels that will be experienced throughout the whole experiment, must be recognized as a particularly potentially hazardous operation. It is often wise to manipulate stock solutions in shielded areas constructed, wherever possible, in a suitable fume cupboard, so that only the hands can enter, and dose rates outside the shield are <7.5 μSv/h. This avoids the need to designate as a controlled area any area other than that within the shield, and of course this may not be necessary if the conditions for exposure to hands only are met.

While manipulating the primary container, the simple principles of reducing time, increasing distance and use of shielding should be used to reduce external dose rates. It is often useful to have a non-active practice run, with a coloured solution perhaps, to identify any possible problems.

8.5 Design of isotope laboratory

The working area should be designed with the aim of reducing the exposure rates to as low as reasonably achievable in mind, and in particular, the need to control contamination on surfaces by choosing surfacing materials which decontaminate easily. In common with other countries, the UK has an organization (the British Standards Institution) which sets standards and testing procedures designed to assure the performance of products. The British Standard Test, BS 4247, enables surfacing materials to be chosen for use in isotope laboratories which are easy to decontaminate. Choosing such surfacing materials and making them as continuous as possible and without difficult-to-clean corners should be given a high priority. The better materials are not obvious. Stainless steel is extensively used in radioactive areas, but it can be difficult to decontaminate, especially when radio-halides are used. Examples of materials which are very easily decontaminated are: unfilled polypropylene, Perspex and paints of the epoxy or polyurethane type. It is always best to ask for certification of the surfacing material involved when laboratory refurbishment takes place or a new piece of equipment is purchased. Strippable coatings and the use of temporary surface coverings can be advantageous, but their use can greatly increase the volume of radioactive waste produced.

Air conditioning which gives five to 30 air changes per hour is a desirable feature, and a fume cupboard with a design specification which will assure the containment of any radioactive gas generated within it is essential. A draught of 0.5 m/sec with the sash at a normal working position is regarded as satisfactory in Britain, but the over-riding priority is that the fume cupboard should 'contain' radioactivity generated in gaseous or volatile form within it.

The laboratory sink, if it is designated for the disposal of radioactive waste, should be given special attention. The joint between bench and sink should be properly sealed to prevent the ingress of radioactivity; there should be no large trap which might collect and accumulate radioactive items or solids; the waste should find its way to the main drain by the most direct and, preferably, discrete route and be carried by high integrity pipework which is marked at intervals with appropriate warning labels.

Where the area is a controlled area there must be provision for washing (a hand wash basin with elbow- or foot-operated taps is required), and somewhere to leave laboratory coats used in the area. To help prevent accidental entry to controlled areas it should require a positive action to enter, for example, the lifting of a physical barrier, the opening of a door or the stepping over of a shoe-change barrier. With attention to these details an ordinary good well appointed laboratory is usually sufficient for most levels of radiotracer work.

9 Personal protection in the radioisotope laboratory

In all laboratories some form of protective clothing is required, a laboratory coat of suitable design and safety spectacles being a bare minimum. The use of latex or vinyl gloves is a useful first device in the prevention of skin contamination. These gloves must be used in a manner which does not permit their being removed and put back on with the original, possibly contaminated, outside in. Reusable ambidextrous gloves should be marked to prevent this, or used only once. Overshoes should be worn if not wearing them would result in the spread of contamination to other areas.

10 Routine monitoring of area

Monitoring the radiation and contamination in the working environment and of the individual on a routine detailed in the Local Rules is required. Most of the effort will, in normal circumstances, be aimed at the environment. The relevant derived limits are, in the case of external dose, and air contamination, derived in a fairly straightforward way. On the one hand the annual limit is divided by the number of working hours in the year, and on the other, with some knowledge of a worker's breathing rate we can calculate the limiting air concentration which would lead to an intake of an annual limit on intake. Here the simplicity ends; the author cannot find sufficient evidence to dispel his fear that confusion exists in the minds of whoever derived the limits for surface contamination.

10.1 External dose measurement

The measurement of dose rate requires the use of a measuring device which is sensitive to the type of radiation being given of by the isotope used and which will give a reading which reflects the true radiation dose. The simplest monitor is a G–M counter with the output fed to a rate meter. The visual indication is of a count rate. The relationship between the count rate and true dose rate is given by calibrating the monitor in conditions of accurately known dose rate. Errors occur in measurement when the energy of the radiation given off by the isotope used differs from that of the source used in calibration. For example, a simple G–M device which has been calibrated using a cobalt-60 source will probably show a reading much higher than the true value when used to measure a dose rate produced by 100 keV X-rays. This response energy dependence is all but eliminated with more elaborate (compensated) G–M devices and in monitors which employ ion chambers as their detector. If a single isotope is used, expense can be avoided by choosing a simple G–M monitor and having it calibrated against standard sources of the same, or similar, emitter. The multi-isotope user will require the more sophisticated, and more expensive, monitor of the compensated G–M or ion chamber type. In either case it is important to be able to detect with confidence dose rates of 7.5 μSv/h and 2 mSv/h since these are important limiting values for routine monitoring.

10.2 Measurement of air contamination

The measurement of activity concentration in air is undertaken by direct reading instrumentation in only a very few circumstances. For example, tritium as a gas, or radon/thoron gases, are sometimes measured by passing air through a chamber containing two electrodes. The resulting ionization current is then a measure of the activity concentration. More often than not a known volume of air is drawn through a trap, which may be, for example, a glass-fibre filter paper or an activated charcoal adsorber, and the trap analysed later in the radio-chemistry laboratory.

10.3 Measurement of surface contamination

Surface contamination measurements are best used as nothing more than an indicator of the overall control and housekeeping in a laboratory. Changes in the normal or usual level of contamination should be investigated. They usually indicate a malpractice or a change in working conditions. Two methods are used to measure surface contamination. The direct method involves observing the response of a suitable detector when held close to the surface, and comparing it with the response of the instrument held in a similar position close to a 'standard of surface contamination'. This method is suitable when the ambient radiation level is low and when the isotope being used may be detected by a detector outside the source. Tritium contamination is not usually measured by the direct method. The indirect method involves swabbing of the area with subsequent measurement of the activity removed.

Monitoring results should be recorded in the manner previously discussed.

10.4 Estimation of individual dose

In the UK, the monitoring of individual dose involves a responsibility under the regulations to keep records of individuals doses for 50 years after the last entry. Laboratories who undertake measurements of activity in samples taken to assess personal dose must be approved by the HSE, as must any laboratory offering a calibration service for measuring instruments.

11 Transport of radioactive material

A particularly hazard-prone event in the use of radioactive material is its transport from one place to another. Hand-held containers slip from wet fingers, glass vials break and contamination results. There is a need to control radiation exposure and contain radioactive material during its movement by hand within, say, the building or plant. More stringent controls are required during wheeled transport off site or shipment by air or water.

The UK legislation that governs the transport of radioactive material is The Radioactive Material (Road Transport) Act 1991 (for the USA see Chapter 12). The Act grants the Secretary of State leave to make such regulations as appear to him to be necessary or expedient

- to prevent any injury to health or any damage to property or to the environment being caused by the transport of radioactive material; and

- to give effect to such international regulations for the safe transport of radio-active material as may from time to time be published by the International Atomic Energy Agency (IAEA).

The relevant publication is the *IAEA Regulations for the Safe Transport of Radio-active Material*, 1985 (amended 1990) (8). These regulations, and the Act, apply to radioactive material with an activity concentration of >70 kBq/kg or, in the UK, such lesser specific activity as may be specified in an order made by the Secretary of State. They cover all operations associated with the movement of radioactive material. This includes the design, fabrication and maintenance of packaging; the preparation, consigning, handling, carriage, storage in transit; and receipt at final destination.

From a practical point of view, the least restrictive transport option is that of shipping radioactive material in 'excepted packages'. The IAEA regulations should be consulted before transporting any radioactive materials. The usual require-ments for excepted packages are shown in the next section.

11.1 Requirements and controls for transport of excepted packages

(1) Other dangerous properties must be taken into account.

(2) Radiation levels at any external point on the surface of the package must not exceed 5 µSv/h.

(3) Non-fixed contamination on any external surface shall not exceed 0.4 Bq cm^{-2} for beta and gamma emitters and alpha emitters of low toxicity, and 0.04 Bq cm^{-2} for all other alpha emitters.

(4) Each package of gross mass exceeding 50 kg must have its permissible gross mass marked on the outside of the package.

(5) The consignor must include in the transport documents with each consign-ment the following information in the order given:

   (a) the United Nations number assigned to the material as specified in Appendix 1 of the regulations, e.g. '2910';

   (b) all items and material transported under the provisions for excepted packages shall be described in the transport document as 'RADIOACTIVE MATERIAL, EXCEPTED PACKAGE' and

   (c) shall include the proper shipping name of the substance or article being transported from the list of United Nations number in Appendix 1 of the regulation, e.g. 'INSTRUMENTS OR ARTICLES', 'LIMITED QUANTITY OF MATERIAL', 'EMPTY PACKAGING', etc.

(6) When an empty package is shipped as an excepted package, the previously displayed labels must not be visible.

(7) Where the radioactive material is enclosed in, or forms part of any instru-ment or other manufactured article with activity not exceeding the item or

package limit in the table above, shipment may be made as an excepted package provided that:

(a) the radiation level at 10 cm from any point on any external surface of any unpacked instrument or article is no greater than 0.1 mSv/h and

(b) each article or instrument (except radioluminescent timepieces or devices) bears the marking 'radioactive'.

(8) Radioactive material in forms other than those above may be transported as excepted packages provided that:

(a) the package retains its contents under conditions likely to be encountered in routine transport; and

(b) the package bears the marking 'radioactive' on an internal surface which warns of the presence of radioactive material on opening.

In addition, the general design requirements for packages must be met. These include ease of decontamination of outer surfaces, resistance to the collection and retention of water, etc. There are also additional requirements where packages are sent by post.

Where the conditions for excepted packages cannot be met, the next best option may be shipment in 'Type A' packages. There are, however, activity limits specified in the IAEA regulations. The design and manufacture of Type A packages is such that they meet certain performance standards of containment under test conditions. The design must be approved and certified by a competent authority. Operating, handling and annual maintenance procedures are required as part of the IAEA quality assurance conditions.

Four categories are assigned to Type A package (Table 9), which depend on the radiation level and 'transport index' (TI). The TI of a package is the maximum dose rate in mSv/h at 1 m from the external surface multiplied by 100.

The summary above covers only some aspects of shipping radioactive material, but must be read in conjunction with the Regulations, which cover many other aspects of control.

12 Disposal of radioactive waste

In the UK the disposal of radioactive waste is controlled by the Radioactive Substances Act 1993. This requires a Certificate of Authorization to be obtained

Table 9 Categories of package

Transport index (TI)	Maximum radiation level at the external surface (mSv/h)	Category
0*	≤0.005	I-WHITE
>0 but ≤1[a]	>0.005 but ≤0.5	II-YELLOW
>1 but ≤10	>0.5 but ≤2	III-YELLOW
>10	>2 but ≤10	III-YELLOW but under exclusive use

[a]If the measured TI is ≤0.05 the value quoted may be as zero.

before radioactive waste is disposed. Although there are some 20 exemption orders made under the Act, the majority of radioactive waste disposed is controlled by the Authorization.

The radioactive waste that is generated falls in the main into either very-low-level waste (VLLW) or low-level waste (LLW). In addition, waste that is radioactive, but of an activity concentration <0.4 Bq/g, does not need to receive special treatment. Liquid scintillation cocktails containing tritium and carbon-14 at activity concentrations <4 Bq/g may also be disposed without reference to the Act.

12.1 Solid waste

The disposal of solid VLLW material is usually permitted via the normal rubbish under what is known as the 'dustbin authorization'. Certain conditions are imposed. They are:

- no dustbin (0.1 m^3) shall contain >400 kBq;
- no single item must contain >40 kBq;
- the disposal must be recorded.

This method of disposal of solid VLLW is by far the most cost-effective way of disposing of this category of waste and must be used whenever possible if high disposal charges are to be avoided.

The disposal of solid LLW is much more complicated, with the use of authorized routes and proper record keeping being essential.

For example, the most commonly used radionuclides are often of relatively low radiotoxicity, and lend themselves to either incineration or disposal as VLLW after a period of decay. Incineration routes are available for the disposal of all radionuclides except alpha emitters and strontium-90, but are very restrictive for radionuclides other than tritium, carbon-14 and iodine-125. The decaying of tritium and carbon-14 is obviously impractical and LLW containing these nuclides can often go for incineration (e.g. for an activity of <4 GBq in total). Iodine-125 and iodine-131 are limited to a smaller activity (e.g. a total of 40 MBq) and all other beta/gamma-emitting nuclides are limited to less again (e.g. 1 MBq). A monthly consignment limit may also be imposed to restrict the rate of disposal via this route.

A store for allowing the shorter nuclides to decay and be disposed as VLLW may be available. A formal quality assurance (QA) plan that will require strict adherence to operational procedures is necessary for this type of storage. It will almost certainly require that wastes be presented for storage in bags containing (wherever possible) single radionuclides whose activity at the date of presentation is realistically estimated. The QA plan will ensure that each bag is traceable and that appropriate records are kept.

12.2 Liquid low-level waste (aqueous and approved liquid scintillation cocktails)

Liquid LLW must normally be disposed of under procedures detailed in a QA plan for the disposal of liquid wastes. For example, at Manchester University, the licensed monthly limit for the radionuclide categories is divided between depart-ments, to produce departmental limits. A fraction of the departmental limit is allocated to each of the department's designated sinks, and a log is attached to each sink containing the sink's limits. The disposer records the activities and names of the radionuclide that are disposed of in any month. The Radiological Protection Supervisor collates the sink data monthly and returns departmental data to a central service where records are kept which demonstrate compliance with the University limits.

Radioactive waste disposal is becoming more and more difficult, and the 'small user' is faced with spiralling direct costs and awesomely complicated admin-istrative procedures. It is in the user's interest that these factors be taken into account at the planning stage of any experiment if he or she and the department are to minimize both.

References

1. International Commission on Radiological Protection (1978–1990). *Limits for Intakes of Radionuclides*. ICRP Publication 30 (8 vols). Pergamon Press, Oxford.
2. Delacroix, D., Guerre, J. P., Leblanc, P. and Hickman, C. (1998). *Radionuclide and Radiation Protection Data Handbook 1998*. Radiation Protection Dosimetry vol. 76, nos 1 and 2. Nuclear Technology Publishing, Ashford, Kent.
3. International Commission on Radiological Protection. (1991). *1990 Recommendations of the International Commission on Radiological Protection*. ICRP Publication 60. Annals of the ICRP, Vol. 21, nos 1–3. Pergamon Press, Oxford.
4. Draper, G. J., Little, M. P., Sorahan, T. *et al.* (1997). *Cancer in the Offspring of Radiation Workers – a Record Linkage Study*. NRPB R298. National Radiological Protection Board, Chilton.
5. International Commission on Radiological Protection. (1977). *Recommendations of the International Commission on Radiological Protection*. Annals of the ICRP, Vol. 1, no. 3. Pergamon Press, Oxford.
6. *Radioactive Substances Act 1993*. Her Majesty's Stationery Office, London.
7. *The Ionising Radiations Regulations 1999*. Statutory Instrument no. 3232. Her Majesty's Stationery Office, London.
8. International Atomic Energy Authority. (1990). *IAEA Regulations for the Safe Transport of Radioactive Material*. Published as IAEA Safety Series no. 6 (1990). IAEA, Vienna.

Chapter 3

Risk assessments for the use of radioisotopes in biology

PETER E. BALLANCE

University of Sussex, Health and Safety Office, Brighton, Sussex BN1 9RJ, UK

1 Introduction

Ever since the early 1950s when radioisotopes began to be used extensively in experimental biology, research group leaders, supervisors and individual researchers instinctively asked themselves what were the hazards from the radioactive materials to be used, what were the risks to health and how these risks should be minimized. In the process of selecting radioisotopes for particular experiments, research workers needed to obtain detailed information on the way radioisotopes decayed. In order to use radioisotopes as tracers for specific elements or compounds in biological systems, they needed to know whether charged particles or electromagnetic radiation (or both) were emitted, and hence how to determine the best choice of radiation detection equipment to be used. From the physical data available they were able to gain some insight into the likely hazards which would be associated with the presence of the radioactive elements. With the well documented knowledge of the effects of ionizing radiation on living organisms, particularly the effects of atomic weapons, they instinctively sought to assess the risks to their health and, in turn, to take steps to minimize these risks.

Using simple hand-held Geiger–Müller radiation detectors, research workers were able to check the effectiveness of any lead shielding they used and to monitor for radioactive contamination on their bench surfaces. In the more efficiently managed laboratories, these procedures for reducing the external dose from penetrating radiation and the internal dose from the uptake of contamination by ingestion or inhalation were written down as laboratory rules. These workers were, thus, identifying hazards and risks to health and establishing safe systems of work to minimize risks. They were under no specific legal obligation to undertake these risk assessments, but as scientists they were taking the responsibility for their own health and that of their students and colleagues. In other words, they were behaving responsibly in accepting their common-law duty of care obligations.

Today's research supervisors face the same fundamental health and safety problems, but with the added requirement to comply with very specific safety and radiation protection legislation. In recent years the whole process of hazard and risk assessment has become much more formal and is seen by the law-enforcing authorities as the most important procedure for reducing workplace accidents. Research supervisors now not only have to deal effectively with the risks to health, but also have to be aware that failure to comply with health and safety legislation may result in the closure of their laboratory or prosecution under criminal law of either their institution or, perhaps worse, themselves! In addition, at all times they have to be ready to receive health and safety law enforcement inspectors and show them their laboratory facilities for handling radionuclides and to provide documentary evidence of compliance with their legal obligations, including the requirement to undertake risk assessments.

Before describing the actual procedure for making hazard and risk assessments, it is first necessary to consider the most important health and safety legislation dealing with the requirement to undertake risk assessments. What follows is based on legislation operating in the UK; however, the same guiding principles apply wherever work with radioisotopes is carried out. A summary of US legislation is provided in Chapter 11.

2 The Health and Safety at Work etc. Act 1974

The main provisions of this excellent legal framework are contained in Sections 2 and 3 of the Act (1). Whereas Section 1 outlines the broad aims of the Act, securing the health, safety and welfare of persons at work, and protecting other persons (including the public in general) against risks arising from the activities of persons at work, Section 2(2) describes the more precise duties of each research establishment. Section 2(2)(a) calls for equipment and services, e.g. electricity, water and heating, to be safe and, above all, requires safe systems of work to be established.

Implied in the last requirement is the need to carry out risk assessments. The essential requirements for producing safe systems of work involve, first, examining the hazards which might be present and, second, giving due consideration to the risks to health which might arise. Only by this systematic approach to proposed work activities is it possible to prepare effective systems of work to minimize the risks to health in the laboratory.

Section 2(2)(b) calls for the safe use, handling, storage and transport of substances, and Section 5 includes a requirement to control emissions to atmosphere, e.g. of volatile chemicals from the discharge stacks associated with fume cupboards and microbiological safety cabinets. Again, compliance with these requirements requires careful assessments of hazards and risks of exposure before the safe systems for transport, storage, handling and use of substances can be put in place.

In Section 2(2)(c), the Health and Safety at Work etc. Act 1974 stresses the need to provide information so that all supervised staff and research students are fully aware of the hazards present and the risks to health. Compliance with this requirement can only be achieved if a sufficiently comprehensive risk assessment has been undertaken.

Within Section 2(2)(c) are the requirements to ensure that training needs are evaluated and that appropriate training provides a necessary level of knowledge, an awareness of the risks to health, and the practical skills in safe working methods. This important section also requires that staff receive the required level of instruction and that they are effectively supervised. This duty again carries an implied requirement for research supervisors to ensure that risks to health have been recognized and are being effectively controlled by all members of their group, including postgraduate research students. Paying attention to the requirements of Section 2 of the Act should ensure that the health and safety of postgraduate students is controlled sufficiently to achieve compliance with Section 3 of the Act.

The Health and Safety at Work Act describes the powers of the enforcing authority (the Health and Safety Executive, or HSE). All researchers and University employees must be fully aware that HSE inspectors may enter laboratories and have the authority to serve Improvement and Prohibition Notices and to bring criminal prosecutions.

An Improvement Notice may list poor standards of equipment or unsafe procedures which need to be remedied. The Notice usually allows a reasonable time for these deficiencies to be rectified. A Prohibition Notice, on the other hand, may in some circumstances require the almost immediate stoppage of particular laboratory procedures. In exceptional cases (e.g. following a serious accident or finding serious deficiencies within the systems of work) a criminal prosecution may be brought. Most prosecutions arise from failure to comply with Sections 2(2)(a), 2(2)(b), 2(2)(c) or 3 of the Health and Safety at Work Act. Although prosecutions are rare, a prosecution should be anticipated whenever an employee or student is seriously injured in a laboratory accident.

3 The Management of Health and Safety at Work Regulations 1999

Regulation 3 of these Regulations (2) requires the employing organizations to make a suitable and sufficient assessment of the risks to the health and safety of their employees and of the risks to the health and safety of persons present who are not employed by the establishment. This last requirement concerns the health and safety of postgraduate research workers and any students who may be working in the laboratory or at fieldwork locations. In addition, it relates to protecting visiting research workers and any members of the public who may be affected by the research activities.

Unlike the more general principles set out in the Health and Safety at Work Act 1974, the Management of Health and Safety at Work Regulations 1999 were produced in response to a European Directive. They therefore tend to be much more detailed in their requirements. In addition they are targeted at individual heads of department, research group heads or individual supervisors of research workers. Biologists using radionuclides need to be aware of their responsibilities: if a serious accident occurs within their laboratory they, as the individual in charge of the injured person, could face prosecution unless they can clearly demonstrate that a suitable and sufficient risk assessment for the work had been undertaken and that the findings of the assessment had been conveyed to all those working in the laboratory. Although the main purpose of the assessment is to establish safe systems of work, the legal requirement of Regulation 3 is to assist the 'employer' to determine what measures need to be taken to comply with the employer's duties under the relevant Health and Safety Acts and Regulations. These will include compliance with the Health and Safety at Work Act 1974 as well as several of the following specific Regulations:

- the Management of Health and Safety at Work Regulations 1999 (requiring risks to health to be effectively managed);

- the Provision and Use of Work Equipment Regulations 1998 (3) (requiring equipment—e.g. electrophoresis equipment or fume cupboards—to be safe);

- the Personal Protective Equipment at Work Regulations 1992 (4) (these are concerned with the provision of eye protection, respiratory protection, etc.);

- the Workplace (Health, Safety and Welfare) Regulations 1992 (5) (these are concerned with laboratory floors not being potentially slippery, with working space and ventilation, etc.);

- the Manual Handling Operations Regulations 1992 (6) (these are concerned with risks of injury through manually lifting materials or equipment—they may be especially relevant when new equipment is being installed, when laboratories are being moved during building refurbishments, or when gas cylinders are being moved).

- the Ionising Radiations Regulations 1999 (7); as well as setting dose limits, these Regulations contain a clear requirement to restrict, so far as is reasonably practicable, the extent to which employees and other persons are exposed to ionizing radiation. To comply with this overall objective, prior risk assessments are required to be undertaken in order to identify the measures to be taken to restrict the radiation exposure;

- the Approved Code of Practice (8) which accompanies the Regulations (7) gives details of the most important matters to be considered in any risk assessment procedure. It is essential that the following matters are given due consideration:

(i)   the nature of the radioisotopes and the quantities to be used;

(ii) the risks to health from radioactive contamination or from penetrating X or γ radiation or from high energy beta particles;

(iii) the proposed systems of work to be followed and whether these are likely to be accompanied by any release of significant radioactive contaminations especially in the form of dusts or vapours (there may also be a risk of high hand doses if intense sources of radiation are to be manipulated);

(iv) the estimated dose rates and possible doses to the workers involved and to other persons who may be affected;

(v) possible accident situations and the steps required to either prevent their occurrence or to limit their consequences.

The prior risk assessment should establish what actions will be needed to limit radiation exposure (e.g. design features which may involve shielding to attenuate penetrating radiation or the provision of a fume cupboard or glove box to contain airborne radioactive contamination). These 'engineering controls' will always need to be supported by a description of the safe systems of work to be followed, either attached to, or included within, the experimental protocol.

- the Control of Substances Hazardous to Health Regulations 1994 (9) (these are concerned with the risks to health from chemicals and from biological agents);

- the Radioactive Substances Act 1993 (10) (although this Act is concerned with protecting the environment, it is advisable to consider this Act as part of any overall risk assessment for work involving radiochemicals. The Act is enforced by the Environment Agency in the UK and their inspectors may bring a prosecution against the establishment for even very minor breaches of any certificates which authorize the use of radioisotopes or their disposal);

- the Electricity at Work Regulations 1989 (11) (these Regulations are designed to ensure no one is able to make contact with live parts of electrical equipment —they are particularly relevant in the case of electrophoresis equipment);

- the Radioactive Material (Road Transport) (Great Britain) Regulations 1996 (12).

In larger establishments Local Rules for the implementation of each of the above Acts and Regulations will have been produced by working groups headed by a research director and comprising researchers, technicians, trades union safety representatives, the Director of Safety Services and the Radiation Protection Adviser. Copies of Local Rules booklets are usually issued to individual researchers. Copies may also be made available on the site computer network.

4 The risk assessment process

4.1 The content of the risk assessment

From the above list of Acts and Regulations it can be seen that the risk assessment involving protection against ionizing radiation may constitute only a small fraction of the overall risk assessment for a particular project.

For each risk assessment, researchers need to examine all the hazards which may be present, including especially the risks to health from biological agents, electricity, highly flammable liquids and gases, toxic chemicals, corrosive chemicals and chemicals which may have long-term effects (e.g. carcinogens and teratogens). In examining all the hazards associated with a research project, it is important not to waste any time on trivial risks but instead to concentrate on those which may pose the greatest risks to health. The assessment should include a consideration of the risks of injury from the common causes of laboratory accidents (e.g. slippery floor surfaces; cuts from glassware; splashes and spillages of material) as well as those events which, although they are rarely likely to happen, can cause devastating injuries and property damage (e.g. explosions or fires due to the use of reactive chemicals or highly flammable liquids). Included in the rarely occurring category are risks of uptake of radiochemicals or carcinogenic agents, risks of significant doses from external radiation, and ingestion or inhalation of harmful biological agents.

The assessment must include the risks to health during routine operations as well as the risks during special or 'one-off' operations. In addition, the assessment should include a consideration of the risks to contractors or maintenance staff who may need to enter the laboratory. The assessment should also consider the arrangements for the disposal of waste materials. Due to the increasingly high costs of disposals, it is advised that not only should the disposal methods be included in the risk assessment, but also that Local Rules should require the means of disposal to be resolved before materials can be ordered or synthesized. Each risk assessment must also indicate the actions to be taken following an accident.

4.2 Responsibility for making risk assessments

The Management of Health and Safety at Work Regulations 1999 require each establishment to have in place an organization for the effective management of the risks to health. Included in this organization should be a clearly defined management structure with clear statements of the responsibilities of research group heads, research supervisors and individual research workers. There should also be competent persons appointed who are able to provide authoritative guidance and to provide sources of information to facilitate the risk assessment procedure. Although directors of research establishments would be expected to ensure that risk assessments are undertaken, it will be individual project supervisors who carry that responsibility in the laboratory. Here they must make judgements as to whether they do each assessment themselves or whether for some work activities they get their postgraduate students to undertake their own risk assessments. This last approach is likely to provide excellent training for new researchers, but it is essential that the assessments undertaken are carefully checked and approved by the supervisors. There will be some critical work activities where the assessments must be undertaken by the supervisors themselves and, in some cases, the work assessment may need to be checked by the research group head, and approval to proceed may be needed at director level. Supervisors who allow work practices to continue knowing that hazard and risk assessments have not been undertaken, may be seen as consenting or conniving at the commission of an offence.

A study of the prosecution of a research supervisor at a British university may clarify some of the questions surrounding the legal responsibilities associated with the undertaking of risk assessments for laboratory work (13). In 1995 a chemistry lecturer was charged under Section 36(1) of the Health and Safety at Work Act 1974 (1) with failure to produce a suitable and sufficient assessment of the risks to the health and safety of a person not employed by the university, and by default causing the university to contravene Regulation 3 of the Management of Health and Safety at Work Regulations 1999 (2). The charge was made because there had been an explosion in which a third-year undergraduate student suffered severe injuries to his left hand. In court, the case against the lecturer collapsed when the student admitted that he *had* been given photocopies of the relevant literature, *had* been warned of the danger and *had* been instructed to follow safety precautions. It was also disclosed that the lecturer had completed a general risk assessment for the project being undertaken.

Although the judge's comments gave the impression that risk assessments can be suitable and sufficient without being recorded, provided that the relevant hazards and risks to health and protective/preventive measures have been identified, it seems reasonable to assume that in completing the university's risk assessment form, and providing photocopied safety information from the litera-ture, the supervisor had in fact provided a suitable and sufficient risk assessment. It was also revealed that the lecturer had given specific written instructions to the student who was shown how to perform the reaction safely by ensuring that no water was present in the reaction mixture. The supervisor had personally supervised the drying of the solvent used. Although this case appears to have established no legal precedents, it does suggest that the HSE may be moving away from prosecuting research and teaching establishments (provided that the institution has established an effective system for the management of health and safety), and instead are prepared to prosecute individual research super-visors who fail to provide clear evidence that they have subjected their work activities to a suitable and sufficient risk assessment.

5 Suggested methods for making risk assessments

There are no hard and fast rules for achieving a suitable and sufficient risk assess-ment, other than the need to undertake a careful and systematic examination of each research project or work activity. It is therefore sensible to examine a range of different approaches, and for research supervisors to choose those which are applicable to their own management arrangements and to the particular projects being undertaken.

5.1 General laboratory hazards

These risks, which may include falls, glassware injuries, electrocution, effects of corrosive chemicals, etc., may be assessed by using the following semi-quantitative approach, in which for each 'hazard' (the potential to cause harm, e.g. a smooth laboratory floor surface) one estimates its risk to health, i.e. the severity of the event multiplied by the probability of it happening.

Table 1 Assessment scores

Hazard severity		Probability of occurrence		
		unlikely 1	possible 2	likely 3
No injury	0	0	0	0
Slight injury	1	1	2	3
Significant injury ($\geq$4 days off work)	2	2	4	6
Severe injury (death or major injury)	3	3	6	9

A system of scoring can then be used to determine what actions are needed to control the risks to health, and the urgency for actions to be taken. An assessment score is obtained by multiplying the severity of the hazard by the probability of its occurrence, as shown in *Table 1*.

Priorities for taking action to control or eliminate risks can then be assigned as follows:

9 take action at once
6 take action as soon as possible
4 take action as soon as practicable
3 take action within a reasonable time
2 low priority for action
1 very low priority
0 take no action

It must be remembered that there will be instances where the risk of injury may appear very low but where the assessment identifies a requirement to comply with safety legislation. In these cases the priority score will need to be increased to fall within the 3–6 range, the final figure being based upon a judgement of the likely consequences of any delay in achieving compliance.

A formal semi-quantitative approach is taken when undertaking assessments for work involving genetically modified organisms. Detailed advice for these assessments is provided by the HSE (14).

5.2 Use of generic risk assessments in biological research and teaching

5.2.1 Experimental work in teaching laboratories

Generic risk assessments to meet the requirements of the Management of Health and Safety at Work Regulations 1999 can be drawn to the attention of students and demonstrators for each teaching laboratory project in the form of an easy-to-read table such as the example shown in *Table 2*.

It is essential that the generic risk assessments are undertaken with great care. Usually for each assessment there will be just one or two points of detail which are absolutely critical to the safety of the experiment; these critical pieces

Table 2 Student experimental work protocol—information about hazards to health[a]

Substance/ apparatus/ process/ equipment	Possible hazards							
	Explosion	Fire	Corrosion	Toxic	Pressure/ vacuum	Electricity	Radioactivity/ radiation	Other (specify)

[a] To complete the assessments, the hazards which may give rise to risks to health are to be ticked and below each tick reference is made to the generic safe procedure(s) to be followed, e.g. LP = laboratory precautions (reference number) to be followed, ASN = attached safety notes. Where appropriate, reference may also be made to specific pages of the establishment's Local Rules.

of information must be drawn to the attention of students by including the necessary warnings and safe procedures within the experimental work schedules. In addition, it is suggested that these warnings should be repeated in notes, in bold type, added to the schedule.

For most teaching projects the hazards and risks to health from the use of very small quantities of radioisotopes will be minimal and the risk assessment will need to stress the risks from the use of non-radioactive chemicals. Teaching protocols may include small quantities of ^{14}C, ^{35}S and ^{33}P and similar low energy beta-emitting radioisotopes; these carry no risk of doses to the skin from beta radiation. Most teaching laboratory work with ^{125}I will involve the use of no more than 40 kBq for immunoassay work. Teaching experiments should never involve iodination reactions where tens of MBq are normally used. When attempts are made to assess doses which may arise from the ingestion, inhalation (or uptake through the skin) of radioactive contamination it is possible to make use of the dose coefficients (SvBq^{-1}) which appear in the International Commission of Radiological Protection (ICRP) publication 68 (15). The dose coefficient represents the committed effective dose per unit intake (SvBq^{-1}), integrated over 50 years, e(50). Annual limits on intake (Bq) can be obtained by dividing the annual average effective dose limit (0.025) by the dose coefficient e(50). *Table 3* shows examples of ICRP 68 e(50) values for some of the radioisotopes commonly used in biological research and teaching.

Table 3 Examples of ICRP 68 dose coefficients for intakes of radioisotopes by workers (based on ICRP 60 recommendations (17), where einh and eing are committed effective dose for inhalation and ingestion respectively, integrated over 50 years.

Radioisotope	Inhalation einh(50) (Sv Bq^{-1})	Ingestion eing(50) (Sv Bq^{-1})
Tritiated water	1.8×10^{-11}	1.8×10^{-11}
Tritiated compounds	4.1×10^{-11}	4.2×10^{-11}
^{14}C	5.8×10^{-10}	5.8×10^{-10}
^{35}S, organic	1.2×10^{-10}	7.7×10^{-10}
^{32}P		2.4×10^{-9}
^{33}P		2.4×10^{-10}
^{125}I	1.4×10^{-8}	1.5×10^{-8}

In the case of teaching experiments involving 40 kBq of ^{125}I, ingestion of as much as 1% of the material (400 Bq) would result in a committed effective dose of only $400 \times 1.5 \times 10^{-8} \times 10^{6}$ μSv, i.e. 6 μSv. This dose may be compared to the dose rate of about 4 μSv/h received by long-distance airline passengers, and the average annual dose, 2.6mSv, received from background radiation (30). In the case of the teaching experiment with 40 kBq of ^{125}I, there would therefore be no need for significant control measures other than to wear gloves to avoid uptake through the skin, to monitor for contamination and to clean up any spillages promptly.

Assessments for other radionuclides used in teaching experiments can also give rise to very low committed effective doses from any uptake of radioactivity. The use of ≤1 MBq of ^{35}S would result in a dose of no more than 7.7 μSv even if 1% of the material were ingested. Similar Good Radiochemical Practice precautions to those outlined for ^{125}I would be sufficient to minimize any uptake of radioisotope. These would include the use of gloves to prevent uptake through the skin, covering working surfaces with Benchcote (polythene-backed paper), working in trays to contain spillages, and, since significant quantities are present, the use of rate meters with hand-held probes to monitor surface contamination. Examination of the e(50) values in *Table 3* shows that radioactive hydrogen, ^{3}H (tritium), is even less hazardous than ^{35}S, having a maximum beta energy of only 18.57 keV, as opposed to the 167.50 keV beta radiation emitted by ^{35}S. An ingestion of 1 MBq of ^{3}H would only result in a committed effective dose of 42 μSv. Due to its very low energy beta emissions, surface monitoring for ^{3}H, with scintillation or Geiger–Müller probes, is impossible. Instead, 'wipe tests' using 50 mm glass microfibre filter discs are required. Each disc is placed in a vial containing approximately 12 ml of scintillation solution and counted in a liquid scintillation spectrometer. The surface concentration of radioactive contamination (Bq cm^{-2}) is estimated by assuming that the filter disc removes 10% of the radioactivity present. For all general radiochemical laboratory work, a derived working level of 4 Bq cm^{-2} is an achievable, practicable and effective standard for the control of radioactive contamination on laboratory surfaces. One-tenth of this surface concentration should be used for the more radiotoxic alpha-emitting radioisotopes. No matter how low the risks appear, it is always good practice to go through the risk assessment procedure and to explain the results of each assessment as a means of reassuring undergraduate students (and their parents!).

Having outlined the very low risks from radioactive materials used in biological teaching, it is essential that careful assessment is made of the much more serious risks to health from non-radioactive chemicals. This assessment is required to comply with the Control of Substances Hazardous to Health Regulations (1994) (9). For teaching laboratories the chemicals used may be listed as suggested in *Table 2*. Alternatively, *Table 4* may be used for the chemicals to be used in each experiment, as follows:

Table 4 Hazards to health from chemicals to be used

Substance	Health hazards			
	Highly flammable	Toxic	Corrosive	Other hazards

Each substance is listed and the relevant hazard columns are ticked. For most experiments the following simple warnings and precautions, set out as basic chemical safety rules, should provide sufficient information to minimize the risks to health.

i. Basic chemical safety rules

Highly flammable materials: to avoid the risk of fire, these materials must be kept away from sources of ignition. For distillations and solvent extractions, flame-proof heating equipment must be used. What may not be obvious is that vapours from some highly flammable liquids may travel several metres along a bench or floor surface to reach a source of ignition. Appropriate fire extinguishers must be available.

Toxic materials: some compounds, e.g. hydrogen cyanide and phenol, may cause rapid death by poisoning. Some, e.g. chlorine and nitrogen dioxide, may be severe respiratory irritants. Some, e.g. carcinogens, mutagens or teratogens, may cause long-term effects. It is essential to work in a glove box or fume cupboard to contain fine dusts, vapours or gases. No mouth pipetting must be undertaken and skin should be protected by gloves. These basic precautions are also applicable to radioactive materials.

Corrosive materials: these can cause chemical burns to the skin or eyes. The wearing of protective gloves and eye protection are both essential. If material does enter the eyes, they should be carefully irrigated with sterile saline solution or tap water. Contaminated clothing must be removed; this is particularly important also for toxic chemicals, e.g. phenol.

5.2.2 Research project risk assessments

Although there are no specific rules for making these assessments, research workers may find it helpful to follow a systematically arranged risk assessment form. It is recommended that research directors who issue hazard and risk assessment forms, remind research supervisors that it is a legal requirement (with criminal law penalties) that suitable and sufficient risk assessments are undertaken before the project is started. They should be advised also to ensure that by completing the risk assessment forms they will have the evidence to prove to inspectors from the enforcing authorities that the assessments have been undertaken and that their findings and safe systems of work have been communicated to every individual involved in the project. In addition to a risk assessment form for the overall research project, there may be a requirement to produce a special risk assessment form if there is a clearly foreseeable risk of death or severe injury or serious breach of safety legislation and the materials or procedures involved are not included within any of the generic risk assessments contained in the establishment's local rules or safe working procedures.

Supervisors should be further advised to issue copies of risk assessments to the individuals under their supervision, and to others whose health and safety may be affected by the work being undertaken. This last point is particularly relevant in shared laboratories or where temporary staff or undergraduate students are working in research laboratories. Copies should also be sent to the director of the establishment so that he or she can demonstrate compliance when being visited by either HSE inspectors or safety staff from within the organization who are undertaking regular audits of the quality and health and safety management.

The following examples of project risk assessments and special hazard and risk assessments (*Protocols 1* and *2*) are based upon risk assessment forms used in the School of Biological Sciences at the University of Sussex.

Protocol 1

Example of a risk assessment form for a research project

SCHOOL OF BIOLOGICAL SCIENCES

Project Hazard and Risk Assessment Document No. .... Date ....

Project title ....

Names of research workers/technicians ....

....

Description of project (include substances, procedures, equipment—use separate sheets if necessary)..

....

Duties of researchers/technicians: to follow Good Laboratory Practice; to read manufacturers' warning labels and safety data sheets for substance to be used; not to use equipment or apparatus unless trained in its use, and unless it conforms to current safety requirements; to work in accordance with relevant University Local Rules, Codes of Practice and Safety Regulations, titles of which are listed overleaf. The list is not exclusive.

Special Hazard Assessments: the supervisor and the above research workers/technicians have together considered the substances and procedures likely to be used in the above project that are not covered by the establishment's Local Rules, Codes of Practice or Safety Guidance Notes. For these substances/procedures, special risk assessment forms have been prepared as follows.

Substances/procedures.. Document no. ....

....

Copies of the Special Risk Assessment document should be attached to this Project Hazard and Risk Assessment document.

One copy of the Project Hazard Assessment, and any associated Special Hazard Assessments, must be kept in a file in the laboratory where the work will be undertaken. A copy should be retained by each of the persons named above and one copy sent to the School Laboratory Director.

Signatures: Date:

Research supervisor... ....

Researcher/technician...................................... ....

Researcher/technician...................................... ....

Protocol 1 continued

Supplementary information: list of policy documents, Local Rules, Codes of Practice and safety guidance notes

- Establishment Safety Policy (and safety management organization)
- Local Rules and Notes of Guidance for the implementation of the Management of Health and Safety at Work Regulations 1999
- Local Rules for Inspection, Monitoring and Auditing of the Management of Health and Safety at Work
- Local Rules for the implementation of the Manual Handling Operations Regulations 1992
- Local Rules for the implementation of the Provision and Use of Work Equipment Regulations 1992
- Notes of Guidance for the implementation of the Workplace (Health, Safety and Welfare) Regulations 1992
- Local Rules for the implementation of the Personal Protective Equipment Regulations 1992
- Local Rules for the implementation of the Control of Substances Hazardous to Health Regulations 1994
- Local Rules for the control of poisons
- Local Rules for the disposal of waste chemicals
- Local Rules for the control and use of carcinogenic, mutagenic and teratogenic substances
- Local Rules for the use of lasers
- Local Rules for the control and use of X-ray equipment
- Local Rules for working with radioactive materials and ionizing radiation
- Local Rules for the safe operation of centrifuges
- Code of Practice for Safe Working with Laboratory Glassware
- Local Rules for the control of biological agents
- Local Rules for work with human and animal blood samples, human and animal blood products and other tissue specimens of human or animal origin
- Local Rules for the implementation of the Genetically Modified Organisms (Contained Use) Regulations 1992, as amended in 1996
- Local Rules for working in establishment buildings outside normal working hours
- Notes of guidance for the implementation of the Electricity at Work Regulations 1989
- Local Rules for the safe operation of electrophoresis apparatus
- High Pressure Safety Code
- Safety guidance notes (these include: storage and use of gas cylinders, testing electrical equipment, explosive chemicals, distillations, use of phenol, hydrogen fluoride, hydrogen cyanide, solvent purification, fire safety, first aid, and reporting of accidents). Those relevant to each research project may be used as generic risk assessments.

Protocol 2

An example of a Special Risk Assessment form

SCHOOL OF BIOLOGICAL SCIENCES

Special Hazard and Risk Assessment Document No. ....   Date ....

Special and exceptional hazards and risks which are beyond the scope of the generic risk assessments in Local Rules and Codes of Practice require the completion of a Special Hazard and Risk Assessment document, a copy of which must be sent to the Laboratory Director. Completion of the assessment document is mandatory before the experimental work is started. The document must be completed by the research supervisors and fully discussed with the research workers and technicians involved. Discussion with the School and the Establishment Safety Adviser may also be necessary.

A copy of this document must be issued to each worker involved and a copy kept in a clearly labelled file in the laboratory. The information in sections (1), (2), (6f) (7), and (8) must be displayed at the work location.

(1) Name of substance or procedure ....

Quantities of hazardous material(s) to be used ....

....

....

(2) Details of special hazard(s)/risk(s) (use extra pages if necessary) ....

....

(3) Personnel involved

Name of supervisor..

Names of research workers/technicians	Signatures	Date
....	....	....
....	....	....

The above workers have received from me notification of special hazards and risks, and instructions on the safe procedures to be used, including the following:

Supervisor's signature.................................   Date ....

(4) Location of work...   Storage locations ....

(5) Dates and times when work will proceed ....

Overnight running permit and emergency contact telephone numbers ....

....

(6) Procedures and precautions

(a) Brief descriptions of experimental procedures (use extra pages if necessary): ....

....

(b) Specify grade of containment facility to be used:

(i) fume cupboard ....

(ii) microbiological safety cabinet ....

Protocol 2 continued

    (c) Protective clothing required ....

    (d) Details of special monitoring (including biological monitoring) required ....

    (e) Waste disposal procedures ....

    (f) Special first aid procedures required ....

(7) Emergency actions in the event of an accident

    (a) Uncontrolled release (spillage/loss of containment) ....

    (b) Fire ....

    (c) Shut-down procedure ....

(8) Actions in the event of a service failure

    (a) Electricity ....

    (b) Water ....

    (c) Other (specify) ....

In making an assessment of special hazards and risks, supervisors and research workers should aim to identify the worst hazards present and the worst things that may go wrong. They should then identify the procedures which must be followed to minimize the risks to health, as well as the actions to be taken in an accident or emergency situation. In some cases the completion of standardized risk assessment documents may not be appropriate. Instead researchers may prefer to devise their own more descriptive style of recording the findings of their assessments. Whatever approach is taken, the HSE in the UK (16) have suggested five steps that should always be completed:

(1) Carefully examine the hazards which may be present, including the materials to be used and procedures to be followed. Consider which could result in either serious injury and/or uncontrolled release of material.

(2) Consider who may suffer harm; do not forget cleaners, maintenance and security staff, and members of the emergency services who may enter the building. It is important also to consider whether members of the public or the environment may be affected.

(3) Determine which hazards may cause a risk to health and whether the risks are high, medium or low, and whether the relevant legal requirements have been met. Examine ways of reducing the risks to health.

(4) Always clearly record the major risks to health and the protective and preventive measures to be taken.

(5) Review hazard and risk assessments at regular intervals, and always if there are major changes to particular research projects. It is good practice for research supervisors, together with their researchers and technicians, to review their assessments at least once a year and to update the signatures at the end of each assessment document.

The descriptive style of writing risk assessments may be appropriate for the use of a particular piece of specialist apparatus. The assessment should describe the process involved, the hazards which may be present, the risks to health and the precautions to be taken to manage effectively the risks to health. Usually there will need to be strict control of quantities of substances present, and the temperature and pressure of any reactions. There must also be a management system in place to ensure that only authorized persons who have been adequately trained are permitted to use the equipment. A system for logging the use of the equipment and the operations involved is also advisable.

6 Hazards and risks to health from the use of radioactive substances and ionizing radiation

Having suggested a systematic approach to research project risk assessments, there remains the problem of integrating assessments of work with ionizing radiation into the all-embracing risk assessments for research projects. Clearly the project and special hazard and risk assessment procedure outlined in Section 5 could be followed. The use of radioactive materials would then be entered on the project risk assessment document together with the advice to follow the establishment's Local Rules for work with radioactive materials and ionizing radiation. Because work with ionizing radiation and radioisotopes requires compliance with specific legislation (e.g., in the UK, the Ionising Radiations Regulations 1999 (7) and the Radioactive Substances Act 1993 (10)), it is better to undertake completely separate hazard and risk assessments for those parts of each research project where radioisotopes are to be used.

Most work with radioisotopes will fall within generally applicable generic hazard and risk assessments which may be included in a Local Rules booklet, or within Local Rules for each laboratory. Regulation 17 of the Ionising Radiations Regulations 1999 requires the preparation of Local Rules. A copy of the Local Rules should be issued to everyone who uses radioactive materials. They may be issued as part of a training package in which potential users are taught basic nuclear physics, interaction of radiation with matter, detection and measurement of radioactivity and radiation, the units and quantities used in radiation protection, the biological effects of radiation, protection standards advised by the ICRP (publication 60) (17) and the European standards for radiological protection which are enshrined in UK legislation. On the more practical side, the training course should explain methods of operational radiation protection, including hazard and risk assessments and methods of minimizing doses from external radiation as well as from internally absorbed radioisotopes (18).

To ensure compliance with the environmental protection requirements of the Radioactive Substances Act 1993, course candidates should be made aware of the controls on quantities of radioisotopes which may be ordered, the need to maintain stock records on a monthly basis and the limits on quantities of radioactive waste which may be either stored to decay or discharged from the laboratory. Disposal routes which may be used may include discharging gases to atmosphere, liquids to drains, and solid waste to dustbins, to landfill or to an incinerator.

The training programme should include the details of Good Radiochemical Laboratory Practice (19) and should include either practical laboratory work with unsealed radioisotopes or a video presentation of contamination and radiation control during the actual radioisotope work which occurs in the establishment. Equipping users of radioisotopes with this level of detailed training will considerably simplify the undertaking of hazard and risk assessments.

Protocol 3

Example of a form for controlling the ordering of radioisotopes, and the risk assessment for their use

APPLICATION TO ORDER RADIONUCLIDES

After completion this form must be sent by the responsible person to the Radiation Protection Supervisor, who forwards it to the Safety Office.

ORDER NUMBER: ....

NAME OF LICENSED RADIATION WORKER: ....

RADIATION LICENCE NO:................   SCHOOL: ....

LABORATORY OR ROOM NO: ....   TEL. NO: ....

HAZARD AND RISK ASSESSMENT:

  – Tick box if work falls within scope of the Local Rules (SSC-48-1)   ☐

  – Tick box if work is covered by SSW1 (^{32}P) or SSW2 (^{125}I)[a]   ☐

  – If special assessment has been prepared, attach copy and tick box   ☐

BRIEF DESCRIPTION OF EXPERIMENT: ....

Tick box if this order is for a sealed source   ☐

Item no.	Code no.	Quantity (MBq)	Radionuclide	Date required

Conversion factors: $1\ mCi$ $=$ $3.7 \times 10^7\ Bq$ $=$ *37 megabecquerels (MBq)*
$1\ mCi$ $=$ $3.7 \times 10^4\ Bq$ $=$ *37 kilobecquerels (kBq)*

I certify that the use or supervision of the use of the radionuclide will be carried out only by either myself or by other competent licensed workers.

Signed: ....

Date: ....

Risk assessment checked and approved by:...
(signature of responsible person)

[a]SSW1 and SSW2 are special systems of work and risk assessments for ^{32}P and ^{125}I.

Although most assessments can simply be described as 'within the scope of the Local Rules', it is recommended that a record is made of each assessment. A simple way of achieving this is to include a requirement to record that the hazards and risks of particular experiments have been assessed, and that they are covered by the generic assessments provided by the Local Rules. This record can easily be included on an 'Application to Order Radioisotopes' form, an example of which is shown in *Protocol 3*.

The hazards and risks associated with the proposed work should be discussed with a desginated responsible person, who will then be in a position to sign the 'Application to Order' form, thereby allowing the material to be ordered, and at the same time confirming to the Radiation Protection Supervisor (RPS) that the risks to health and the safe system to be followed are covered by the Local Rules document. This procedure will also allow the responsible person or the RPS to check that the researcher attempting to order the radioisotope has obtained a licence to work with radioactive materials and has completed the required course of training.

Some work with radioisotopes may require procedures which are beyond the general generic assessments provided in the Local Rules. These experiments, e.g. ^{125}I iodinations, ^{32}P DNA labelling protocols or sequencing work, will require special hazard and risk assessments.

7 Special hazard and risk assessment for ^{125}I

7.1 The hazard

Radioactive iodine decays by electron capture, emitting 35.49 keV gamma photons from 6.7% of the disintegrations. Also produced are 27.20, 27.40 and 31.00 keV X-rays which arise from the electron cloud of the ^{125}Te daughter (20).

7.2 Risks to health and radiation protection procedures to be followed

Iodine entering the body through inhalation, ingestion or absorption through the skin tends to accumulate mainly in the thyroid, where it has an effective half-life of 120 days (20). At low doses the risk to health is thyroid cancer (probability 8×10^{-4} Sv^{-1}). A massive uptake of ^{125}I could cause loss of thyroid function if dose rates exceed 500 mSv/year. The main risk to health comes from the inhalation of iodine vapour, which may be released during iodination reactions. These reactions must therefore be undertaken in a fume cupboard. Disposable PVC or latex gloves will minimize skin absorption. Three or four hours after each iodination, and early the following day, researchers are advised to use a thin sodium iodide crystal probe and rate meter (e.g. from Mini Instruments, Type 5-42B; see Appendix) to check the extent of any ^{125}I accumulation in their thyroid (the detector is placed on the skin at the position of the tie knot). The committed equivalent dose ($H_T(50)$) is approximately 1.1 μSv/Bq. An indicated 1 c.p.s. above background with a Mini 5-42B probe represents $\sim$20 Bq in the thyroid. A very high count rate, e.g. 250 c.p.s., would imply an $H_T(50)$ of 5.5 mSv (20).

Researchers finding thyroid count rates of twice background or more should report their findings to their RPS and Radiation Protection Adviser (RPA) for further, more accurate measurements to be made.

Examination of the dose coefficients (*Table 3*) for a selection of the radio-isotopes which are routinely used in biological research (15) shows that ^{125}I is much more radiotoxic than ^{32}P, ^{35}S or ^{3}H. It is therefore essential that contamination is controlled carefully. The effectiveness of the techniques used to control contamination can be monitored by making use of the Mini 5-42B probe or by using a much cheaper, low energy X-ray-sensitive Geiger–Müller detector. If significant uptake of ^{125}I is suspected, the dose to the thyroid can be minimized by taking ~100 mg of potassium iodide per day up to a limit of 1.5 g, under supervision.

External radiation from ^{125}I is also a problem (40 MBq at 10 cm gives an equivalent dose rate of approximately 160 μSv/h, increasing to 16 mSv/h at 1 cm). Since the Approved Code of Practice (8) which accompanies the Ionising Radiations Regulations 1999 requires that radioactive materials should not be directly manipulated in the hand, it is necessary to shield the microcentrifuge tubes in which iodination reactions take place. This can be achieved by locating the tube in a block of lead-loaded Perspex (see *Figure 1*). The block can then be held in the fingers whilst undertaking the iodination.

Lead-loaded Perspex material can also be used as a vertically mounted screen to shield the column which is used to separate the ^{125}I-labelled protein from the free ^{125}I. Alternatively, the column can be shielded by placing it inside a length of copper pipe. This reduces the intensity of the low energy X- and gamma-radiation by a factor of 10^3.

Although at distances >50 cm external dose rates are relatively low, it is still

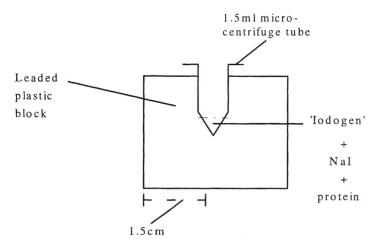

Figure 1 Lead-loaded Perspex holder for microcentrifuge tubes.

good practice to keep solutions behind simple shields (e.g. a vertical piece of blockboard backed with thin lead sheet). Keeping solutions inside the fume cupboard is also effective since the toughened glass of the sash will effectively attenuate the low-energy radiation. A square of lead-backed blockboard should be placed on top of any refrigerator which is used to store stock solutions of ^{125}I. The double thickness of the sheet steel sides and door of the refrigerator do not require additional shielding. With iodination work, provided all the above dose reduction techniques are in place, it is not necessary to issue research workers with personal dosimeters. However, for new work with >40 MBq of ^{125}I, it is advisable to issue fingertip and body thermoluminescent dosimeters for the first few months of the project. Initially these can be checked after one or two operations have been completed, so that if higher than expected doses are found, then the work methods can be quickly changed before significant doses of external radiation have been received.

7.3 Elimination of the hazard

For some ^{125}I labelling techniques (e.g. western blots), it is possible to use non-radioactive detection systems. Not only does an enhanced chemiluminescence (ECL) label give greater sensitivity (1 pg) than ^{125}I (1 ng), but with ECL it is possible to re-wash and re-expose the blot (R. Roberts, Nycomed Amersham plc, personal communication). The ECL marker also gives experimental information more quickly: ECL can give a superb signal after several minutes, whereas ^{125}I even after 18 h of exposure may still only just provide a useful trace on the film.

8 Special hazard and risk assessment for DNA labelling/sequencing work with ^{32}P

The hazards from ^{32}P include high energy (1.71 MeV) beta radiation and the associated *Bremsstrahlung* X-rays, which may become significant when large quantities of stock solutions are present. The risks to health can occur from the external beta radiation, which may deliver massive doses to the skin and from contamination entering the body where the ^{32}P becomes deposited mainly in bone and hence may induce cancer of the bone marrow and the surface of the bone. The protective and preventive measures for reducing the risks to health are described below.

8.1 Minimizing uptakes

The fact that ^{32}P decays by emitting very high energy beta particles, which can be measured easily with very simple portable Geiger–Müller or scintillation probes, and which travel about 30–50 cm in air, enables the presence of even minute quantities of contamination to be detected and the affected areas cleaned promptly. Most probes indicate 10 c.p.s. for a derived working level of 4 Bq cm^{-2}. Even the Mini Instrument 5-42B beryllium window sodium iodide probe will detect ^{32}P contamination, but with only half the efficiency of the probes equipped with Geiger–Müller tubes. Due to the ease of detecting spillages or spread of contamination, there have been very few recorded incidents involving the uptake of ^{32}P. Two recent examples (21), however, involved contamination being deliberately introduced into the working environment. Urine samples and whole body *Bremsstrahlung* measurements indicated uptakes of 11–22 MBq which resulted in calculated committed effective doses corresponding to ~2 mSv/MBq. For the control of radioactive contamination, the importance of systematic regular monitoring can never be overemphasized.

8.2 Assessment of risks and control of external radiation from ^{32}P

High doses of beta radiation can cause skin desquamation (shedding of surface layers), ulceration and epithelial necrosis (death of epithelial cells). These deterministic effects, however, only occur if dose rates exceed 400 mSv/h and if the threshold absorbed dose (~50 Gy) is exceeded. The Ionising Radiations Regulations 1999 dose limit is 500 mSv/year to the skin, averaged over 1 cm^2. It is extremely unlikely that these doses would ever be obtained in biological research laboratories. The most significant remaining risk is the induction of skin cancer (17).

The following represents a systematic approach to reducing exposures from external radiation from ^{32}P (or any other high-energy beta-emitting radioisotope).

- Plan the experimental work and identify what may be critical sources of radiation. It may be useful to carry out a 'dry run' using non-radioactive solutions.

- Quantify the hazard by:
  - calculations
  - thermoluminescent dosimeter measurements
  - monitoring, to find any unforeseen sources of radiation.

- Use protective procedures to minimize exposure, taking special precautions with the 'critical sources' which have been identified.

- Check the effectiveness of the protective measures by wearing thermoluminescent fingertip dosimeters and, if necessary, body or head dosimeters.

8.2.1 Identifying possible critical sources of exposure

This will involve a careful check on the following (but not exclusive) list of examples:

- stock solutions and stock vials;
- solutions in microcentrifuge tubes;
- material on small plastic columns;
- extended sources, e.g. membrane labelling solutions in hybridization tubes or bags;
- harvested cells;
- small drops of contamination on skin or clothing.

8.2.2 Examples of rapid dose rate assessment calculations (22)

Consider a researcher who may handle 40 microcentrifuge tubes (each containing 2 MBq) every week. What fingertip dose would she receive if she held each tube for 30 sec? Assume the radioactive solution is 1 cm from the skin.

Using the following approximate formula for high-energy beta emitters gives the equivalent dose rate at 30 cm:

$$\text{radiation dose } (\mu\text{Sv/h}) \text{ at 30 cm} = 54 \times (\text{radioactivity in MBq})$$
$$= (54 \times 2) \text{ i.e. } 108 \ \mu\text{Sv/h}$$

The inverse square law will provide the dose rate at 1 cm,

i.e. $108 \times 30^2 \ \mu\text{Sv/h}$
$= 97.2 \text{ mSv in 60 min}$
$= 0.81 \text{ mSv in 30 sec}$

Therefore 40 tubes would give 32.4 mSv/week, i.e. 1620 mSv/year. This equivalent dose is above the skin dose limit of 500 mSv/year specified in the Ionising Radiations Regulations 1999, and is more than the three tenths of the dose limit which would trigger a researcher being designated as a 'classified person' required to undergo medical examinations and obtain their radiation dosimeters from an HSE 'Approved Laboratory'.

To confirm these initial calculations, thermoluminescent dosimeters may be used to make direct measurements of the beta radiation emanating from critical sources.

Ballance *et al.* (22) have published measurements comparing surface dose rates from stock vials, solutions being dispensed from a hypodermic syringe and solutions in microcentrifuge tubes (*Table 5*).

The dose rates from *Bremsstrahlung* X-rays can be calculated (23) using the simplified formula:

$$\text{dose rate from } Bremsstrahlung \text{ (in } \mu\text{Sv/h}) \text{ at any distance } r \text{ in cm} = \frac{\text{MBq} \times 4.2}{r^2}$$

For even 100 MBq ^{32}P, the *Bremsstrahlung* only gives a dose rate of 0.2 μSv/h at 50 cm. Shielding will be unnecessary except where large quantities of ^{32}P are being used. The peak *Bremsstrahlung* energy is 56 keV, and hence lead sheet or leaded Perspex screens are needed to provide effective shielding. Although lead and similar high atomic number elements increase the *Bremsstrahlung* yield, they are also extremely effective in absorbing the low-energy photons produced.

Table 5 Use of thermoluminescent dosimeter to identify critical sources of beta radiation

Source	Activity (MBq)	Dose rate (μSv min^{-1})
Stock vials	74	54–175
Vial in Perspex shield	74	0.1–0.3
Syringe	1	1.76
Microcentrifuge tubes (base)	1	1024–2000

Calculations of doses to the skin from splashes of contamination (22) show that very high doses may be received unless research workers monitor their skin surfaces frequently and remove promptly any radioactivity found on their skin. Considering the fluence rate of beta particles entering the skin from a 1 cm^2 area contaminated by 1 MBq, Ballance et al. (22) have shown that the dose rate to the skin could be as high as 1.82 Sv/h. This high value is derived from the formula:

$$D = \Phi \mu_a \bar{E} \text{ MeV g}^{-1} \text{ sec}^{-1}$$

where D is the dose rate, Φ is the fluence rate (0.5×10^6) of the beta particles cm^2/sec, μ_a is the beta absorption coefficient (9.3 cm^2 g^{-1}) at the maximum beta energy of 1.71 MeV, $\bar{E}$ is the average beta energy (0.68 MeV) and 1 MeV is 1.6×10^{-13} J.

Thus $D = 0.5 \times 10^6 \times 9.3 \times 0.68 \times 1.6 \times 10^{-10} \times 3600$ J kg/h
   = 1.82 Gy/h
   = 1.82 Sv/h (radiation weighting factor 1 for ^{32}P beta particles)

Failure to remove contamination until the afternoon of the following day (20 h) would have exposed the skin to a dose of 36.4 Sv. This level of exposure is close to the threshold for deterministic effects on the skin (17). Even a few kilobecquerels on the skin could, if not detected and removed promptly, result in tens of millisieverts being received each time, with the real possibility that the 500 mSv/year dose limit could be exceeded unless better standards of contamination control are exercised.

8.2.3 Protective and preventive procedures to minimize beta radiation doses

Doses to the face and front of the body can be reduced by working behind a 1 cm thick Perspex screen. These can be made up in-house, although elegant screened workstations are available from commercial suppliers. Working behind a screen can also minimize the risk of contamination of the face and eyes when dispensing from stock solutions and when opening newly arrived transport containers. In the case of the latter, leakage of solution should always be assumed to have occurred until wipe tests have verified that this is not the case.

Apart from splashes on the skin, calculations and measurements show that the manipulation of ^{32}P in thin-walled plastic microcentrifuge tubes is the most critical phase of the experimental protocols. In this case dose reduction can be achieved remarkably well by only handling the tubes when they are located within 1 cm thick Perspex blocks (MacGregor blocks).

When tubes are removed from the shield blocks they must never be handled directly with the fingers. Instead, researchers should use remote handling forceps, 'Cee-Vee' type reachers (22) or similar devices (*Figure 2*). Screw-topped microcentrifuge tubes are much easier to open and close using remote handling tools than the 'flip-top' type. The latter are more likely to lead to droplets of contamination being ejected onto gloves and working surfaces.

A programme of fingertip dosimetry needs to be in operation to check the

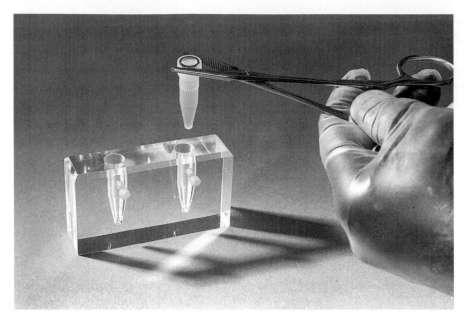

Figure 2 A Perspex block should be used to hold and shield a radioactive source.

effectiveness of the protection procedures. Flexible lithium fluoride (LiF) dosi-meters, either enclosed in pouches in the index fingers of two-way stretch nylon gloves, or directly held over the inner face of the fingertip, give a reasonably accurate measurement of skin dose if microcentrifuge tubes are inadvertently handled. If ring dosimeters or wrist dosimeters are used, doses to the finger and thumb holding the tubes are likely to be underestimated by a factor of 40. This is poor radiological protection practice. At the University of Sussex, before MacGregor blocks and handling tools were used, fingertip doses ranged from 5 to 34 mSv/month. Once tubes were held in the hand in shield blocks, and when remote handling devices were used, doses fell to <0.5 mSv/month. In recent years this has fallen to an average of 22 µSv/month. The dosimetry results also support the decision not to designate the researchers as 'classified persons'; no medical examinations are therefore required. To support systems of work aimed at keeping doses as low as is reasonably practicable, Regulation 8(3) of the Ionis-ing Radiations Regulations 1999 requires employers to establish locally enforced dose constraints, each of which would trigger a detailed investigation of the cause. It is suggested that a dose rate of 1000 µSv per calendar quarter may be an appropriate constraint for hand doses arising from work with ^{32}P.

8.2.4 Further strategies for reducing external radiation dose for DNA labelling/sequencing

i. Changing the radioactive marker

It may be possible to use the more expensive ^{33}P labelled nucleotides instead of ^{32}P. Nucleotides labelled with ^{35}S may also be an alternative. Both emit much lower energy beta radiation (^{35}S, 167.5 keV; ^{33}P, 249 keV) than ^{32}P (maximum energy 1.71 MeV) (23). By using ^{35}S or ^{33}P, the hazard from high-energy beta radiation is eliminated. An additional advantage is that equivalent doses to bone from any uptake into the body would be significantly lower following an uptake of ^{33}P. Dose coefficients for the ingestion of ^{32}P and ^{33}P (*Table 3*) indicate the much lower risk to health from an uptake of ^{33}P.

For DNA sequencing ^{35}S is excellent. Although it can be used for alpha labelling of nucleotides, –OH groups are replaced by –SH groups, so the group may not be recognized, and hence ^{35}S is unsuitable for gamma (end) labelling (see Chapter 6). However, ^{33}P-labelled nucleotides are natural species and are perfect for end labelling (R. Roberts, Nycomed Amersham plc, personal communication). Both ^{35}S and ^{33}P produce superb autoradiographs with better resolution than is possible with ^{32}P. To obtain sufficient detail, ^{35}S autoradiographs can take ≤48 h, compared with only 7 h for ^{33}P (24).

ii. Using non-radioactive markers

For labelling DNA of <200 base pairs (bp), the use of ^{32}P is still the method of choice where the greatest sensitivity (0.01 pg) is required. For >200 bp, ECL detection systems give reasonable sensitivity (0.5 pg) with the added advantages of prolonged storage time and speed of detection. Where a 2 day exposure might be needed for ^{32}P, the same quality of autoradiograph could be obtained in 1 h using ECL (R. Roberts, Nycomed Amersham plc, personal communication). Typical exposure times are 10 h for ^{32}P and 10 min for ECL. This can represent an enormous gain in research productivity. In addition, whereas the physical half-lives of ^{32}P and ^{33}P are 14.27 days and 25.4 days respectively (23), ECL probes can be stored for ≤12 months. For ^{32}P the original probe has to be stripped off. With ECL it is possible to re-probe repeatedly. As well as needing less time to obtain a satisfactory signal, the non-radioactive labelling systems also provide much shorter experimental protocols. Whereas radioactive protocols can take 8–16 h, fluorescence labelling protocols are usually 8 h, whilst the protocols using dioxetane can be completed in 2–3 h (R. Roberts, Nycomed Amersham plc, personal communication). The 'Alkaline Phosphatase Direct' system recently introduced by Nycomed Amersham plc reduces experimental protocols to just 50 min. The use of these highly sensitive non-radioactive labelling techniques represents the ultimate in radiological protection. In recent years the cost of radioactive waste disposal has become a major component of the research budget of many institutions. The switch to non-radioactive labelling therefore has an additional economic benefit as well as minimizing risks to the health of thousands of research workers.

9 Risk assessments for gamma-emitting radionuclides

Researchers can use one of the simple formulae which are available (22,23) to make rapid assessments of the risks to health from gamma-emitting radioisotopes used in biological research (e.g. [86]Rb, [42]K, [22]Na, used in plant physiology; ^{51}Cr, ^{59}Fe, [57]Co, used in metabolic studies). It is important to remember that many gamma-emitting radioisotopes also emit high-energy beta radiation. Following the risk assessment procedure for ^{32}P described in Section 8, and applying similar preventive and protective measures, will provide the necessary degree of protection from risks to health from the beta radiation emitted by the beta/gamma-emitting radioisotopes.

Biological research workers will usually require a rapid estimate to be made of the gamma dose rate at various distances from their stock solutions or from any concentrated sources (e.g. harvested plant material or collected samples from animal tissues). For these rapid assessment calculations it is usually possible to treat the radioactive material as a point source.

The following formula gives the approximate equivalent gamma dose rate at any distance, d metres from the source:

$$\text{dose rate } (\mu Sv/h) \approx \frac{135 \text{ E GBq}}{d^2}$$

where E is the total gamma energy per disintegration.

Use of the inverse square law formula, $R_1(d_1^2) = R_2(d_2^2)$ (where R_1 is the dose rate at distance d_1), allows dose rates to be calculated at any distance from the source. To provide more accurate equivalent dose rates it may be possible to use measured specific gamma ray constant values ($\mu Sv/h$ in air at 1 m from 1 MBq) which can be found in *The Radiochemical Manual* (23) or similar source books.

Taking [22]Na as an example, and assuming a plant physiology researcher wishes to dispense from a 7.4 MBq stock solution, the dose rate at 1 m quoted by *The Radiochemical Manual* is 0.3 $\mu Sv/h$ per MBq. At 1 m, 7.4 MBq would give 2.22 $\mu Sv/h$. At 10 cm this dose rate would increase to 222 $\mu Sv/h$ and, if the vial were handled (1 cm) it would give fingertip dose rates in excess of 22 mSv/h. It is immediately clear from these rapid calculations that the vial must only be held using a remote handling device, never in the fingers. Pipetting of samples should be done using a shielded pipette, and stock solutions should only be stored behind lead bricks. Information on thicknesses of shielding materials for a range of gamma photon energies are given in *The Radiochemical Manual* (23) and the *Handbook of Radiological Protection* (25).

In addition to keeping the amount of radioactive material as small as practicable and using shielding, gamma dose rates can be further reduced by keeping as far away as practicable from stock solutions and labelled experimental material, and by reducing exposure times by careful planning of the work. Depending upon the gamma dose rates, it may be necessary for individual researchers to be supplied with personal (whole body) dosimeters, especially where it is impracticable to shield large numbers of experimental plants or animals. Where radioactive animals are to be released into the environment, the risk assessments will need to foresee virtually every eventuality. For example, if radioactive badgers are released, consideration should be given to calculating doses to a child who may pick up and carry home a dead animal. In this example the risks to health of children accidentally eating sandwiches contaminated by radioactive badger faeces may also need to be calculated before the project can be approved by the Environment Agency!

10 Risk assessments for environmental protection

In very small establishments, individual researchers rather than professional health physicists may be required to deal with enforcing officers from the Environment Agency and to negotiate the amounts of radioactivity which may be discharged into the environment as radioactive waste. In order to comply with the UK environmental protection legislation (the Radioactive Substances Act 1993) a Certificate of Authorization for the storage and disposal of radioactive waste has to be obtained. This necessitates an examination of the various routes which may be available for disposal of the waste. Solid waste might be stored several years until it has decayed, an excellent choice for ^{32}P, ^{33}P, ^{35}S or ^{125}I. This route has the advantage of having no impact on the environment whatsoever and no doses to any members of the public. The only problem is the cost of the storage facility, which also needs to be secure. Low-level waste may be incinerated or buried at a landfill site. The former releases radioactive gases into the air; leachates from the latter may contaminate ground water. Incineration may also produce radioactive ash which then goes into a landfill site, a costly, energy-consuming procedure which may have no advantage whatsoever over direct landfill disposal. Aqueous soluble waste, including biodegradable liquid scintillation solutions, may be discharged to foul drainage systems and thence to river or sea. Non-aqueous flammable solvents are best incinerated. Volatile radioactive waste may be discharged to the atmosphere from the above-roof fume cupboard effluent stacks. For researchers given responsibility for making environmental risk assessments, the UK Association of University Radiation Protection Officers (AURPO) has produced helpful guidance (26). Excellent guidance notes also accompany the form RSA1 issued by the Environment Agency, and which has to be completed when applying for the authorization to discharge the waste. For the amounts of radioactivity handled by most research and teaching establishments, these environmental impact assessments can be made on the basis of simple worst-case calculations.

Taking each load delivered to the landfill, and assuming that the gamma emitters present are a point source 2 m below the surface, the annual dose to any site operators can be calculated (assuming they spend no more than 5 min each day near the buried waste). To determine the dose to anyone drinking 'run-off' water from a stream near the site, it can be assumed that all the rain in a year leaches all the deposited radioisotopes in that year into the water reaching local streams or ditches. Assuming a person drinks 0.8 m^3 of the water each year and that all the dissolved radioactivity is retained in the body, then, for each radionuclide, an annual committed equivalent dose can be determined from the dose coefficients (Sv Bq^{-1}) in ICRP publication 68 (15).

An assessment of the doses to persons catching and eating fish from a local river can be made by first assuming all the radioisotopes deposited in 1 year are leached into the river. The annual volume flow of the river gives the radioactive concentration. Research literature will give the concentration factor, water to fish, and an estimate of typical annual quantities of fish caught and eaten by riverbank fishermen (the critical group) will provide an estimate of the activity ingested by each member of the group. Applying the appropriate dose coefficient from ICRP publication 68 will yield the annual doses received. Similar calculations will give the doses to persons catching fish where a foul sewer carrying radioactive waste discharges into either the river or the sea.

Assessments of doses to persons downwind of incinerators or roof discharge points from fume cupboards require the calculation of downwind concentrations at various distances from the point at which the discharge plume reaches the point of interest (27). Taking an incinerator as an example and knowing the amount of radioisotope (e.g. ^{35}S) in the feed drums and the burn time (e.g. 1.5 h), the release rate in Bq/sec can be determined. Knowing that 2000 h at the derived air concentration (Bq/m^3) gives the annual average effective dose limit (20 mSv), 2000 h at the downwind concentration will normally be found to result in a committed effective dose of 10–30 μSv. This figure is reduced by a factor of ~1000 when the exposure times are taken into account (e.g. assuming a person spends 10 min each day standing in the plume, and the incinerator is only operated 20 days each year).

Before the Environment Agency will grant a Certificate of Authorization for any waste storage or discharge, they will require full details of the environmental risk assessment calculations. Their stated aim is to ensure that annual doses to members of the public are <20 μSv. It is expected that this environmental objective would be easily achieved by most biological research establishments.

11 Risks to the health of biological research workers in relation to other risks

It is clear from the excellent review by Rees (28) that there are significant risks to health if researchers pay too much attention to getting research results and not enough attention to radiological protection. Rees sees the most significant risks arising from a failure to carry out Good Radiochemical Laboratory Practice, especially if they fail to control radioactive contamination by not monitoring their work surfaces and themselves throughout every working day. In this case researchers may suffer high effective doses due to uptake of contamination. Intakes may not be detected due to the short half-life of the isotope used. The risk of high skin doses also must never be overlooked and in this chapter emphasis has been made on the need to avoid handling radioactive solutions in microcentrifuge tubes and to check the effectiveness of the safe systems of work being followed by the use of fingertip dosimeters. The use of ring-type dosimeters may seriously underestimate the skin dose by a factor of 40 and thereby give a false sense of security.

In comparing the risks of death from ionizing radiation with the risk of deaths due to physical injuries or the toxic effects of chemicals, the HSE (29) noted that, whereas deaths due to releases of toxic chemicals, explosions or transport accidents, or failures of hydroelectric dams tend to be instantaneous, deaths due to ionizing radiation usually occur several years after the irradiation has occurred. These late deaths, due to cancer induction following either a protracted or a short period of exposure, represent the loss of ~15 years of life. Deaths due to vehicle accidents and other accidental causes result in the loss of ~35 years. In the following discussions this difference has been ignored, and hence the risks of death from radiation have been overestimated.

Whilst recognizing the disparity between late deaths due to radiation and short-term deaths from transportation or workplace accidents, the HSE (29) have provided information (*Table 6*) which allows comparisons to be made between the annual risks of death from ionizing radiation and those from a range of other work activities.

The National Radiological Protection Board, in a 1993 review of radiation exposure of the UK population (30), have enabled further comparisons to be made (*Table 7*) between the annual doses received by research laboratory workers and other groups of workers exposed to ionizing radiation. Examination of both tables demonstrates the very low health risks from the use of radioisotopes in research laboratories.

It must be remembered, however, that the risks shown in *Table 3* represent probabilities, and laboratory workers should never lose sight of the fact that unless they work to very high standards of radiation protection they face the real possibility of high committed doses from internally deposited radioisotopes. In

Table 6 UK annual risks of death for 1989 (29)

Activity	Risk as annual experiences	Deaths per million
All causes, population average	1 in 87	11490
All cancer deaths, population average	1 in 374	2880
Deep sea fishing	1 in 750	1340
Extraction of oil and gas	1 in 990	1011
Factory and office workers receiving 15 mSv year^{-1} (from radon)	1 in 1333	750
Mining	1 in 3900	254
Road accidents, population average	1 in 10204	98
Construction work	1 in 10200	98
Agriculture	1 in 10500	74
Nuclear power station workers (1 mSv year^{-1})	1 in 20000	50
All manufacturing	1 in 53000	19
Service industries	1 in 150000	6.6
Research, education (0.1 mSv year^{-1})[a]	1 in 200000	5
Death by lightning	1 in 10^7	0.1

[a] From reference 30.

Table 7 Comparisons between annual doses received by university researchers and other groups exposed to ionizing radiation (30)

Group exposed	Average annual dose (μSv)
2500 workers in buildings with high concentrations of radon gas	15 000
Miners	
non-coal mines	4500
deep coal mines	600
Air crew (4 μSv h^{-1})	2000
Industrial radiographers	800
Medical, dental and veterinary staff	100
Research/university staff and postgraduates[a]	100

[a] Excluding skin/extremity doses.

the case of ^{32}P, unless the external beta radiation is strictly controlled, annual dose limits to the skin may be exceeded. Failure to monitor external surfaces of the body might also result in spectacularly high doses to the skin. Poor standards of radiation or contamination control not only increase the risk to health but are likely to result in enforcement action being taken against either the establishment or the individual supervising the research.

References

1. *Health and Safety at Work etc. Act 1974*, Chapter 37. Her Majesty's Stationery Office, London.
2. Health and Safety Commission. (1999). *The Management of Health and Safety at Work Regulations 1999.* Her Majesty's Stationery Office, London
3. *Health and Safety, The Provision and Use of Work Equipment Regulations 1998.* SI 1998, no. 2806. Her Majesty's Stationery Office, London.
4. Health and Safety Executive. (1992). *Personal Protective Equipment at Work, Personal Protective Equipment at Work Regulations 1992, Guidance on Regulations.* L25. Her Majesty's Stationery Office, London.
5. Health and Safety Commission. (1992). *Workplace Health, Safety and Welfare, Workplace (Health, Safety and Welfare) Regulations 1992, Approved Code of Practice and Guidance.* L24. Her Majesty's Stationery Office, London.
6. Health and Safety Executive. (1992). *Manual Handling, Manual Handling Operations Regulations 1992, Guidance on Regulations.* L23. Her Majesty's Stationery Office, London.
7. *Health and Safety, The Ionising Radiations Regulations 1999.* SI 1999 no. 3232. Her Majesty's Stationery Office, London.
8. Ionising Radiations Regulations 1999, Approved Code of Practice, Draft. http://www.hse.gov.uk/hthdir/irracop.htm
9. Health and Safety Commission. (1995). *General COSHH ACOP (Control of Substances Hazardous to Health) and Carcinogens ACOP (Control of carcinogenic substances) and Biological Agents ACOP (Control of biological agents), Control of Substances Hazardous to Health Regulations 1994, Approved Codes of Practice.* L5, HSE Books, Sudbury, Suffolk.
10. *Radioactive Substances Act 1993*, Chapter 12. Her Majesty's Stationery Office, London.
11. Health and Safety Executive. (1989). *Memorandum of Guidance on the Electricity at Work Regulations 1989.* Health and Safety series booklet HS(R)25. Her Majesty's Stationery Office, London.

12. *Atomic Energy and Radioactive Substances, The Radioactive Material (Road Transport)(Great Britain) Regulations 1996.* SI 1996 no. 1350. Her Majesty's Stationery Office, London.

13. Mordell, L. (1995). Court room drama turns to farce. *Chemistry in Britain*, August 1995, 591.

14. Health and Safety Executive. (1993). *Revised ACGM Guidelines on the Risk Assessment of Operations Involving the Contained Use of Genetically Modified Micro-organisms.* New ACGM/HSE/DOE, Note 7. Health and Safety Executive Advisory Committee on Genetic Modification Secretariat, London.

15. International Commission on Radiological Protection. (1994). *Dose Coefficients for Intakes of Radionuclides by Workers.* ICRP publication no. 68 [Replacement of ICRP 61]. *Annals of the ICRP*, Vol. 24, no. 4. Pergamon Press, Oxford.

16. Health and Safety Executive. (1998). *5 steps to risk assessment, Case Studies.* HSG183. HSE Books, Sudbury, Suffolk.

17. International Commission on Radiological Protection. (1991) *1990 Recommendations of the International Commission on Radiological Protection.* ICRP Publication no. 60. Annals of the ICRP, Vol. 21, nos 1–3. Pergamon Press, Oxford.

18. Connor, K.J. and McLintock, I.S. (1994). *Radiation Protection Handbook for Laboratory Workers.* HHSC Handbook no. 14. H and H Scientific Consultants Ltd, Leeds.

19. Scott-Wood, G. (1997). *A Practical Approach to the Use of Radiation in Molecular Biology.* HHSC Handbook No. 22. H and H Scientific Consultants Ltd, Leeds.

20. Prime, D. (1985). *Health Physics Aspects of Radioiodines.* (Association of University Radiation Protection Officers), Occupational Hygiene Monograph no. 13 (Science Reviews). H and H Scientific Consultants Ltd, Leeds.

21. Special Session—P-32 Incidents at MIT and NIH, report of oral presentations at the Forty-First Annual Meeting of the Health Physics Society, July 1996. *Health Physics*, **70**(6), Suppl., S18–19. 1996.

22. Ballance, P.E., Day, L.R. and Morgan, J. (1992). *Phosphorus-32: Practical Radiation Protection.* H and H Scientific Consultants, Leeds.

23. Longworth, G. (ed.) (1998). *The Radiochemical Manual.* AE Technology plc, Harwell, Oxfordshire.

24. Zagursky, R.J., Conway, P.S. and Kashdan, M.A. (1991). *Biotechniques*, **11**, 36.

25. Association of University Radiation Protection Officers. (1990). *Handbook of Radiological Protection, Part I: Data.* Reprinted in 1990 for members of the Association of University Radiation Protection Officers from the original published by HMSO in 1971.

26. Baker, J. (ed.) (1990). *AURPO Guidance Note No. 3, Completion of RSA3 Forms.* Association of University Radiation Protection Officers, UK.

27. Bryant, P.M. (1964). *Methods of Estimation of the Dispersion of Windborne Material and Data to Assist in their Application.* AHSB (RP) R42. Authority Health and Safety Branch, AERE Harwell, Oxfordshire. [British Library Document Supply Centre.]

28. Rees, B. (1996). *Health Physics*, **70**, 639.

29. Health and Safety Executive. (1988). *The Tolerability of Risk from Nuclear Power Stations.* Her Majesty's Stationery Office, London.

30. Hughes, J.S. and O'Riordan, M.C. (1993). *Radiation Exposure of the UK Population—1993 Review.* NRPB-R263. National Radiological Protection Board, Chilton. Her Majesty's Stationery Office, London.

Chapter 4

Radioisotope detection using X-ray film

RON A. LASKEY

MRC Cancer Cell Unit, Hutchison MRC Research Centre, Hills Road, Cambridge CB2 2XZ, UK

1 Introduction

During the last three decades gel electrophoresis has steadily replaced other methods for analytical fractionation of macromolecules. This chapter describes and discusses methods of detecting radioisotopes within flat samples such as acrylamide or agarose gels, nitrocellulose filters, thin layer chromatography plates and paper chromatograms. For all of these samples X-ray film provides a convenient non-destructive means of isotope detection which combines high sensitivity with high resolution. Alternative storage phosphor technology is described in Chapter 5. Autoradiography of microscopy slides is described in Chapter 9.

For some applications such as detection of ^{14}C or ^{35}S on the surface of thin layer chromatograms direct autoradiography on X-ray film is ideally matched to the radioactive emissions. However, for the great majority of combinations of isotope and sample, sensitivity can be greatly improved by converting emitted β particles or γ rays to light, using either X-ray intensifying screens for high energy emitters such as ^{32}P, or organic scintillators for weak β emitters such as ^3H, ^{14}C or ^{35}S. Although sensitivity can be increased in this way, conversion of ionizing radiation to light has several compensating disadvantages. First, it decreases resolution by causing secondary scattering. Second, it requires exposure at low temperature ($-70\,^{\circ}C$) because the response of film to light is fundamentally different from its response to ionizing radiation for reasons which are explained in Section 7. Third, this different response of film to light means that, unlike direct autoradiography, methods which convert to light are not quantitative unless the film is pre-exposed to an instantaneous flash of light to by-pass a reversible stage of image formation (see Sections 6.3 and 7).

This chapter starts with a guide to selection of the most appropriate method for a range of isotopes and applications. It then describes procedures for individual methods and evaluates their merits and drawbacks. These sections are followed by several general considerations such as choice of materials and exposure conditions. Finally the principles which underlie these detection methods are summarized in the belief that this information can help the user to avoid pitfalls.

RON A. LASKEY

2 Choosing a class of detection method: autoradiography versus fluorography or intensifying screens

Table 1 summarizes the suitability of various detection methods for different combinations of radioisotope and sample type. *Table 2* summarizes their sensitivities.

Table 1 Choosing the appropriate method

Isotope	Sample	
^3H (0.0186 MeV)	Acrylamide gels Agarose gels Paper chromatograms Thin layer chromatograms Nitrocellulose filters	Fluorography (Section 5)
^{14}C or ^{35}S (0.156 or 0.167 MeV)	Acrylamide gels Agarose	Fluorography for maximum sensitivity (Section 5) or direct autoradiography for extreme resolution (Section 3)
	Thin layer chromatograms Paper chromatograms Nitrocellulose filters	Direct autoradiography (Section 3). [There may be slight enhancement by fluorography (Section 5)]
^{32}P (1.79 MeV) or any γ-emitting isotope	All types of flat sample	Intensifying screen for maximum sensitivity (Section 4) or direct autoradiography for maximum resolution (Section 3)

Fluorography or intensifying screens require 'screen-type' X-ray film exposed at −70°C (Section 6) and only pre-flashed film gives quantitative images and maximum sensitivity (Sections 6 and 7). In contrast, direct autoradiography does not require 'screen-type' film, exposure at −70°C or pre-flashed film for quantitative accuracy or sensitivity (Sections 2 and 7).

2.1 Direct autoradiography gives high resolution but limited sensitivity

For direct autoradiography an X-ray film is held as close as possible to the sample in a light-tight container and exposed at any convenient temperature (see below for details). This simple method gives optimum resolution with moderate sensitivity for isotopes with emission energies equal to or greater than ^{14}C. It produces quantitative images in which the absorbance of the film image is directly proportional to the amount of radioactivity for all film absorbances up to 1.5.

2.2 Fluorography increases sensitivity for weak β emitters

Direct autoradiography is ideally suited to unquenched emissions from ^{14}C or ^{35}S. This situation is achieved when these isotopes are on the surface of a thin layer chromatography plate. However, when these or weaker β emitters such as ^3H are located within the lattices of acrylamide or agarose gels or similar samples, then the β particles are absorbed within the sample ('quenched') and fail to reach the film.

The resolution to this problem is impregnation of the sample with an organic

64

Table 2 Sensitivities of film detection methods for commonly used radioisotopes

Isotope	Method	Detection limit d.p.m./cm² for 24 h	Relative performance compared with direct autoradiography
^{3}H (0.0186 MeV)	Direct autoradiography (Section 3)	$> 8 \times 10^6$	1
	Fluorography using PPO (Section 5)	8000	>1000
^{14}C or ^{35}S (0.156 or 0.167 MeV respectively)	Direct autoradiography (Section 3)	6000	1
	Fluorography using PPO	400	15
^{32}P (1.79 MeV)	Direct autoradiography (Section 3)	525	1
	Intensifying screen (Section 4)	50	10.5
^{125}I (0.035 MeV γ rays, plus X-rays and electrons)	Direct autoradiography (Section 3)	1600	1
	Intensifying screen (Section 4)	100	16

Data from refs. 1, 3, and 4.

scintillator. This procedure is called fluorography. Beta particles from the sample excite the scintillator producing ultraviolet light which is not absorbed by the sample, but which can escape to produce a *photographic* image on the film. Several fluorographic reagents and procedures have been described. They differ in sensitivity, image quality, preparation time, cost and convenience. Their relative merits are described in Section 5. Because a photographic image is produced, rather than an autoradiographic image, films must be exposed at $-70\,°$C. They must also be pre-exposed if quantitative images are required, or if extreme sensitivity is required in long exposures (Sections 6.3 and 7).

2.3 Intensifying screens increase sensitivity for high energy β particles or γ rays (1)

Gamma rays or high energy β particles such as those from ^{32}P pass through and beyond an X-ray film, so that most of their emission energy is not recorded by the film but wasted. This problem can be overcome by placing the film between the sample and a high density fluorescent 'intensifying screen'. Emissions which pass through and beyond the film excite the screen causing it to emit UV light which superimposes a photographic image over the autoradiographic image. As with fluorography, the production of a *photographic* image from the very low intensities of light involved requires exposure at $-70\,°$C. Furthermore, contrary to widespread misunderstandings, images obtained using intensifying screens are not quantitative or fully sensitive to small amounts of radioactivity unless the reversible stage of image formation is by-passed by pre-exposing the film to a brief flash of light (see Section 7).

Although intensifying screens greatly increase the sensitivity of detection for ^{32}P- or γ-emitting isotopes (*Table 1*) they result in decreased resolution. This is because both the primary emissions and the secondary emissions of light diverge from the source. For many purposes the slight loss of resolution is unimportant. However, for uses such as DNA sequencing by the old gel method, the loss of resolution becomes a serious disadvantage and direct autoradiography should be considered as an alternative.

3 Direct autoradiography

Direct autoradiography without scintillators or intensifying screens is the method of choice when spatial resolution is more important than absolute sensitivity. In addition there are also rare examples of circumstances when an X-ray film can capture most of the emissions from one face of the sample. For example, ^{35}S or ^{14}C emissions from a thin layer chromatography plate are essentially unquenched and escape from the surface of the plate. Therefore direct autoradiography is roughly as efficient as fluorography for these samples (2). However, this would not be true of weaker β-emitters such as ^{3}H or for more highly quenched samples such as ^{14}C or ^{35}S in polyacrylamide gels. For each of these cases fluorography is substantially more efficient (2–4).

Any type of X-ray film can be used for direct autoradiography, but maximum efficiency is obtained when a 'direct' type of film (as opposed to a 'screen-type' film) is used. 'Direct' film types such as Kodak BioMax MR have a high silver halide content to maximize their absorption efficiency. In the case of ^{3}H, self-absorption within the sample is a very severe problem, but provided that this is not excessive, then direct autoradiography of ^{3}H can be performed using a special purpose film such as Hyperfilm-^{3}H (Amersham Pharmacia Biotech) which lacks the 'anti-scratch' plastic coating used on most films.

Essentially the same procedure for direct autoradiography is applicable to all kinds of flat, dry sample, though the choice of film for optimal sensitivity varies as described above and in Section 6. For wet samples such as acrylamide gels, ^{32}P can be detected efficiently without drying, but resolution is increased by drying, because the thickness of the sample determines the image resolution. However, in the case of weaker emissions such as ^{14}C, ^{35}S, or ^{3}H, drying greatly increases sensitivity as well as resolution, by decreasing self-absorption within the sample. Procedures for drying gels are discussed in Section 6.4.

3.1 Procedure for direct autoradiography

The sample to be exposed should be dry, though in the case of ^{32}P wet samples can also be exposed at the expense of resolution. In the case of ^{32}P the sample may be covered by thin plastic film such as Saran or Clingfilm, but this should not be used for direct autoradiography of weaker β emitters such as ^{14}C, ^{35}S or ^{3}H as they are effectively absorbed by it. To enable the exposed film to be oriented correctly, mark opposite corners of the sample with different symbols using 'radioactive ink'. This is usually ink to which waste radioisotope has been added. For this purpose isotopes with long half-lives (e.g. ^{14}C) are most convenient but others are satisfactory. After exposure orient and align the film correctly by placing the exposed symbols on the film over the ink marks on the sample. Remove a sheet of X-ray film (selected as described above) from the box and place it over the sample in a dark room with minimum intensity dark red safe lights. Then firmly clamp it to the sample to achieve the closest possible contact. The most convenient way to achieve this is by using a medical X-ray cassette obtainable from most medical X-ray film manufacturers. A cheaper but less convenient alternative is to clamp the film and sample between glass plates using spring clips and to expose in a light-tight envelope.

Films can be exposed at ambient temperature for direct autoradiography. There is no gain in sensitivity by exposing at lower temperatures. Nor is there any advantage in pre-exposing the film. Exposure at $-70\,°C$ can have advantages when wet samples such as acrylamide gels are exposed since it can prevent diffusion of solute molecules. When ^{32}P or γ emitting isotopes such as ^{125}I or ^{131}I are used, care should be taken to keep samples and exposures away from film stocks to avoid fogging. Similarly, stacking exposures together can lead to 'phantom' images from adjacent samples. One way of minimizing this problem is to use X-ray intensifying screens as barriers either between cassettes or inside cassettes at ambient temperature. If exposed at ambient temperature, the screen will not contribute to the film image and so will not decrease the resolution.

After exposure remove and process the films in a dark room following the manufacturer's instructions. Most types of X-ray film can be processed either by hand or using an automatic film processor. However, films which lack protective anti-scratch layers such as Hyperfilm-^{3}H are not suitable for automatic processing but should only be processed by hand.

4 Use of X-ray intensifying screens to increase sensitivity for ^{32}P and γ-ray emitters

High-energy β particles such as those emitted by ^{32}P, or γ rays such as those emitted by ^{125}I or ^{131}I, are too penetrative to be absorbed efficiently by X-ray film. Consequently, most of their energy is wasted during direct autoradiography as it passes through and beyond the film.

This problem can be overcome by converting any emissions which pass beyond the film into light. The light produced passes back through the film, superimposing a photographic image over the autoradiographic image. This conversion is achieved routinely in medical radiography by placing a high density fluorescent screen behind the film. Calcium tungstate is used widely though other fluors are discussed below. Radiation which passes beyond the film is absorbed

a
10 hours
without
screen

b
1 hour
without
screen

c
1 hour with
screen −70°C
(pre-flashed)

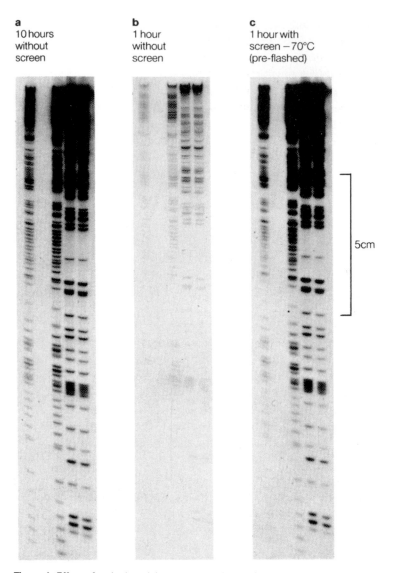

5cm

Figure 1 Effect of a single calcium tungstate intensifying screen on sensitivity and resolution of [32]P. Note that the screen increases sensitivity by approximately tenfold (compare b with c) but decreases resolution (compare c with a). All exposures were recorded on Fuji RX film. The gel containing ^{32}P-labelled DNA fragments was kindly provided by Dr S. E. Kearsey.

more efficiently by the dense screen, producing light which generates the secondary photographic image.

Although this procedure is used routinely in medical radiography, at least one additional step is required for its use in the longer exposures required for radio-isotope detection (1). This arises because the initial step of photographic image formation by light is reversible as explained in Section 7. However, this problem can be partly overcome by exposing the film at −70°C (1,5) and completely overcome by combining low temperature exposure with pre-exposure of the film to a brief flash of light to bypass the reversible stage of image formation. This latter step is essential if image quantitation is required (1,6).

Figure 1 shows that the advantage of using X-ray intensifying screens is increased sensitivity, but that there is a disadvantage of decreased resolution.

To obtain the increased sensitivity which intensifying screens offer it is necessary to use a 'screen type' X-ray film (see Section 6.1). The spectral sensitivity of these films is matched to the emission spectra of intensifying screens. Many manufacturers make suitable films such as Kodak, Fuji, Agfa, or Amersham. In general the 'fastest' (most sensitive) films are suitable, but these also give the most grainy images and they have the shortest shelf-lives. Most 'screen-type' films are sensitive to UV or blue light. However some are sensitized to green emissions from rare-earth intensifying screens. These are not recommended for use with other types of screen and there are problems with rare-earth screens as described below.

There is a wide choice of intensifying screens available (see Section 6.1), but some of the most sensitive screens are not suitable for long exposures at $-70\,^{\circ}$C, especially when pre-exposed film is used, because they contain low levels of endogenous radioisotopes which cause the film to blacken in the absence of a sample. This problem was observed for screens consisting of europium-activated barium fluorochloride or terbium-activated rare earth oxysulfides. It should be noted that this problem is not encountered when these products are used in medical radiography since much shorter exposures are used at room temperature. The problem arises because sensitivity to the light produced from trace amounts of radioactivity is far greater at $-70\,^{\circ}$C. This point is the key to the success of the intensifying screen method and is explained further in Section 7.

Protocol 1

Procedure for indirect autoradiography using intensifying screen (1)

Equipment and reagents
- Clingfilm or Saran Wrap
- X-ray film
- intensifying screen
- X-ray cassette

Method
1. Work in a dark room with dark red illumination.
2. If the sample is wet cover it with Clingfilm or Saran Wrap. Gels can be dried as described in Section 6.5.
3. Select only 'screen-type' X-ray film such as Kodak X-OMAT AR, Amersham Hyperfilm-MP, Fuji Super RX or Agfa Cronex 10T (see Section 6.1).

4 For quantitative accuracy or maximum sensitivity for small amounts of radio-activity, pre-expose the film to an instantaneous flash of light. See Section 6.3 for pre-flashing procedure and Section 7 for an explanation of its effect.

5 Mark two corners of the sample with different symbols using radioactive ink (see Section 3.1). For wet samples this can be achieved by writing on adhesive labels stuck to the Clingfilm cover.

6 Place one sheet of X-ray film (pre-flashed if required) over the sample.

7 Select an X-ray intensifying screen according to the criteria described in Sections 4 and 6.1. Clean and dry the screen if necessary.

8 Place the intensifying screen, face down, on top of the film so that the film is en-closed between the screen and the sample.

9 Clamp sample, film and screen tightly together, ideally in a medical X-ray cassette (see Section 3.1 for alternatives).

10 Expose at $-70\,^{\circ}$C. If the film has been correctly pre-flashed, exposure can also be performed at ambient room temperature, but at only 50% the efficiency obtained at $-70\,^{\circ}$C. Without pre-flashing exposure at $-70\,^{\circ}$C is essential to achieve the benefit of intensifying screens.

11 For translucent samples such as dry gels on filter paper, sensitivity can be increased further by placing a second screen outside the sample so that the order is screen 1, gel, film, screen 2 (5). Although this can increase sensitivity up to twofold, it sub-stantially decreases resolution. This point should be remembered when using a cassette which already contains two screens for a high resolution sample.

12 After exposure, develop the film according to the manufacturer's instructions.

5 Fluorography of weak β emitters

The efficiency of detection of radioisotopes which emit weak β particles (e.g. ^{3}H, ^{14}C or ^{35}S) is limited by the extent of absorption of the β particle in the sample. The extent of this problem depends both on the isotope's emission energy and on the extent to which the radioactive molecules are buried within the sample. For example, the very weak emissions from ^{3}H are essentially unable to escape from within the matrix of an acrylamide gel, and are therefore most unsuitable for direct autoradiography, whereas the stronger emissions from ^{14}C or ^{35}S can be detected efficiently by direct autoradiography of thin-layer plates (2), but not of acrylamide gels (3), because most of the emission energy is absorbed within the gel and therefore unable to reach the film.

Fluorography is designed to overcome this problem, by impregnating the sample with an organic scintillator. This is excited by β particles from the iso-topically labelled sample so that it emits UV light which can escape more freely from the sample to expose the film. Because the energy of an emitted β particle is divided into smaller quanta of light, it is necessary to expose film at $-70\,°C$ to obtain a fluorographic image (7,8). In addition, a quantitative response of the film and maximum sensitivity for small amounts of radioactivity are only obtained when the reversible stage of image formation is bypassed by pre-exposing the film to a brief flash of light (4). The physical principles underlying this effect are discussed in Section 7, together with a cautionary illustration of how easily this effect is underestimated.

The conversion of β particles to light has additional consequences. First, it makes it necessary to use a screen-type X-ray film such as Kodak X-OMAT AR, Kodak BioMax MS, Fuji Super RX or Amersham Hyperfilm-MP in order to achieve maximum sensitivity. Second, it decreases resolution because in addition to the dispersed spread of β particles, the secondary light emissions are also dispersed.

Most of the applications of fluorography concern acrylamide gels and several procedures exist to achieve this. However, fluorographic procedures have also been devised for several other types of sample such as agarose gels, thin-layer plates, paper, or nitrocellulose sheets. To simplify the choice of method the description of fluorographic procedures is divided. Section 5.1 considers general points which apply to a wide range of procedures. Section 5.2 considers the various procedures for fluorography of acrylamide gels and section 5.3 considers fluorography of other types of sample.

5.1 General technical procedures for fluorography of acrylamide gels by all methods

Protocol 2

Technical procedures for fluorography of acrylamide gels

Equipment and reagents

- 7–10% acetic acid
- ethanol or methanol
- Method described in detailed protocols

Method

1 Unstained gels can be fixed by soaking in 7–10% acetic acid, or 7% acetic acid, 10% methanol, 83% H_2O for 15–30 min. This minimizes diffusion of bands during fluoro-graphic impregnation. Diffusion of protein bands during impregnation is usually negligible, but smaller molecules such as oligonucleotides generated in DNA sequencing reactions could diffuse significantly.

2 Gels which have been stained in Coomassie blue can be processed directly for fluoro-graphy as the blue stain does not absorb fluorographic emissions too severely. However, most fluorographic procedures remove stain, so stained gels should be photographed before fluorography.

Gels which have been stained in ethidium bromide must have the stain removed before fluorography, because ethidium bromide absorbs fluorographic emissions

Protocol 2 continued

efficiently. This is easily achieved by soaking the gel in ethanol or methanol for 30 min before proceeding. Similarly, silver staining quenches fluorographic emissions, so silver should be removed, such as by immersion in photographic fixer before proceeding.

3 Acrylamide gels which have been dried must be rehydrated by soaking in water for approximately 1 h before proceeding with fluorographic impregnation.

4 After fluorographic impregnation by one of the methods described in Section 5.2, gels should ideally be dried before exposure as described in Section 6.4. Although this is not essential for fluorography, drying the gel increases both sensitivity and resolution. In addition, it prevents diffusion of labelled bands.

5 After drying, the orientation of the gel should be marked using radio-active ink (ink to which waste radioisotopes has been added), marking two corners with different symbols. This will allow the relative orientations of the film and the gel to be determined by super-position.

6 Fluorographic exposures require 'screen-type' X-ray film such as Kodak X-OmatAR, Kodak BioMax MS, Amersham Hyperfilm-MP, Fuji Super RX, etc. (see Section 6.1). There are sensitized to the wavelengths of light emitted by intensifying screens and, coincidentally, fluorographic scintillators.

7 For fluorographic exposure films should be held in close contact with the dried gel. This is achieved most conveniently by enclosing them in an X-ray film cassette. Note that it is not necessary to include an intensifying screen, but nor is it necessary to remove it if one is present.

8 Fluorographic exposures must be performed at $-70\,^{\circ}$C or below to overcome the reversible stage of latent image formation in the film. An exception is that film which has been hypersensitized by pre-exposure to a flash of light (see next paragraph and Sections 6.3 and 7) can be exposed at room temperature with approximately 50% the efficiency which would be obtained at $-70\,^{\circ}$C.

9 To achieve maximum sensitivity and quantitative accuracy of fluorography, film should be hypersensitized by pre-exposure to an instantaneous flash of light as described in Section 6.3 and explained in Section 7.

10 After exposure film should be processed in a dark room with a safelight according to the manufacturer's instructions.

5.2 Procedures for fluorography of acrylamide gels

The various methods which are available for fluorography of acrylamide gels differ in cost, convenience, time, efficiency and image quality. They are summarized in *Table 3*. One method uses sodium salicylate (9), two use the organic scintillator PPO (2,5-diphenyloxazole), and at least one of the commercial methods uses a derivative of PPO which is more soluble in water. The commercial prod-

Table 3 Methods for fluorography of acrylamide gels

	Method	Advantages	Disadvantages
1.	Commercial (e.g. Amplify[a], Enlightning[b] or Enhance)	Convenience and speed	Relatively expensive (see note concerning effect of gel drying on sensitivity in Section 5.2.1)
2.	PPO in DMSO (3)	Sensitivity. Cost. Image quality	Tedious procedure. Potential hazard
3.	PPO in acetic acid (10)	Sensitivity. Cost. Image quality	Potential hazard
4.	Sodium salicylate (9)	Sensitivity. Cost. Speed	Grainy image. Potential hazard

[a]Amplify is a registered trade mark of Amersham Pharmacia Biotech.
[b]Enlightning and Enhance are registered trade marks of New England Nuclear.

ucts are the most convenient to use, but they are also relatively expensive. In spite of commercial claims to the contrary, the author receives repeated comments from users that the original method (3) which uses a solution of PPO in dimethyl sulfoxide (DMSO) yields the best combination of sensitivity and resolution. The two disadvantages of this method are the long procedure time and the potential hazard posed by the skin-penetrating properties of DMSO. An alternative method has been described which uses glacial acetic acid in place of DMSO as the solvent for PPO (10). This offers some saving of time over DMSO, but in other respects the relative merits of these methods are essentially similar. The final method which is described below uses an aqueous solution of sodium salicylate (9).

When comparing claimed sensitivity and impregnation times of these various methods, it is important to note that time can be saved at the expense of sensitivity and that some claims have been based on comparisons with suboptimal versions of alternative methods.

5.2.1 Procedures for impregnating acrylamide gels with commercial products such as Amplify, Enlightning or Enhance

Precise details for impregnating gels with these materials are provided by the manufacturers together with the products.

The fastest of these impregnation protocols are those for Amplify (Amersham Pharmacia Biotech) and Enlightning (New England Nuclear). For these the gel is simply soaked in the product for 15–30 min and then dried under vacuum. Excess product can be re-used for further gels. Although these procedures are extremely convenient, their sensitivity can be affected by the gel drying procedure, as it is possible to suck some of the scintillant back out of the gel under vacuum. Methods which precipitate the scintillant before drying, such as the longer procedure for Enhance (New England Nuclear) or the PPO methods in *Protocols 3* and *4*, avoid this problem.

Protocol 3

Procedure for impregnating acrylamide gels with PPO using dimethyl sulfoxide (3)

Regents

- dimethyl sulfoxide (DMSO)
- 10% (u/v) ethanol in water
- 22% (w/v) 2,5-diphenyloxazole (PPO) in DMSO

Method

1   Before fluorographic impregnation, unstained gels can be fixed. Dried gels can be rehydrated as described in Section 5.1.

2   Soak the gel in approximately 20 volumes of DMSO for 30 min. Use a fume-hood and wear rubber gloves to prevent skin contact with DMSO for this and later stages. Swirl the gel gently in a sealed container such as a plastic food storage box.

3   Repeat Step 2 with fresh DMSO to remove all water from the gel.

4   Keep both stocks of DMSO separately for reuse in the same sequence with other gels.

5   Soak the gel for 3 h in 4 volumes of a 22% (w/v) solution of PPO in DMSO. Use a sealed container to prevent water absorption and agitate the gel gently. Three hours is the optimum time, but this can be decreased significantly with slight loss of resolution (see ref. 3 for details).

6   Soak the gel in water for 1 h to remove DMSO and to precipitate PPO in the gel. This step decreases drying time and ensures high resolution by forming very small crystals of PPO in the gel. In addition it prevents removal of the scintillator from the gel by suction during drying, a problem which affects efficiency of some fluorographic methods. Gels will shrink during immersion in DMSO, but re-swell during the water step.

7   Dry the gel and expose it to pre-flashed screen-type X-ray film as described in Section 5.1.

8   To recover excess PPO from solutions in DMSO, pour the solution into 3 volumes of 10% (v/v) ethanol. Filter after 10 min, wash with water and air-dry (or dry under vacuum) for reuse. Use of 10% ethanol can be replaced by water, but the PPO crystal size will be smaller, so that filtration, rinsing and drying will take longer.

9   The optimum concentration of PPO varies with the acrylamide concentration of the gel. The procedure described is optimal for gels which contain more than 10% acrylamide. 10% PPO should be used for gels which contain less than 5% acrylamide and 15% PPO should be used for gels which contain between 5% and 10% acrylamide.

Protocol 4

Procedure for impregnating acrylamide gels with PPO using acetic acid (10)

Reagents
- acetic acid
- 20% (w/v) PPO in acetic acid

Method

1 Soak the gel in undiluted acetic acid for 5 min.

2 Soak the gel in 20% (w/v) PPO in acetic acid for 1.5 h. Gentle agitation in a sealed container is recommended.

3 Soak gel in water for 30 min.

4 Dry and expose the gel to pre-flashed screen-type X-ray film at $-70\,^{\circ}$C as described in Section 5.1.

5 Recover excess PPO for reuse by addition of five volumes of water. Collect the PPO crystals by filtration, rinse and air-dry.

Protocol 5

Procedure for impregnating acrylamide gels with sodium salicylate (9)

Reagent
- 1 M sodium salicylate, pH 5–7

Method

1 Before impregnation gels may be stained or fixed as described in Section 5.1, but gels fixed in acid should be soaked in water for 30 min to prevent precipitation of salicylic acid.

2 Soak the gel in 10 vols of 1 M sodium salicylate, pH 5–7 for 30 min.

3 Dry and expose the gel to pre-flashed screen-type X-ray film at $-70\,^{\circ}$C as described in Section 5.1.

Excess salicylate solution can be reused, but prolonged storage causes brown discoloration presumably due to oxidation. To avoid this salicylate can be recovered as salicylic acid by precipitation with equimolar HCl followed by washing with water and drying overnight on a Buchner funnel with suction.

Salicylate is toxic and like DMSO it is readily absorbed through the skin. Therefore wear gloves when handling gels in salicylate. Consult local regulations concerning the handling of these reagents.

5.3 Fluorography of other samples excluding acrylamide gels

Fluorography is suitable for many other types of sample including agarose gels, thin layer plates, nitrocellulose filters or paper. Although impregnation procedures differ from those used for acrylamide gels, the principles of film choice and exposure conditions are the same as those described in Section 5.1.

5.3.1 Fluorography of agarose gels

Although the principles of fluorography are similar for agarose and acrylamide gels, there are four important additional considerations for agarose. First, agarose gels are more fragile and they break easily in response to mechanical agitation. Second, they are frequently stained with ethidium bromide which must be removed or it will quench fluorographic emissions. The ethanol (or methanol) procedures described here remove ethidium bromide from nucleic acids in the gel. Third, agarose gels melt when heated; therefore extra care is necessary when drying them. Fourth, scintillant is easily sucked out of the gel during drying. This makes it desirable to precipitate the scintillant in the gel before drying.

Procedures for fluorography of agarose gels

Agarose gels can be impregnated with scintillator using Amplify (Amersham Pharmacia Biotech), but fluorographic efficiency varies with the level of vacuum used to dry the gel as this may suck the scintillant out of the gel, and not just its solvent. Alternatively, the PPO in acetic acid method (10) described in *Protocol 4* can be used. However, in the author's experience, PPO in ethanol provides an efficient method (6) which also removes ethidium bromide which is bound to nucleic acids in the gel.

The recommended procedure (6) is as follows:

Protocol 6

Fluorography of agarose gels

Reagents

- 3% (w/v) PPO in absolute ethanol or methanol
- 100% ethanol or methanol

Method

1. Soak the gel in approximately 10 volumes of 100% ethanol or methanol (95% ethanol is unsatisfactory) in a sealed container to prevent water absorption. Agitate very gently, taking care not to break the gel.

2. Transfer the gel to fresh ethanol for a further 30 min to remove water completely.

3. Retain these two ethanol stocks in sealed bottles for re-use in the same sequence with future gels.

4. Transfer the gel to 4 volumes of a 3% w/v solution of PPO in absolute ethanol (or

Protocol 6 continued

methanol) for approximately 3 h. If the gel appears white at this stage due to pre-cipitation of PPO, return it to ethanol to remove residual water, before continuing in 3% PPO in ethanol.

5   Transfer the gel to water to precipitate PPO in the gel. This prevents suction of PPO from the gel during drying and ensures that very small crystals of PPO are formed. Large crystals reflect fluorographic emissions distorting the image.

6   Dry the gel under vacuum essentially as described in Section 6.5 except that the gel must not be overheated or it will melt before drying. This will be revealed as a very blurred image when compared to the ethidium bromide stain.

7   Expose the dried film to pre-flashed screen-type X-ray film at $-70\,^{\circ}$C as described in Section 5.1.

Any of the immersion steps can be extended overnight as nucleic acids will not diffuse during this procedure.

5.3.2 Fluorography of paper chromatograms (11)

Protocol 7

Fluorography of paper chromatograms

Equipment and reagents

- 7% (w/v) PPO in ether (or alternative solvent such as ethanol or acetone)
- ultraviolet light source

Method

1   Soak the paper in a 7% (w/v) solution of PPO in ether.

2   Dry the paper in air and expose to pre-flashed screen-type X-ray film as described in Section 5.1.

3   Uniformity of impregnation should be checked by observing under ultraviolet illumination to ensure a uniform level of fluorescence.

5.3.3 Fluorography of thin-layer chromatograms (2)

As explained earlier, fluorography of ^{14}C and ^{35}S on thin-layer chromatograms is not much more efficient than direct autoradiography, though there is a large gain in efficiency by fluorography of ^{3}H (2). For the most widely used method, simply pour a 7% (w/v) solution of PPO in ether rapidly over the dried chromato-gram (2). Alternatively, stand the dried plate in a 7% solution of PPO in acetone in a chromatography tank and allow the solution to ascend (4). To detect molecules which are soluble in ether or acetone, but not in water; stand the dried plate in Amplify (Amersham Pharmacia Biotech) and allow it to ascend the plate.

A further increase in sensitivity has been reported for the following more complex procedure (12).

Protocol 8

Fluorography of thin-layer chromatograms

Reagents

- 2-methylnaphthalene
- PPO

Method

1 Melt 2-methylnaphthalene in a water bath (melting point 34–36 °C) and add 0.4% (w/v) PPO.

2 Pour this mixture into a warm Pyrex dish on a hot-plate in a fume cupboard and immerse the dried thin-layer plate until soaked (usually less than 1 min).

3 Drain the plate and allow the methyl naphthalene to solidify before exposure.

When the plates treated by any of the above procedures are dry expose them to pre-flashed screen-type X-ray film at −70 °C as described in Section 5.1.

5.3.4 Fluorography of nitrocellulose filters (13)

Protein or nucleic acid blots on nitrocellulose filters can be revealed conveniently by fluorography. Although the method was described for ^{3}H, ^{14}C or ^{35}S (13), it is not clear from the literature how much it increases sensitivity for ^{14}C and ^{35}S. By analogy with thin-layer plates the gain for these two isotopes compared to direct autoradiography may be much less for nitrocellulose than for gels.

Protocol 9

Fluorography of nitrocellulose

Reagents

- 20% (w/v) PPO in toluene or ether

Method

1 It is essential that the filter to be impregnated must be completely dry.

2 Soak the dried filter in 20% (w/v) PPO in toluene or ether.

3 Air-dry and check impregnation is complete by uniform fluorescence under ultra-violet illumination. If fluorescence is patchy re-dry the sample thoroughly and repeat the impregnation procedure.

4 Expose to preflashed screen-type X-ray film at −70 °C as described in Section 5.1.

---

6 General technical considerations for radioisotope detection by X-ray film

The sections which follow briefly survey general points which may apply to many of the individual methods described above. For this reason they also repeat several points made earlier in the text.

6.1 Choice of films and screens

The techniques described in this chapter can all be performed with medical X-ray film, but it is important to realize that this is available in two types 'direct' and 'screen-type', neither of which is suitable for all of the methods described here.

Although direct autoradiography without scintillators or intensifying screens can be performed on most types of film, it is most efficient when 'direct' film types are used such as Kodak BioMax MR. These have high contents of silver halide which increase absorption efficiency and hence sensitivity. Furthermore direct autoradiography of ^{3}H requires film which lacks the anti-scratch plastic coating, e.g. Hyperfilm-^{3}H from Amersham.

In contrast, only screen-type film should be used to record from X-ray intensifying screens or fluorography as its absorbance spectrum is matched to the emission spectrum of intensifying screens and fortuitously to the emission spectrum of scintillators like PPO. It should be noted that some films are specifically designed to be sensitive to the green light emitted from lanthanum oxysulfide intensifying screens. We have found this type of film to be less satisfactory for radioisotope detection (1).

Within these constraints a wide choice of films is available, though it should be noted that the most highly sensitive films give the grainiest images and therefore relatively lower resolution. Kodak X-OMAT AR and Amersham Hyperfilm-MP both yield good results and are widely used with intensifying screens or fluorography. Fuji Super RX and Cronex 10T (Agfa) are also suitable.

A wide range of X-ray intensifying screens is also available from manufacturers such as Amersham, Fuji, Kodak or CAWO. In general screens made of calcium tungstate (e.g. Amersham Hyperscreen or CAWO) give the best combination of high resolution and low background for radioisotope detection. Screens containing either europium-activated barium fluorochloride or terbium-activated rare-earth oxysulfides (lanthanum, gadolinium or yttrium) may offer greater sensitivity, but with decreased resolution and greatly increased background. Although these problems are not revealed during medical radiography, for which the screens are designed, they become acute in long exposures when the reversible stage of image formation is manipulated by exposing at $-70\,^\circ$C or by pre-exposure of the film. The screens which have been most satisfactory for radioisotope detection in the author's experience were Amersham Hyperscreen or CAWO (CAWO Photo-chemische Fabrik, Schrobenhausen FRG), though Kodak BioMax MS now offers improved sensitivity over earlier screens. A further increase in sensitivity has been introduced by Kodak Transcreen technology, arranged as: sample-screen-film-reflector. Note also that Chapter 5 describes an efficient and fully quantitative alternative to X-ray films, namely storage phosphor technology.

6.2 Conditions for exposure and processing of film

All films should be handled in dark rooms with dark red safelights. Avoid exposing film to luminous clocks or dials. The shelf-life of sensitive films can be prolonged by storage at 4 °C in sealed bags. It is important to store film stocks well away from radioisotopes.

For maximum resolution and sensitivity it is important that films are clamped in the closest possible contact with the sample and with the screen when a screen is used. The most convenient means of achieving this is a medical X-ray cassette, but glass plates can be clamped together in a light-tight envelope as an alternative. Avoid using adhesive tape because its removal generates light which produces artefactual images.

There is no advantage in exposing direct autoradiographs at −70 °C. Room temperature is just as efficient. However, fluorographs and images from intensifying screens require exposure at −70 °C because the response of film to light is a multi-hit process as explained in Section 7. The exact temperature optimum has not been reported, but it appears to be between −40 °C and −90 °C and probably close to −70 °C or −80 °C. If a low temperature freezer is not available, then intensifying screens or fluorography can be performed at room temperature, provided that the film has been correctly pre-exposed as described in Section 6.3.

When cassettes are stacked in a freezer, films may acquire 'phantom' images from adjacent cassettes. This is a problem for ^{32}P and a severe problem for γ emitters such as ^{125}I or ^{131}I. However, this problem can be minimized by ensuring that there is an intensifying screen between consecutive films in the freezer as screens absorb radiation much more efficiently than film.

Before developing films which have been exposed at −70 °C, they should be allowed to warm. This prevents condensation from forming on films and causing them to stick to the rollers of processing machines.

Development procedures should follow the film manufacturer's instructions. Small bench-top film processors are widely available and extremely convenient if there are many samples to process.

6.3 Pre-exposure of film to hypersensitize it for use with intensifying screens or fluorography

Once ionizing radiation has been converted to light by a fluorographic scintillator or an intensifying screen, the kinetics of the film response become fundamentally different from the response to β particles or γ rays (see Section 7). In summary a back reaction occurs which cancels the latent image produced by earlier photons. Exposing films at −70 °C slows this back reaction. Pre-exposing the film to a flash of light bypasses the back reaction (4). Hence it increases sensitivity for *small* amounts of radioactivity, though not for large amounts, and in particular it allows quantitation of images because all photons contribute equally to the image of pre-flashed film. Pre-exposure has no effect when large amounts of radioactivity are used in short exposures or when direct autoradiography is performed without intensifying screens or scintillators (1,4).

For pre-exposure to hypersensitize a film it is essential that the flash should be single and short, in the order of 1 msec. This can be achieved by attenuating the output from a photographic flash gun. Longer flashes only increase the background fog level of film without hypersensitizing it.

Provided that the flash is of the order of 1 msec, then the increase in background fog level can be used as a convenient index to monitor hypersensitivity. The background absorbance should be increased to between 0.15 and 0.2 (A_{540}) above the absorbance of untreated (but developed) film.

The intensity of the flash from a photographic flash gun can be attenuated by wavelength filtration. Thus orange filters (Wratten numbers 21 or 22) taped to the units decrease the output to approximately the correct level. Further adjustments can be made by adding neutral density filters, or by varying the aperture in an opaque mask or by varying the distance of the flash unit from the film. Trial exposures can be made on a single film by changing the position on the film of a clear window in an opaque mask. The fog levels achieved can be measured using a densitometer or by placing pieces of the film in a spectrophotometer.

It is important to note that the film will only yield a linear response to the amount of radioactivity when the fog level has been raised between 0.1 and 0.2 (A_{540}) above that of untreated film. Whereas unflashed film under-represents small amounts of radioactivity, film which has been pre-flashed to densities above 0.2 over-represents small amounts of radioactivity (*Figure 2*).

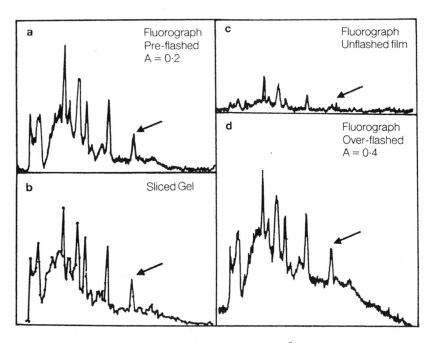

Figure 2 Effects of varied pre-exposure levels on images of ^{3}H distribution obtained by fluorography. Note that the arrowed peak is lost when unflashed film is used and exaggerated when overflashed film is used. (Data from ref. 3.)

6.4 Procedures for drying acrylamide or agarose gels

It is not essential to dry gels before exposing them to X-ray film for any of the methods described in this chapter. Nevertheless, drying the gel increases both sensitivity and resolution. In addition, it prevents diffusion of labelled bands. Commercial gel dryers are available from several manufacturers (such as Bio-Rad). They are designed to heat the gel on a porous polythene support pad, under vacuum. The gel should be placed on a sheet of filter paper such as Whatman 3mm before placing it on the gel dryer. It should then be covered with cling film before drying.

Agarose gels will melt if they are heated excessively before most of their water has been removed; therefore heating should be delayed when drying agarose gels.

If acrylamide gels crack during drying it may help to lower the bisacrylamide concentration for future gels towards the ratio:

$$\% \text{ bisacrylamide} = \frac{1.3}{\% \text{ acrylamide}}$$

Gels made by this formula do not crack when dried under continuous vacuum (14), but 20% acrylamide is required to give good protein resolution with this formula.

7 The underlying principles of radioisotope detection by X-ray film

Photographic emulsions are composed of silver halide crystals (grains of the film) each of which behaves independently. To produce a developable image each silver halide crystal requires several photons of visible light ($\sim$5 in average emulsions), each of which produces an atom of metallic silver. These then catalyse the reduction of the entire silver halide crystal by the developer.

A single hit by a β particle or γ ray can produce hundreds of silver atoms rendering the grain fully developable. Hence direct autoradiography is a linear 'single hit' process in which all emissions are recorded equally until the film is saturated.

However, once the ionizing radiation is converted to multiple photons of light by an intensifying screen or fluorographic scintillant, the response of film is fundamentally different. Each photon produces only a single atom of silver. Although two or more silver atoms in a silver halide crystal are stable, a single silver atom is unstable and it reverts to a silver ion with a half-life of about 1 sec at room temperature. We have suggested previously (1,4,15) that this is the reason why exposure at $-70\,^{\circ}$C is necessary for the low light intensities produced by fluorography and intensifying screens. Lowering the temperature slows the thermal reversion of the single silver atom, increasing the time available to capture a second photon and thus produce a stable pair of silver atoms.

The probability of a second photon being captured by a grain before the first silver atom has reverted is greater for large amounts of radioactivity, and hence higher photon flux, than for small amounts. Hence small amounts of radio-activity are under-represented for both fluorography and intensifying screens, even when exposed at $-70\,^{\circ}$C (*Figure 2*). Pre-exposing film to an instantaneous flash of light overcomes this problem because it provides many of the grains of the film with a stable pair of silver atoms. Thereafter each photon which arrives has an equal chance of contributing to the growth of the latent image. Consequently, correctly pre-flashed film responds linearly to the amount of radio-activity from intensifying screens and fluorographs. For the same reason pre-exposure largely (but not completely) bypasses the need to expose at $-70\,^{\circ}$C.

There is confusion in the literature over the need to pre-flash in order to obtain maximum sensitivity from intensifying screens. This has arisen because the effect of pre-exposure is negligible when tested using large amounts of radioactivity in short exposures. Only *small* amounts of radioactivity are under-represented when radiation is converted to light and there is no effect for large amounts. The practical consequence of this confusion is that serious errors arise when faint bands are compared with dark bands in long exposures of 1 week or more using intensifying screens with untreated film. Under these conditions errors of eight-fold have been observed, using ^{32}P and intensifying screens for peak comparisons on unflashed film. Hence the purpose of this final section is to stress the importance of understanding the underlying principles to ensure success with the 'practical approach'.

Acknowledgements

I am grateful to Amersham Pharmacia Biotech for permission to reproduce illustrations from Booklet 23 *Efficient Detection of Biomolecules by Autoradiography, Fluorography or Chemiluminescence* (Laskey 1993, see ref. 15) and Bronwen Harvey for reading the manuscript.

References

1. Laskey, R. A. and Mills, A. D. (1977). *FEBS Lett.* **82**, 314.
2. Randerath, K. (1970). *Anal. Biochem.* **34**, 188.
3. Bonner, W. M. and Laskey, R. A. (1974). *Eur. J. Biochem.* **46**, 83.
4. Laskey, R. A. and Mills, A. D. (1975). *Eur. J. Biochem,* **56**, 335.
5. Swanstrom, R. and Shank, P. R. (1978). *Anal. Biochem.* **86**, 184.
6. Laskey, R. A. (1980). *Methods Enzymol.* **65**, 363.
7. Luthi, U. and Waser, P. G. (1965). *Nature, Lond.* **205**, 1190.
8. Koren, J. F., Melo, T. B. and Prydz, S. (1970). *J. Chromatog.* **46**, 129.
9. Chamberlain, J. P. (1979). *Anal. Biochem.* **209**, 281.
10. Skinner, M. K. and Griswold, M. D. (1983). *Biochem. J.* **209**, 281.
11. Shine, J., Dalgarno, L. and Hunt, J. A. (1974). *Anal. Biochem.* **59**, 360.
12. Bonner, W. M. and Stedman, J. D. (1978). *Anal. Biochem,* **89**, 247.
13. Southern, E. M. (1975). *J. Mol. Biol.* **98**, 503.
14. Blatter, D. P., Garner, F., Van Slyke, K. and Bradley, A. (1972). *J. Chromatog.* **64**, 147.
15. Laskey, R. A. (1993). *Efficient Detection of Biomolecules by Autoradiography, Fluorography or Chemiluminescence.* Review Booklet 23: 2nd edn. Amersham Pharmacia Biotech.

Chapter 5
The scintillation counter

CHARLES L. DODSON
Beckman Coulter Inc., Fullerton, California, USA

1 Introduction

A radioactive atom or molecule emits small particles or photons or both. Monitoring radioactivity depends upon the detection of such emissions, counting either the number of particles or the number of photons. No method has been discovered to monitor radioactively emitted particles directly. Rather, we monitor the emitted photons or the effects produced by the particles interacting with selected matter. An instrument to measure radioactivity requires a sample, means of detection and means of recording. In experiments conducted early in the nineteenth century, radioactivity was monitored by dark-adapted human eyes counting scintillations produced by solid detectors and lead sulfide, and recorded by human hand. This chapter describes machines that produce the same end result by automation.

Numerous radionuclides exist. Many occur naturally; some, artificially. Biology is concerned with three applications of radionuclides: monitoring physiological processes, therapeutic treatment and monitoring potentially harmful environmental processes. This chapter focuses on the measurement of radionuclides used for the first application. Although the number of molecules present in living forms is very large, a major percentage is composed of a small number of different atoms. Therefore, the number of radionuclides applied to monitoring physiological processes is small and consists of the following: ^{3}H, ^{14}C, ^{35}S, ^{32}P, ^{33}P, ^{125}I, ^{131}I, ^{45}Ca and a few others.

Gamma counters monitor X- and gamma-ray emissions. Liquid scintillation counters monitor alpha and beta particle emissions. Some nuclides can be measured only with gamma counters, some with liquid scintillation counters and some with either. Before discussing these two counters, two fundamental concepts for any measurement of radioactivity are explained.

2 Counting efficiency and counting error

The purpose of measuring radioactivity is to determine either the quantity of radioactivity in a given sample or that radioactivity is present. A detector capable of monitoring each nuclear emission would be ideal. For several reasons all

Table 1 Notation

Type of disintegration	Units
Actual	disintegrations per minute (d.p.m.)
	disintegrations per second (d.p.s.)
	(1 d.p.s. = 1 Becquerel (Bq))
Measured	counts per minute (c.p.m.)
	counts per second (c.p.s.) ≡ 1 Bq

emissions are not counted experimentally. Some emissions have energy below the threshold of the detector. Some miss the detector. Some are absorbed before they arrive at the detector. Only a fraction of the original emissions is counted. Depending on the nuclide, sample preparation and detector, the fraction of emissions counted can be 1% or 99%. The true or actual number of nuclear emissions occurring is expressed as disintegrations per unit time, D, and the number measured is expressed as counts per unit time, C. Actual units are shown in *Table 1*. Becquerel is the preferred unit, but instrument manufacturers report d.p.m. and c.p.m. as well.

The fraction, f, of the true disintegrations, d.p.s., experimentally measured, c.p.s., is given by $f =$ c.p.s./d.p.s. If that fraction is converted to a percentage, then counting efficiency, E, is defined as

$$E = 100 \text{ c.p.s./d.p.s.}$$

Rearrangement of this equation in terms of d.p.s. provides

$$\text{d.p.s.} = 100 \text{ c.p.s.}/E$$

Because c.p.s. is always measured, d.p.s. can be obtained from this equation if the counting efficiency is known. The significance of counting efficiency will be discussed in detail later.

3 Counting error

Any number measured experimentally is subject to error. If the same sample is measured numerous times or numerous 'identical' samples are measured, the set of numbers obtained from all measurements provide a mean, N_M, with a standard deviation, σ. Assuming the measurement device possesses no inherent bias, the measurement is a random process described by a Gaussian or normal error distribution, illustrated in *Figure 1*. Radioactivity measurements are no different. But in addition, radioactive decay itself is a random process described by a Poisson distribution. For a reasonable number of counts, a Gaussian distribution approximates a Poisson distribution very well.

The standard deviation, σ, of N observed counts can be derived rigorously and is $\sqrt{N}$. One standard deviation, σ, corresponds to 0.683 of the area under the Gaussian distribution shown in *Figure 1*. This means that 68.3% of all possible observations are taken into account, or that the measurement of a single value has a 0.683 probability of lying in the reported σ range.

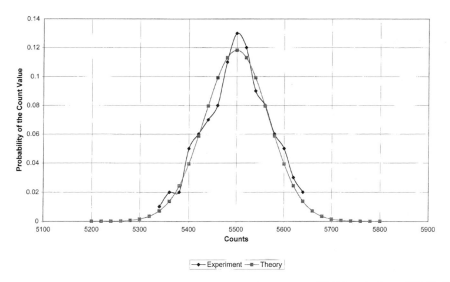

Figure 1 A Gaussian distribution with a mean count from theory of 5500 and a $2\sigma = 135$ ($2\sigma\%$ = 2.5); from 110 measurements of a sample with mean of 5510 and $2\sigma = 133$ ($2\sigma\% = 2.5$).

Reporting a measurement with a 2σ interval corresponds to 95.5% of the area under a Gaussian curve or provides a 95.5% confidence interval. Typically, radio-activity is reported with 2σ error intervals or $2/\sqrt{N}$. The result of a measurement of N_M is reported as $N \pm 2/\sqrt{N}$.

An alternative way of reporting the error range is per cent relative error or $2\sigma\%$ error defined as follows:

$$2\sigma\% - 2(100)\sqrt{N}/N = 200/\sqrt{N} = 200/\sqrt{(Ct)}$$

where C is reported as c.p.s. and t is count time in seconds. *Tables 2* and *3* illustrate these concepts with examples.

4 Origin and nature of nuclear emissions

Unstable or radioactive atoms emit a variety of particles and photons. A variety of processes underlie these emissions. *Beta decay*, β^-, is a two-particle emission

Table 2 Count times for 1 and $2\sigma\%$ confidence intervals for 500 c.p.s. sample

	Count time (s)			Counts	Counts
C.p.s.	1σ	2σ	$\pm\sigma\%$ error	(1σ) 68.3%	(2σ) 95.5%
500	500	2000	0.2	250 000	1 000 000
500	80	320	0.5	40 000	160 000
500	20	80	1.0	10 000	40 000
500	5.0	20	2.0	2500	10 000
500	0.8	3.2	5.0	400	1600
500	0.2	0.8	0.4	100	400

Table 3 1 and 2σ% error for 500 c.p.s. sample with equal count times

C.p.s.	Count time(s)	±σ% error	±2σ% error	Counts
10	30	5.8	11.5	300
100	30	1.8	3.7	3000
1000	30	0.6	1.2	30 000

process. An unstable nucleus yields an electron, β^-, and an anti-neutrino, $\bar{\nu}$, to produce a different nuclide with an atomic number one larger than the initial nuclide. No change in mass number occurs.

$$^A_Z X \rightarrow {}^A_{Z+1} Y + \beta^- + \bar{\nu}$$

For example,

$$^{14}_6 C \rightarrow {}^{14}_7 N + \beta^- + \bar{\nu}$$

Alpha decay is a one-particle emission process. An unstable nucleus emits a helium 2+ ion to produce a different nuclide with four fewer atomic mass units and two fewer protons than the initial nuclide.

$$^A_Z X \rightarrow {}^{A-4}_{Z-2} Y + {}^4_2 He^{2+}$$

For example,

$$^{241}_{95} Am \rightarrow {}^{237}_{93} Np + {}^4_2 He^{2+}$$

Electron capture refers to the nuclear capture of an electron from an inner shell orbital. Electrons from higher orbitals cascade into the vacancy and release X-ray energy. The released X-ray energy equals the difference between the initial and final electron energy states of the atom. In the nucleus a proton interacts with the captured electron to produce a neutron, thereby decreasing the atomic number of the nucleus and releasing a neutrino. Alternatively, the X-ray energy may be transferred to an extranuclear electron causing it to be released. Such electrons are called *Auger electrons*. ^{55}Fe illustrates both processes:

$^{55}_{26} Fe + e^-$ (K shell) $\rightarrow {}^{55}_{25} Mn + 6.5$ keV X-ray *or* 5.9 keV X-ray + Auger e$^-$ (90% of all emissions)

$+ e^-$ (L shell) $\rightarrow {}^{55}_{25} Mn + 0.77$ keV X-ray *or* 0.69 keV X-ray + Auger e$^-$ (10% of all emissions)

An internal conversion process involves a cooperative or coupling interaction between an excited nucleus and orbital electrons. Excess atomic energy is released by emission of a *conversion electron*, its accompanying kinetic energy and sometimes release of gamma radiation as well.

A process that leaves a nucleus in an excited state is followed by gamma emission. This includes some alpha and beta decay processes and the conversion electron process. Frequently, gamma emission occurs singularly.

Other emissions do occur. But here only alpha particles, electrons, positrons and gamma photons are of interest. Note that the names for various electron and photon emissions are intended to describe their origins. X-rays and gamma rays are photons originating from extranuclear and nuclear changes respectively. Beta particles, Auger electrons and conversion electrons are electrons originating from the nucleus, extranuclear electrons and a process involving both the nuclear and extranuclear structures of an atom. Regardless of these origins, the only qualitatively different emissions for understanding the nature of required detectors are photons and electrons. Alpha particles, He^{2+}, have a much larger specific ionization than beta particles. More ions are formed per unit pathlength. Therefore, a smaller percentage of the particle's original energy is converted into photons. A rough rule of thumb is that an alpha particle produces one-tenth as many photons as a beta particle. This means that a 5 MeV alpha particle is detected as a 0.5 MeV particle on a 2 MeV 'beta particle scale'.

5 Interaction of decay species with matter

The origin and nature of nuclear emissions are described in Chapter 2. The interactions of charged particles, electrons and alpha particles, with matter transfers energy to the primary atoms or molecules involved or to those of the products of the initial reaction. Charged particles make a 'track' as they pass through matter. A large number of molecules along the track will be excited. The track's length is determined by the size and energy of the ionizing radiation, as shown by *Figure 2*. In the solid state the mean pathlength for gamma rays, electrons and alpha particles respectively is 1–10, 0.1 and 0.001 cm. Such energy transfer may break chemical bonds and cause single or multiple ionizations or electron excitation. Ionization produces charged pairs, an electron and a cation, which forms the basis for detection in both gaseous and solid states. Electron excitation energy can be released through electronic, vibrational and rotational internal conversion processes within an atom or molecule, ultimately causing them to translate faster. A very sensitive thermometer could measure such a temperature increase. Alternatively, the excited electron may release energy by emission of a photon. Depending on the nature of the state of excitation, the released photon may be

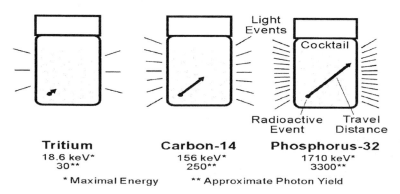

Figure 2 Particle energy and travel distance.

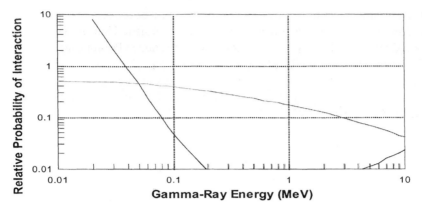

Figure 3 Relative probability of gamma-ray interaction with matter by photoelectric effect, Compton effect and pair production.

an X-ray, gamma ray or of visible or longer wavelength. Such *fluorescence* is an external release of energy and returns the excited chemical species to its electronic ground state. Emission of a photon is the energy transfer basis for detection of electrons and alpha particles with a liquid scintillation counter, the subject of this discussion. Other instrumentation is available to monitor other energy transfer processes in the gaseous and solid states.

Gamma radiation reacts with matter by three processes: photoelectric effect, Compton effect and pair production. *Figure 3* describes the relative probabilities of these three interactions and the nature of their dependence on energy. The photoelectric effect dominates at low energies, up to about 0.02 MeV. Compton and photoelectric effects occur in the 0.02–0.1 MeV range. From 0.1 to 3 MeV, the Compton effect dominates. Pair production refers to the creation of an electron and a positron, has a threshold of 1.02 MeV and is small in the organic media concerned here.

The photoelectric effect is an energy absorption process. A gamma ray is absorbed by an atom which produces an electron with kinetic energy (E_K) equal to the energy of the gamma ray (E_γ) minus the binding energy of the electron (E_B):

$$E_K = E_\gamma - E_B$$

The Compton effect is a scattering process. A photon, $h\nu_1$, incident to the scattering material produces a recoil or Compton electron and a scattered photon with energy $h\nu_2$; where h is Planck's constant and ν the frequency of the photon. A Compton electron can have a range of energies dependent upon the angles of incidence and scatter. Its maximum energy, E_{max}, is given by

$$E_{max} = \frac{2E_\gamma}{0.511 + 2E_\gamma}$$

where E_γ is the energy of the incident gamma ray.

In each of these interactions of gamma rays with matter, an electron is produced. This electron provides the detection of the gamma emitter. Recall that charged particles are detected by the photons they produce. Photons are detected by electrons, and electrons by photons.

6 The liquid scintillation counter

6.1 Introduction

As noted previously, electrons with three different histories are emitted from radionuclides: beta emission from the nucleus, Auger and conversion electrons from the extranuclear electrons. Such nuclides are referred to frequently as β^- emitters, Auger emitters and conversion electron emitters. All three types of radionuclide can be detected and counted with liquid scintillation counters (LSCs). In addition, recall that nuclides thought of primarily as gamma ray emitters also emit Auger and conversion electrons. In such cases, gamma-emitting nuclides can be detected and counted with *particle counting* instruments such as LSCs. Historically the liquid scintillation counter was developed to detect low energy beta particles. Some still refer to the LSC as a beta counter, but an LSC can detect alpha, beta and gamma emitters provided the latter also emits an electron. *Table 4* provides examples of each of these types, their corresponding energies and counting efficiencies.

Table 4 Nuclides detected by liquid scintillation counters (LSCs)

Nuclide	Emission	Energy (keV)	LSC efficiency (%)	γ efficiency (%)	Half-life
^{3}H	β	18.6	65		12.26 years
^{14}C	β	156	96		5730 years
^{32}P	β	1710	98		14.28 days
^{33}P	β	248	97		24.4 years
^{35}S	β	167	96		87.9 days
^{36}Cl	β	714	97		3.05×10^5 years
^{45}Ca	β	252	97		165 days
^{125}I	γ	35, 70		80	59.6 days
^{125}I	conversion e$^-$	5, 30	80		59.6 days
^{57}Co	e$^-$	7, 13, 115, 129	70		271.1 days
^{57}Co	γ	14, 122, 136, 692		75	271.1 days
^{22}Na	β^+	545	95		2.602 years
^{22}Na	γ	1274		55	2.602 years
^{55}Fe	Auger e$^-$	0.6, 5.2	(35?)		2.72 years
^{59}Fe	β	273, 465, 1570	95		44.51 days
^{59}Fe	γ	143, 192, 1095, 1292		25	44.51 days
^{241}Am	β	5486	100		432 years

CHARLES L. DODSON

6.2 The scintillation cocktail

A radioactive sample is presented to the LSC in a vial containing a scintillation solution or cocktail. In its simplest form, a liquid scintillation cocktail contains a solvent, S, and a molecule capable of scintillating or fluorescing, F_1. The solvent molecule has two roles: to dissolve the sample and to transfer energy to the fluorescing solute. The scintillator's role is to indicate the presence of the particle emitted from the radionuclide. Frequently, a secondary fluorescing solute, F_2, is part of the cocktail. After a beta particle is emitted from a nucleus, it travels within the scintillator solution colliding predominantly with solvent molecules because almost all the molecules in the solution are solvent molecules. Some of the kinetic energy of the beta particle is transferred to solvent molecules on each collision until the beta particle either loses its energy or is accepted by another chemical species. Energy transfers among all components in the scintillator solution as described by the six steps in *Figure 4*.

1. $\beta^- + S_1 \rightarrow \beta^- + S_1^*$
2. $S_1^* + S_2 \rightarrow S_1 + S_2^*$
3. $S_2^* + F_1 \rightarrow S_2 + F_1^*$
4. $F_1^* \rightarrow F_1 + h\nu$
5. $F_1^* + F_2 \rightarrow F_1 + F_2^*$
6. $F_2^* \rightarrow F_2 + h\nu$

Figure 4 The primary energy transfer steps within a scintillator solution.

A beta particle excites a solvent molecule's electrons, shown by an asterisk in *Figure 4*. Such excitation can cause ionization, bond breakage or excitation of internal states. An electronically excited solvent molecule most likely collides with a second solvent molecule (step 2), exciting it with the original molecule returning to the electronic ground state. An excited solvent molecule may also excite a primary fluor molecule, step 3, which can emit a photon and return to the ground state, step 4. Alternatively, the excited primary fluor molecule can excite a secondary fluor, step 5, which releases a photon, step 6. Photons produced by fluorescence are detected by photomultiplier tubes in the scintillation counter.

Let us examine the nature of an elementary scintillation cocktail in more detail. *Table 5* is a short list of the numerous molecules used as the solvent. A good solvent should have four properties:

(1) It should dissolve the samples. The radiolabelled molecule in most biological samples is organic, although the sample medium may be aqueous. Benzene and its methyl derivatives are excellent for dissolving organic samples, although other possibilities exist. Solubilizers are added to elementary cocktails to improve aqueous solubility.

(2) The emission wavelength of the solvent molecule should overlap the absorption band of the primary fluor as much as possible. Maximum overlap enhances the efficiency of the energy exchange between the two.

(3) Solvent molecules should not present a health hazard.

92

Table 5 Solvents for scintillation cocktails

Molecule	Relative scintillation	Quantum yield	Fluorescence decay time (nsec)	Emission wavelength	B.P. (°C) at 760 mm	Flashpoint (°C)
Benzene	85	0.06	29	283	80.1	
Toluene	100	0.14	34	285	111	4
p-Xylene	110	0.34	30	291	138	27
m-Xylene		0.14	31	289	139	
o-Xylene		0.16	32	289	144	
Pseudocumene[a]	112	0.33	27.2	293	169	51
PXE						156
DIN						148
Dioxane	65	0.03	2.1	247	12	12

[a] 1,2,4-Trimethyl benzene.

Abbreviations: PXE, phenylxylyethane; DIN, diisopropyl naphthalene.

(4) They should have good optical values for scintillation efficiency, quantum yield and appropriate lifetime.

Quantum yield measures the number of photons emitted for each photon-producing excitation. The best possible result is to emit one photon for each absorbed photon. Relative scintillation yield measures the number of photons emitted in a reference cocktail, with the yield for the solvent toluene arbitrarily set at 100. Fluorescence decay time measures the length of time the molecule remains in its excited state. Ideally this should be very short. 1,2,4-Trimethyl-benzene, also named pseudocumene, is considered one of the best solvents because its optical properties are excellent and its vapour pressure is very small. The latter property contributes to health safety. More recently, phenylxylyl-ethane (PXE) has become recognized as an excellent solvent because of its optical and safety characteristics, especially its high flashpoint and biodegradability.

Note that the relative scintillation of the better solvents is 12% above that of toluene, that the best quantum efficiency is about one-third, that decay times are tens of nanoseconds and the emission wavelengths are about the same.

Table 6 provides a short list of primary fluorescent molecules and their properties. *Table 7* is an analogous list of secondary fluorescent molecules. Quantum yields are high, decay times short and relative scintillation yields high. Primary and secondary solutes degrade only slightly the signal passed on by the solvent.

In summary, an electron released from an unstable nuclide collides with and excites solvent molecules which return to the ground state upon excitation of primary solute molecules which, in turn, repeats the process with secondary solute molecules. The wavelength of the emitted light is 400–425 nm. Such photons are transparent to glass and plastic vials and readily detected by photomultiplier tubes.

Routinely used cocktails contain a fourth ingredient, an emulsifier, so that aqueous samples can be dissolved. Numerous emulsifiers, cationic, anionic, non-

Table 6 Primary fluorescent molecules for scintillation cocktails

Solute	Fluorescence peak (nm)	Decay time, ι (nsec)	Quantum yield, Φ	Relative scintillation yield	Optimum concentration (g/l)
p-Terphenyl	340	1.0	0.77	101	5
PPO	370	1.4	0.83	112	3–7
PBD	370	1–1.4	0.83	155	8–10
t-Butyl-PBD	385	1.0	0.69	153	12
PBBO	397			125	4.2
BIBUQ	380	0.9	0.93	160	24

Abbreviations: PPO, 2,5 diphenyloxazole; PBD, 2-phenyl-5-(4″-biphenylyl)-1,3,4-oxadiazole; t-butyl-PBD, t-butyl-2-phenyl-5-(4″-biphenylyl)-1,3,4-oxadiazole; PBBO, 2-(4-biphenylyl)-6-phenylbenzoxazole; BIBUQ, 4-4″-bis-(2-butyloctyloxy)-p-quarterphenyl.

Table 7 Secondary fluorescent molecules for scintillation cocktails

Solute	Fluorescence peak (nm)	Decay time, ι (nsec)	Quantum yield, Φ	Optimum concentration (g/l)
POPOP	415	1.5	0.93	0.05–0.2
Dimethyl POPOP	427	1.5	0.93	0.1–0.5
α-NPO	400	2.1	0.94	0.05–0.2
bis-MSB	425	1.35	0.94	1.5
BBOT	440	1.1	0.74	1.1

Abbreviations: POPOP, 1,4-bis-(o-methyl-5-phenyloxazolyl)-benzene; dimethyl POPOP, dimethyl 1,4-bis-(o-methyl-5-phenyloxazolyl)-benzene; α-NPO, 2-(1-naphthyl)-5-phenyloxazole; bis-MSB, p-bis-(o-methylstyryl)-benzene; BBOT, 2,5-bis-2-(5-t-butyl-benzoxazolyl)-thiophene.

ionic, pure and mixed, are available optimized for very specific applications. Triton X-100 has some general use.

6.3 Beta particle spectrum

A spectrum of an emitted beta particle observed by an LSC is illustrated by *Figure 5*. Note that the spectrum is broad compared with that of a conversion electron or alpha particle because all beta particles are not emitted with the same energy. Total energy for beta decay from the same nuclide is constant, but the total is divided between beta and anti-neutrino particles. Beta particle energies are reported as the maximum energy, E_{max}, a beta particle can have. ^{14}C has an E_{max} of 156 keV, but ^{14}C beta particles are emitted over the entire 156 keV range. Each beta particle initiates the energy exchange processes described in *Figure 4*. The end result is the emission of 425 nm photons from the cocktail. Low and high energy beta particles can be distinguished because those with high energy produce more photons than those with low energy. In other words, each beta particle decay event within a scintillation solution produces a 'burst' of photons all with the same wavelength. The number of photons in a burst depends upon

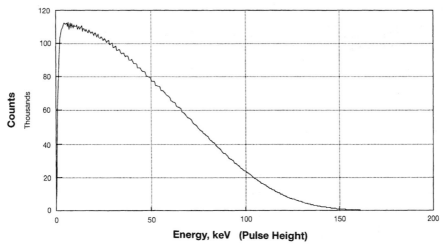

Figure 5 Liquid scintillation spectrum of ^{14}C.

the original energy of the beta particle. Each burst is complete within a few nanoseconds because each burst is produced by one electron traveling with about 90% of the speed of light. Photon bursts are also described as having a pulse height where 'height' refers to the intensity or 'brightness' of the pulse. The x-axis on an liquid scintillation spectrum measures pulse height. The y-axis counts the number of pulses observed for each pulse height. Because pulse height is proportional to beta particle energy, the x-axis may also be labelled in energy.

If an LSC were perfectly efficient, then the number of 'bursts' observed, c.p.m., would equal the number of beta particles emitted, the d.p.m., and the brightness of the burst would provide the energy of the emitted particle.

6.4 The liquid scintillation counter

Figure 6 is a schematic drawing of the major components in an LSC. First consider an overview of the system. The photomultiplier tubes (PMTs) detect photon bursts from the cocktail. Most all the remainder of the system 'processes' the detected signal. A computer serves three roles: control of the instrument, interaction with the user and output of information. Actual output can be a printer, monitor or memory storage device.

The elements of a PMT are shown in *Figure 7*. Photons absorbed by the semi-transparent photocathode provide energy sufficient to release an electron, the photoelectric effect. On average one electron is released for three photons impinging on the photocathode in current PMTs. Because of shaped electrical fields inside the PMT, electrons are focused on to the first anode which has a voltage applied to it. The applied anode voltage combined with the energy of the impinging electron cause the release of three or four electrons. One initial electron at the first anode produces three or four electrons. Amplification has occurred.

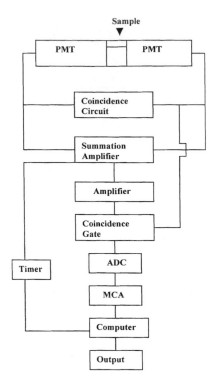

Figure 6 The major components of a liquid scintillation counter.

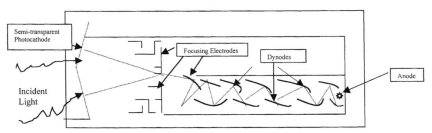

Figure 7 Elements of a photomultiplier tube.

PMTs used in LSCs contain ten to 12 dynode stages, each of which produces three or four electrons for each impinging electron. Consequently, the total amplification process produces about 10^6 electrons for each original electron ejected from the photocathode.

Notice that two PMTs are in the system. Two tubes combined with appropriate reflecting surfaces around the sample vial detect all photons leaving the vial regardless of their direction, an important property for improving observation of low activity samples and accuracy at all count levels. (The vial cap and elevator are also photon reflective.) Since all photons produced in one 'burst' event are not detected by one PMT, the total amplitude is obtained by summing the signals from both PMTs. The summed or total signal intensity is proportional to the energy of the emitted radioactive particle. A second amplifier increases the amplitude of the summed signal, which is then analysed by the coincidence circuitry. The second amplifier may provide either linear or logarithmic amplification.

6.4.1 Coincidence counting

A number of different sources contribute to background events in liquid scintillation counting. Background events produce either a single photon or many photons as discussed for nuclear decay. Two sources for single photon production are noise events within a PMT and chemical reactions, which produce chemiluminescence. Neither of these is wanted because they do not occur from the sample of interest. Such 'noise' events and a true radioactive burst can be distinguished. A single photon event is observed by only one PMT, but a photon burst contains several photons travelling towards each PMT.

Coincidence circuitry monitors the second PMT for ~20 nsec after the first PMT detects a photon. If the second PMT detects no photon, then that event is not counted by the system because only one photon was produced. Consequently, coincidence circuitry eliminates many single photon events being counted and thereby reduces this background source of an LSC markedly. The number of counts passing through such coincidence circuitry is given by $2R_1R_2\iota$, where R_1 and R_2 are the noise count rates in each PMT and ι is the active time (or resolving time) for the coincidence gate. For example, if PMT noise is 1200 c.p.s. and the gate time is 20 nsec, then only 0.058 c.p.s. are reported. A recent study in the author's laboratory of 822 modern PMTs gave a mean noise level of 285 c.p.s., which translates to 0.0032 c.p.s., a most acceptable error.

6.4.2 Analogue–digital converter (ADC)

This electronics component performs the necessary function of converting the amplitude of the original analogue signal into a digital signal so that the digital electronics which follow the ADC in the circuit can process it.

6.4.3 Multichannel analyser (MCA)

An MCA is a device consisting of a number of storage channels. For example, an MCA might have 4096 channels or more. An LSC is designed to monitor an energy range of ~2000 keV. With 4096 or 8192 channels available, each channel monitors ~0.5 or 0.25 keV of energy, respectively. Signal intensity or pulse height is determined by the MCA and that pulse is registered or stored in the channel corresponding to its intensity. Consequently, over time these channel-distributed counts correspond to the spectrum of the sample being monitored. Such a spectrum is a histogram of counts, illustrated by *Figure 8*, a spectrum of ^{3}H and ^{14}C.

Modern LSCs may present a spectrum that is directly proportional to energy or to the logarithm of energy. *Figures 8* and *9* illustrate these two presentations. ^{32}P was omitted from the linear scale plot because the nuclide tail is sufficiently long that the peaks of the spectra are not clearly presented.

In summary, a liquid scintillation cocktail detects particles emitted from an

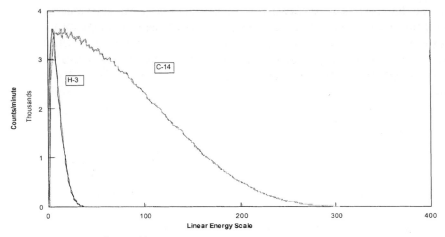

Figure 8 Spectra of ^{3}H and ^{14}C on a linear energy scale.

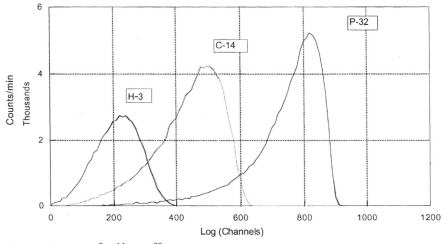

Figure 9 Spectra of ^{3}H, ^{14}C and ^{32}P on a logarithmic scale.

unstable nucleus. These particles interact with components in the cocktail to produce a burst of ~425 nm photons. That photon burst generates photo-electrons upon detection by two PMTs. Both PMT signals are summed, amplified and analysed. Under ideal circumstances each emitted particle produces one photon burst. The intensity of the burst corresponds to the energy of the emitted particle. Analysis then provides a spectrum of the nuclear emission and a count of all observed decayed particles. This spectrum represents the energy distribution of all beta particles, for example, detected by the instrument.

6.4.4 Effects that increase observed count rates

(i) Static discharge

Electric charges may become attached to sample vials. Plastic vials moving frequently within the sample compartment of a counter on a day when humidity is low enhance this effect. If a discharge occurs while the sample is being counted, an extraneous count is registered. Such events produce single photons rather than a short nanosecond burst of photons per event. Consequently, coincidence circuitry prevents some counts being recorded. Nevertheless, if discharges release large numbers of electrons, some will pass through the coincidence circuitry.

(ii) Luminescence

On occasion, chemicals used to dissolve some samples in the liquid scintillation cocktail cause chemical reactions that create molecules in excited electronic states. A single photon is released when the excited molecule returns to the ground electronic state. Such a result is called a 'singles' event by comparison with the nuclear decay process that produces many photons per event. Such chemiluminescence will increase the observed count rate of a sample above the actual sample radioactivity. Chemical kinetics determines the level of luminescence. Reactions may be completed in minutes, hours or days. Therefore, one can observe a decrease in the apparent count rate if such a sample is counted repetitively. Primary causes of chemiluminescence are use of oxidizing agents such as hydrogen peroxide (H_2O_2), basic media and solubilizing agents. Addition of 70 µl of glacial acetic acid to 10 ml of cocktail quickly reduces 'singles' events. Alternatively, 7 ml glacial acetic acid added to a litre of cocktail is more convenient for handling many samples routinely. More detail is available in Fox (1) and Peng (2). A luminescence pulse height spectrum is small, confined within ~7 keV. If energetic nuclides are being counted, then luminescence can be ignored by setting the sample count window to >8 keV. Many modern LSCs can monitor luminescence and remove the excessive counts. Figure 10 illustrates the nature of the luminescence time decay and its effective elimination by a monitoring instrument.

Extraneous photons may also arise from photoluminescence, the excitation of sample vials by exposure to light. Consequently one should avoid having sample preparation stations adjacent to outside windows, other sources of bright light or radioactivity.

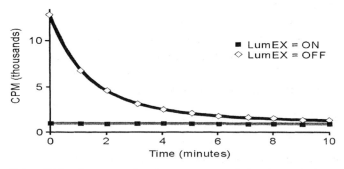

Figure 10 Luminescence decay curves, uncorrected and corrected.

6.4.5 Effects that decrease observed count rates

(i) Quench: adsorption, chemical and colour

If each particle or photon emitted from a radionuclide were counted by the system, then counting efficiency would be 100%: d.p.m. values would equal the observed c.p.m. values. However, even in the optimum case, efficiency is not a perfect 100%. Any effect or process that decreases the counting efficiency below 100% is called a *quenching process*. Three general types of quench have been recognized: chemical, colour and absorption. These names derive from interferences with the basic energy transfer processes in the liquid scintillation cocktail.

Recall that an emitted particle, P, transfers energy to a solvent molecule, S, which transfers energy to a primary solute fluor, F_1, which excites a secondary solute fluor, F_2, which emits a photon, hv. A particle can be blocked from transferring energy to a solvent molecule by any solid material, say filter paper, fibre or powder, lying between the two. *Absorption quench* describes this process. If an electronically excited solvent molecule does not transfer energy to a solute fluor, then *chemical quench* has occurred. Such quench can be caused by low pH. Excess H_3O^+ ions can attach to the fluor, changing its absorption or emission characteristics. Energy absorption or emission is prevented. The last energy transfer step is detection of emitted photons by the PMTs. If photons are emitted by the fluor but do not reach the PMT, then no detection occurs. Photons do not reach the PMTs because a chemical absorbed them. *Colour quench* has occurred. Emitted photons are in the blue region of the visible spectrum. Red, green and yellow molecules are efficient absorbers of photons with blue wavelengths. Such solutions will evidence colour quench.

Schematically, these quench processes are described in *Figure 11*.

Normal Process	Absorption Quench	Chemical Quench	Colour Quench
$P + S \rightarrow P + S^*$	$P \parallel S \nrightarrow P + S^*$	$P + S \rightarrow P + S^*$	$P + S \rightarrow P + S^*$
$S^* + F \rightarrow S + F^*$		$S^* + F \nrightarrow S + F^*$	$S^* + F \rightarrow S + F^*$
$F^* \rightarrow F + hv$			$F^* \rightarrow F + hv$
$hv + PMT \rightarrow$ detection			$hv \parallel PMT \nrightarrow$ detection

Figure 11 Characterization of the basic quench processes.

6.4.6 Quench affects LSC spectra

Although each of the quench types decreases a sample's count rate, each affects a sample's spectrum differently. Chemical and colour quench shift the spectrum toward lower pulse height, as shown by *Figure 12*. Notice that the area under the spectrum decreases as the spectrum 'moves towards the left'. This measures the loss in total counts as quench increases. Second, quench reduces the number of photons detected normally in any single 'photon burst'. Therefore the intensity or height of a pulse is decreased and the spectrum appears to move to the left. Note also the relation between spectrum shift and the counting efficiency shown in *Figure 12*.

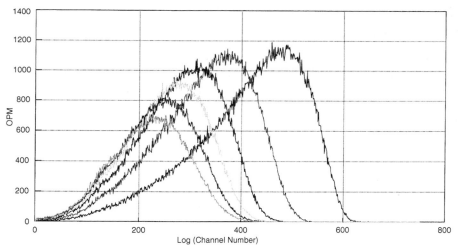

Figure 12 General effect of chemical and colour quench on liquid scintillation spectrum.

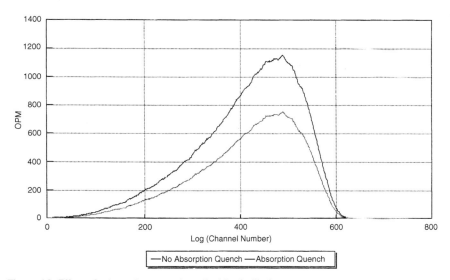

Figure 13 Effect of adsorption quench on liquid scintillation spectrum.

On the other hand, absorption quench uniformly decreases the number of counts all across the spectrum, as illustrated by *Figure 13*. No spectral shift occurs, but the area under the curve decreases. Some writers do not refer to this process as quench because quench monitors cannot correct for it. Here quench refers to any process, correctable or not, that reduces count rate.

6.4.7 Is quench correction required?

Radioactively tagged samples are counted for comparison with other samples. Did a positive or spurious effect take place? If so, what is the quantitative value of the effect? Direct comparison of two samples is acceptable only if all samples have been counted with the same efficiency, in which case a c.p.m. measurement is required. If the counting efficiency of two samples requiring comparison differs, then the counting efficiency of each must be determined so that d.p.m. values are available according to d.p.m. = c.p.m./E. In different words, the samples are comparable because each has been converted to the activity it would possess under conditions of 100% counting efficiency.

6.4.8 Quench monitors

Quench monitors have been developed based upon their ability to follow the spectral shift described above. Generally speaking, two types of quench monitor exist. One type depends only on properties of the sample spectrum itself and is called an *internal monitor*. The second depends upon irradiation of the sample by an external gamma source that produces a Compton spectrum of the sample matrix. This type is called an *external monitor*. Commercial counters present a number of different internal and external monitors. Consequently, examples of each type are introduced here, rather than describing them all.

(i) Internal quench monitors

Because the sample spectrum itself shifts to lower energy as quench increases, that shift is the basis of a quench monitor. One approach is the definition of a fixed reference axis on the spectral graph across which the spectrum shifts. Another is a definition of a spectrum's centre, which moves across the graph. The first is called sample channels ratio (SCR) and the second an isotope's centre number (IC#) or spectral index of the sample (SIS). *Figure 14* illustrates SCR and *Figure 15* the IC# or SIS.

Figure 14 shows unquenched and quenched spectra of a sample. A thick bar positioned at $x = 400$ defines a fixed axis that divides each spectrum into two parts: 0–400 and 400–600. SCR measures the count rates, A and B, of a spectrum on each side of the fixed axis and calculates the ratio A/B. As quench increases

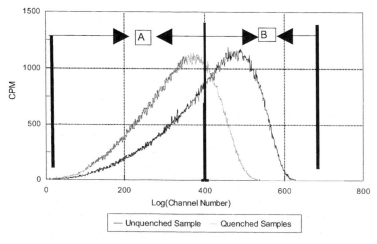

Figure 14 Internal quench monitor: SCR.

102

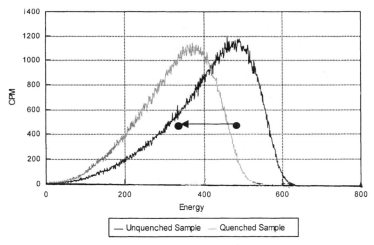

Figure 15 Internal quench monitor: IC# or SIS.

and the spectrum shifts towards the left, the SCR value changes because the areas of the two sections of the spectrum change.

Figure 15 again shows unquenched and quenched spectra of a sample. A circle marks the 'centre of counts' of each spectrum, a number that corresponds to its position along the *x*-axis. As quench shifts the spectrum leftwards, the centre of the spectrum, IC#, shifts with it. In other words, the IC# monitors that shift and therefore monitors quench.

Three difficulties arise with internal quench monitors. If the activity of the sample is small, a long count time is required to obtain sufficient sample counts for a precise value of the monitor. Second, internal quench monitors are specific for each nuclide. Third, they cannot be used for multinuclide samples. External quench monitors solve all three limitations.

(ii) External quench monitors

External quench monitors require a gamma-radiating source external to the sample. Gamma radiation of the sample matrix generates Compton electrons that are subject to exactly the same quench effects as the sample itself. A Compton spectrum shifts towards lower energy and decreases in count rate just as a radiolabelled sample spectrum does. Count time required to obtain sufficient counts for a precise value of the quench monitor is comparatively short.

Several nuclides have been used as external sources, but currently [133]Ba and ^{137}Cs predominate. [226]Ra and [152]Eu have been used also. *Figure 16* provides Compton spectra of three external standards. Within LSCs external gamma sources are sealed and stored in lead containers. A standard is positioned automatically adjacent to a sample vial and radiates it with gamma rays that produce the Compton spectra shown by *Figure 16*. Compton electrons with largest kinetic energy produce the most intense photon burst. Compton electrons are produced with increasing energy as the angle between the incoming gamma ray and the electron is decreased. Maximum energy corresponds to zero angle.

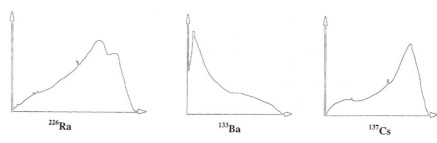

Figure 16 Compton spectra produced by ^{226}Ra, ^{133}Ba and ^{137}Cs.

Instrument manufacturers have adopted different definitions for quench parameters. Either an integrated Compton spectrum or a specific property of the spectrum is used to define a quench parameter.

6.4.9 Specific quench monitors

(i) Spectrum analysis

Packard Instruments uses transformed spectral index of the external standard (tSIE) as an external quench monitor. The Compton spectrum produced by ^{133}Ba is integrated cumulatively and a selected linear portion is extrapolated to the energy axis. This x-axis intercept defines the tSIE, which has a value of 1000 for unquenched samples and decreases in value as counting efficiency decreases.

Beckman Coulter Inc. uses Horrocks' number as an external quench monitor. The Compton edge, the sharp decrease on the right side of the ^{137}Cs spectrum in *Figure 16*, is analysed. An inflection point on the edge, determined from the second derivative of the spectrum, has an energy value of 476 keV for an unquenched sample. A Horrocks' number of 0 is assigned for an unquenched sample. As quench increases, the Compton spectrum shifts towards lower energy and the inflection point moves across channels of the MCA. The difference in channel number between the unquenched and quenched positions of the inflection point is the Horrocks' number, which increases as counting efficiency decreases.

LKB (Wallac) counters also use ^{133}Ba as an external standard. 'Standard quench parameter' (External), (SQP(E))is the quench monitor. The endpoint of the Compton spectrum is used, where endpoint is defined as that point corresponding to 99% of the total Compton spectrum.

6.4.10 Quench monitors: historical note

At one time quench was monitored by internal standards and by external standards channels ratio (ESCR). Rarely are these used now. An *internal standard* method requires counting a sample, removing the vial cap, adding a standard (known d.p.m.) to the sample and recounting. Counts above that observed from the sample were produced by the standard. That excess compared with the standard's d.p.m. values provided the counting efficiency, which was assumed equal to the sample's. This method, believed the most accurate by many, is rarely used today because reasonable automation of the procedure is next to impossible.

Pipetting error and sample loss upon removal of the vial cap are also inherent sources of error.

ESCR is the external standard equivalent of SCR, giving the ratio of the count rates from two areas of interest in the spectrum generated by the external standard. ESCR is rarely used today because gamma radiation of the vial produces spectral contributions to the lower energy region that cannot be eliminated and thereby causes the monitor to be in error. tSIE, Horrocks' number and SQP(E) avoid this problem because they use the upper portion of the Compton spectrum.

6.4.11 Quench curves

A quench curve is a plot of the counting efficiency of a specific radionuclide as a function of a given quench monitor. *Figures 17* and *18* provide examples for Horrocks' number and tSIE, respectively. *Figure 17* displays essentially the entire quench curve domain. Typically ^{3}H is the lowest-energy beta emitter and ^{32}P the highest-energy emitter incorporated as biological tags. Therefore, quench curves for all other nuclides fall somewhere between these two.

Once quench curve information is available, obtaining a value of the quench

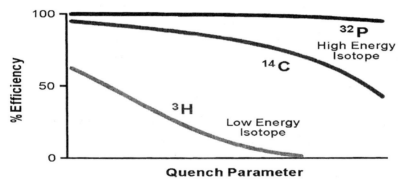

Figure 17 Quench curves with per cent efficiency as a function of Horrocks' number (H#).

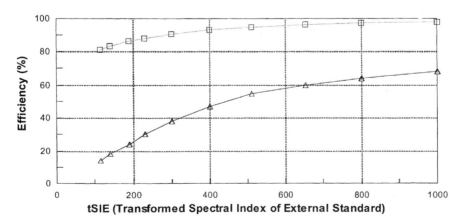

Figure 18 Quench curves as a function of tSIE.

monitor for an unknown allows the quench curve to provide the counting efficiency. Then the formula d.p.m. = 100 × c.p.m./counting efficiency is applied.

How is a quench curve obtained? A quench curve is composed of six to ten samples of the desired nuclide. Each sample contains exactly the same activity. Each successive sample contains successively more quenching agent, which decreases the counting efficiency. Such standards can be purchased or prepared for ^{3}H and ^{14}C. Quench curve standards for all other nuclides must be prepared by the user. A general protocol for preparing such standards follows.

Protocol 1

Preparation of quench standards

Equipment and reagents

- A calibrated sample of the required nuclide
- Scintillation vials
- Scintillation cocktail
- Quenching agents: either that present in experimental samples or nitromethane

Method

1 Obtain a calibrated sample of the desired nuclide from instrument or chemical suppliers.[a]

2 Add the same activity to each of 12 glass vials (100 000–200 000 d.p.m. gives a reasonable count time).

3 Add the same cocktail volume to each of the 12 vials: 5 ml for small vials or 10 ml for standard vials.

4 Cap the vials and count them in an LSC. Each vial must have the same activity. Counting all vials at this stage ensures that pipetting error has been monitored. Eliminate vials that differ by >1% from the mean of the set. Depending on your pipetting technique, preparation of more than 12 vials may be necessary.

5 Add varying amounts of a quenching agent to vials 2–10. Ideally the quenching agent should match that expected for the unknown samples. However, nitromethane has been found generally useful. As an example, 10 μl successive increments for vials 2–10 gives counting efficiencies for tritium over the range 60–5%. No quenching agent is added to the first vial.

6 Standards should be counted so that a 2σ% of 0.5 is obtained.

[a] ^{3}H and ^{14}C quench standards are available from instrument suppliers, Amersham and others in either 20 or 6 ml vials

Modern LSCs have automated routines for counting standards used for establishing quench curves. Manufacturer's recommendations should be consulted.

6.4.12 Quench curve preparation and use on commercial LSCs

Automation is a major advantage of commercial LSCs. Although details differ for different manufacturers, all current LSCs provide for some degree of automation for counting quench curve standards, fitting the resulting curve mathematically, providing statistics on the precision of the curve, counting unknown samples and determining their counting efficiencies and calculating the d.p.m.

Curve fitting takes the general form of efficiency, E, as a function of the quench parameter, Q,

$$E = f(\Sigma C_i Q^i),$$

where C_i are coefficients for an ith order polynomial, frequently third order, or

$$E = C_0 + C_1 Q^1 + C_2 Q^2 + C_3 Q^3$$

Fitting may be by least squares, an averaging of the data, or by smoothed spline functions that pass through all data points. If the fit is by least squares, then a statistical report describes the precision of the fitted curve for each of the data points.

Counting efficiency of unknown samples is computed from the mathematical curve fit. Accuracy of the d.p.m. results depend upon several factors:

(1) Commercial counters require that the quench parameter be calibrated. For example, tSIE is set equal to 1000 for an unquenched sample, and Horrocks' number to 0. The manufacturer's procedure must be followed.

(2) A quench curve is valid only for a specific nuclide and on the same counter originally used.

(3) Counting efficiencies outside the range of the original quench curve are suspect. Commercial LSCs usually provide a numerical result but flag the answer if it is outside the quench range.

On older counters without spectral analysis capability (no SQP(E), Horrocks' number or tSIE), quench agent and the volume and composition of the cocktail should be equivalent for unknowns and quench standards. Also a 'plastic vial effect' presents a problem for some older counters. The effect arises because liquid scintillant migrates into the vial and produces an additional Compton spectrum upon radiation by the external standard. Such additional counts interfere with the computation of the sample's counting efficiency.

7 Advanced topics

Current LSCs with one sample detector provide many capabilities beyond the basic ones discussed thus far. Some of these will be outlined here. In-depth discussions can be found in the liquid scintillation literature (3) and manufacturers' manuals and technical bulletins.

7.1 Multi-label counting

Preparation of appropriate quench curves and counting unknowns with two or three radionuclide labels have been automated. Two differences arise for the dual-label case compared with the single label. First, four quench curves must be determined. Second, as quench increases and spectra shift towards lower energy,

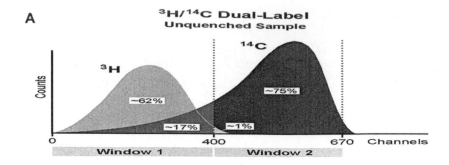

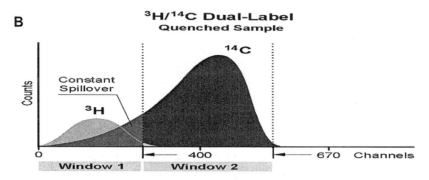

Figure 19 Dual label of (A) ^{14}C and ^{3}H unquenched and (B) quenched.

the two counting windows must track this shift accurately. Both are illustrated by *Figure 19A* and *B*. *Figure 19A* shows the relationship of unquenched samples of ^{3}H and ^{14}C with each other and their respective count windows. *Figure 19B* gives the same information for quenched samples of the two nuclides. Recall that a beta spectrum extends from 0.0 keV to its maximum energy. Therefore, it is impossible that count windows contain only one of the two nuclides. The total count rates observed in each window are composed of contributions from both nuclides. ^{3}H has a quench curve for windows A and B, the lower and upper windows as shown by *Figure 19*. Likewise, ^{14}C spills into window A and has a quench curve for each window. C.p.m. values in the two windows can be represented by the following two equations, where E is efficiency.

$$c.p.m._A = E_{1A}d.p.m._1 + E_{2A}d.p.m._2$$

$$c.p.m._B = E_{1B}d.p.m._1 + E_{2B}d.p.m._2$$

These equations can be solved for the two unknowns, d.p.m.$_1$ and d.p.m.$_2$, the actual activities of ^{3}H and ^{14}C.

$$d.p.m._1 = \frac{c.p.m._1E_{2B} - c.p.m._2E_{2A}}{E_{1A}E_{2B} - E_{1B}E_{2A}}$$

$$d.p.m._2 = \frac{c.p.m._2E_{1A} - c.p.m._1E_{1B}}{E_{1A}E_{2B} - E_{1B}E_{2A}}$$

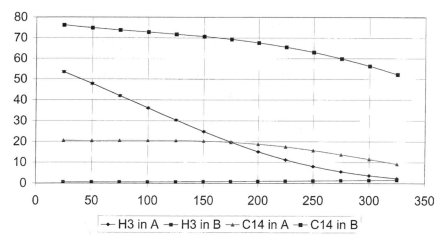

Figure 20 Quench curves for the dual-label pair ^3H and ^{14}C in windows A and B.

The c.p.m. are determined by the instrument for each unknown sample and each of the efficiencies is obtained from one of the appropriate four quench curves. *Figure 20* illustrates the general nature of the four quench curves required for dual-label counting. The example shown is for ^3H and ^{14}C. These vary in detail depending upon the specific quench monitor.

As the spectrum shifts with increasing quench, instruments are designed to track that shift so that the change in spill of each nuclide into the other's window is minimized. Observe the initial constancy and small decline in efficiency for each nuclide in the other nuclide's primary window in *Figure 20*. The details are different for each manufacturer, each of which also uses different terminology to describe automatic tracking. Beckman Coulter call their window tracking system 'automatic quench compensation' (AQC), LKB call theirs 'automatic window setting' (AWS) and Packard instruments call their 'automatic efficiency control' (AEC).

Automated triple-label d.p.m. recovery is available also. Nine quench curves must be prepared and three count windows defined and tracked properly. Additional information is available from manufacturers and the open scientific literature.

For both dual- and triple-label d.p.m. recovery, standards for the quench curves must be prepared as discussed previously (*Protocol 1*). Quench standards are counted automatically by the counter which also mathematically fits and stores all relevant quench curves for future use by the instrument.

7.2 Colour quench detection and correction

Chemical and colour quench curves do not superimpose. Consequently, it is not possible to use chemical quench curves to correct for colour quench samples. A colour-quenched sample with the same value of the external quench monitor as a chemically quenched sample has a smaller counting efficiency, as shown by

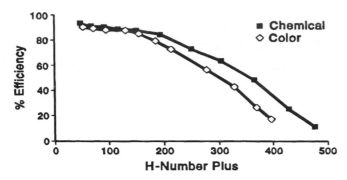

Figure 21 Relationship between colour and chemical quench curves.

Figure 21. Note that at small quench values the two curves essentially super-impose, but differ increasingly below a counting efficiency of ~80% for ^{14}C. Different coloured absorbers exhibit different colour quench curves as well. Some current LSC instrumentation has the capability to detect and correct for coloured samples automatically. Some counters will also colour correct multilabelled samples.

7.3 D.p.m. correction without quench curves

A technique, efficiency extrapolation, has been developed that provides d.p.m. results for single-label samples of beta or alpha emitters without the need for preparing quench curves. This is most helpful for unusual nuclides, especially if such nuclides are not required routinely. Efficiency extrapolation is independent of cocktail, sample volume, vial size and composition, and chemical quench. A set of samples containing different nuclides can be counted under a single program, unlike the use of quench curves which require the sole appropriate nuclide. Half-life correction is available analogous with quench curve usage. The concept depends upon the inherent shape of the relevant LSC spectra and appropriate calibration of the LSC for this technique (4).

Protocol 2

Efficiency extrapolation calibration and sample counting

Equipment and reagents

- A counter capable of carrying out efficiency extrapolation
- Unknown single labelled samples

- One known standard of each nuclide, in scintillation vials with cocktail of choice

Method

1   Enter d.p.m. and date of preparation of the ^{14}C and ^{3}H standards into the counter's 'set-up' page.

Protocol 2 continued

2 Place the standards into the efficiency extrapolation calibration rack.

3 Place single-labelled samples in any nuclide order in racks following the calibration rack and count.

7.4 Sample monitoring

Two sample monitors are available to help avoid erroneous results. The origin of luminescence 'singles' events is discussed in Section 6.4.4. Because one photon is produced per luminescence event, LSCs that monitor all light events both in and out of coincidence from each PMT separately can detect 'single' events during a sample count and subtract them from a sample's observed total count rate. The printed result provides both the corrected count rate and the percentage of the uncorrected count rate that are single photon counts.

Second, some samples will develop two phases after sitting for some time at room temperature or after being placed into the counting compartment of an LSC. Samples with two phases provide erroneous results. Organic and aqueous phases are counted with different efficiencies. Depending on the volumes of the two phases, quantities of the unknown may be distributed differently in two otherwise identical samples. Consequently, two-phased samples should not be accepted for analysis. A *two-phase monitor* detects such a condition and flags such samples accordingly.

8 Special counting situations

8.1 Čerenkov counting

A charged particle passing through a medium causes polarization of the molecules along its path. When the polarized molecules return to their initial state, photons are emitted. Observation of these photons depends on a threshold effect. If the velocity of the charged particle is greater than c/n (the speed of light in a vacuum divided by the refractive index of the medium), then Čerenkov radiation can be observed. The threshold energy dependence on the refractive index of the medium is plotted in *Figure 22* according to the following equation:

$$E \text{ (in keV)} = 511 \left(\frac{n}{(n^2 - 1)^{1/2}} - 1 \right)$$

For water, $n = 1.333$, giving a threshold energy of about 262 keV. Beta particle energies of ^{3}H, ^{14}C and ^{35}S are too small to produce Čerenkov radiation in water, but about half of the emitted beta particles from ^{32}P exceed the threshold for water and can be Čerenkov counted. *Figure 23* compares the pulse height distribution of ^{32}P counted in a cocktail and by the Čerenkov technique. Note the difference required for the count windows because the efficiency of Čerenkov counting is smaller.

CHARLES L. DODSON

Table 8 Advantages and disadvantages of Čerenkov counting

Advantages	Disadvantages
No cocktail required	Limited nuclides in water
No chemical quench	Colour quench
Sample can be recovered	Less efficient than liquid scintillation counting
No scintillation fluid waste	Sample volume dependence

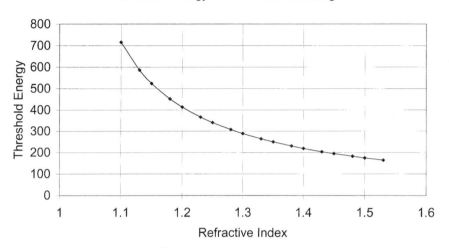

Figure 22 Threshold energy for Č erenkov counting.

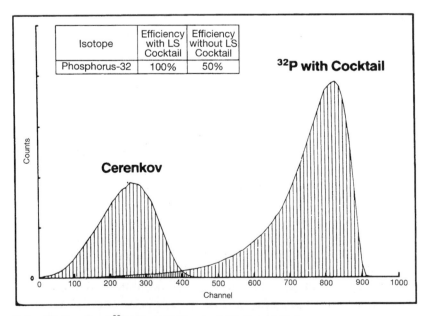

Figure 23 Spectra of ^{32}P with and without scintillation cocktail.

112

8.2 Low-level counting

Background events can be divided into those that are quenchable, detected by the scintillation cocktail, and those that are unquenchable. Quenchable events have long lifetimes and can be distinguished and removed as such. About half of all background events are quenchable and about two-thirds of those lie in the tritium region of the spectrum. Tritium samples with activities as low as 0.17 c.p.s. can be counted. For statistical reasons, 'low-level counting' should be used if sample activity falls below 6 c.p.s. (A rule of thumb for the quantitative lower limit for a counter is given by $8(B/t)^{1/2}$ where B is the measured background in c.p.m. and t is the count time in minutes.) However, the disadvantage is that much longer count times are required to obtain high precision results. Although removal of quenchable background improves primarily the tritium region of the spectrum, some benefit occurs for the ^{14}C region.

9 Solid scintillators and programmable coincidence gates

Solid scintillators have been used for decades to detect radionuclide emissions. Solid organic scintillators are well-known detectors of gamma rays. During the past decade, inorganic solid scintillators have been developed for use on LSCs. The variety of components is similar to that for liquid cocktails. For brevity, consider ytterium silicate doped with cerium, Ce. Energy is transferred from a beta particle, for example, to Yt, analogous with the energy-transfer role of a liquid scintillation cocktail's solvent. Energy is then transferred from Yt to Ce, analogous with a liquid scintillation cocktail's fluorescent solute. Ce returns to the ground state and releases a photon at 395 nm, a wavelength ideal for current PMTs.

Solid scintillators have advantages and disadvantages compared with liquid scintillators and compared with filter paper counting. *Figure 24* demonstrates that the light output is greater than that for liquid scintillators counting the same sample. *Figure 25* shows that the lifetime of the Ce excited state is longer than that of fluorescent solutes in the liquid case. Larger pulse height means that counting efficiency is improved, especially for low-energy radionuclides such as tritium. A sample must be counted in a wider window to monitor the additional photons. The second property requires an increase in coincidence gate time from 20 to ~100 nsec. Incorporation of both properties improves counting efficiency of tritium by a factor of 2 compared with filter paper counting. (Xtalscint is the name of one such commercial product.) Advantages and disadvantages are summarized in *Table 9*, assuming that the correct counting window and coincidence gate time are used. Solid scintillators are available in the form of sheets and filters with various diameters for use in microtitre plates or vial caps. They may be attached to porous or non-porous surfaces for solution and filtration applications. Use of 200 μl samples without liquid scintillation cocktail combined with the sample preparation throughput conveniences of cell harvesters and microtitre plates are useful for DNA hybridization, transport, metabolism, receptor assays and others.

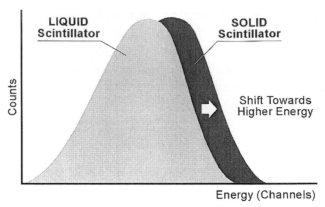

Figure 24 Light output for liquid and solid scintillators.

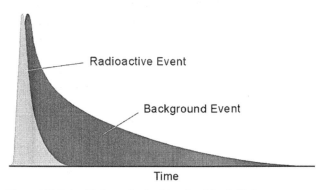

Figure 25 Pulse lifetimes for liquid and solid scintillators.

Table 9 Advantages and disadvantages of solid scintillators

Advantages	Disadvantages
Higher counting efficiency	Mainly for ^{3}H; less for other nuclides
No chemical quench	No multi-labelled counting
Low background	Dissolve in concentrated acids
Chemically inert to aqueous buffers and bases	Samples must be dry
Safety and storage	External quench monitors cannot be used

no vapour pressure
   not absorbed through skin
   does not burn
   requires about 1/300 storage disposal volume
	Cannot use volatile samples

200 μl samples	

10 Sample preparation

Best results in liquid scintillation counting depend on careful sample preparation, the object being to produce a homogeneous mixture between the sample and cocktail whether liquid or solid. The physical state of the sample, its volume and chemical environment are basic considerations for the selections of sample treatment and cocktail. An extensive literature, including entire books, exists on the preparation of samples for liquid scintillation counting (1,2). Here, an outline is given of the primary concerns.

10.1 Cocktail properties

Cocktails, regardless of manufacturer and cost, fall into one of six types: organic, multi-purpose, high sample capacity (which includes extremes of pH and ionic strength), filter counting, flow counting and special applications. For all types, stability (both in time and with temperature) is important. Cocktail manufacturers provide most of the information needed for choosing the best cocktail for a given sample in the form of phase diagrams, which are plots of temperature in terms of per cent sample load. An example is given by *Figure 26*. Per cent sample load is the per cent of the total volume provided by the sample. Six phase diagrams in *Figure 26* describe different chemical environments, starting with water at the top and concluding with a sucrose solution. The light regions of the diagrams are labelled single-phase; dark regions are two-phase. Consequently, for six chemical environments, one can deduce the volume of sample needed for the system to remain single-phase upon addition of 10 ml of cocktail over a temperature range of 10–30 °C. Ready-safe (Beckman Coulter, Inc) cocktail is multi-purpose, providing reasonable sample holding, up to 12–20% at 20 °C, over a range of sample types. In addition it has low vapour pressure, high flashpoint and good biodegradability.

If a larger sample handling capacity is needed, consider the phase diagrams in *Figure 27*. For all chemical conditions given except sucrose, this high-capacity cocktail accepts ≤50% sample loading. Also notice the two-phase regions indicating per cent sample loads not available for use.

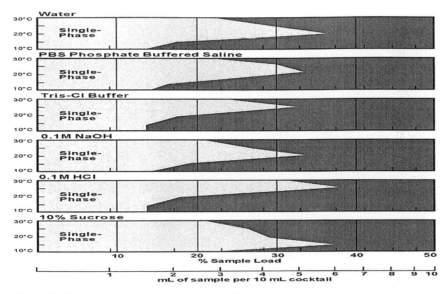

Figure 26 Phase diagram for multi-purpose cocktail.

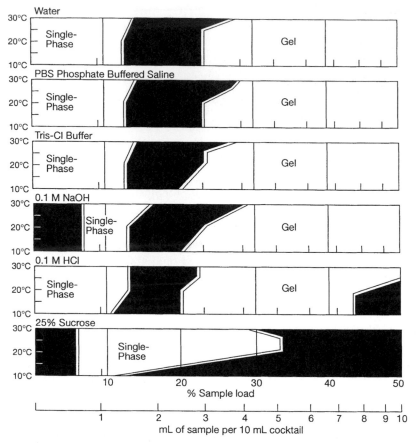

Figure 27 Phase diagrams for cocktail with high sample capacity.

10.2 Sample types

10.2.1 Organically soluble

Samples soluble in organic aromatic solvents should be added to cocktails without emulsifiers. These cocktails contain only solvent and fluorescent solutes. They are commercially available or can be prepared in the laboratory.

Protocol 3

Preparation of general-purpose organic cocktail

Equipment and reagents

- 1 litre volumetric flask
- 2,5-Diphenyloxazole
- *p-bis*(*o*-methylstyryl)benzene
- Pseudocumene

Protocol 3 continued

Method

1   Place 5.0 g of 2.5-diphenyloxazole in a 1 litre volumetric flask.

2   Add 0.125 g p-bis(o-methylstyryl)benzene to the volumetric flask.

3   Dilute with pseudocumene to 1 litre while stirring.

10.2.2 Aqueous (including inorganic) samples

This type requires emulsifiers in the cocktail. As previously noted, the specific emulsifier or mixture of emulsifiers distinguishes otherwise similar cocktails. Both cocktails in *Figures 26* and *27* contain emulsifiers. Such phase diagrams provide the information needed to choose a suitable cocktail. Older literature on liquid scintillation counting contains a variety of recipes for aqueous sample cocktails. These formulations tended to be dangerous to the environment or the user and had limited sample loading capability.

10.2.3 Biological fluids and tissues

Solubilization by quaternary amines or combustion are the standard treatments of biological fluids and tissue samples. Commercial quaternary amines are available from Packard as Soluene-350, from Beckman Coulter as BTS-450 and from Nycomed Amersham as NCS. The numbers after the first two indicate the number of milligrams one millilitre of the solubilizer can treat. Two typical protocols follow.

Protocol 4

Preparation of tissue samples

Equipment and reagents

- Experimental tissue
- Solubilizer
- Hydrogen peroxide
- 40°C oven
- Glass scintillation vials
- Liquid scintillation cocktail

Method

1   Add 2 ml of solubilizer to <200 mg of tissue.

2   Incubate for 2–4 h at 40°C or until solubilized.

3   Remove colour, if present, by adding 0.5 ml 30% H_2O_2.

4   Add 10–15 ml of a cocktail containing no emulsifiers.

5   Chemiluminescence that results from use of quaternary amines can be reduced rapidly by addition of glacial acetic acid, 7 ml per liter of cocktail or 70 μl per 10 ml of cocktail.

Protocol 5

Liquid scintillation counting of plasma or serum

Materials and reagents

- Labelled plasma or serum samples to be counted
- Scintillation cocktail for aqueous samples
- Scintillation cocktail for non-aqueous samples

- Tissue solubilizer
- 40 °C oven
- Glass scintillation vials

Method

1   Add 1 ml of plasma or serum to 10–15 ml aqueous scintillation cocktail.

2   Shake vigorously until clear. If not clear, go to step **4**.

3   Count the sample.[a]

4   Add 0.25 ml plasma or serum to 0.75 ml of tissue solubilizer.

5   Incubate for 1 h at 40 °C.

6   Add 8–10 ml of non-aqueous cocktail (e.g. see *Protocol 3*) and count.

[a] The expected ^3H efficiency is 35–45%.

Ultrasonic agitation aids the solution process markedly. Commercial quaternary amines dissolve coarsely ground tissues and tissue homogenates.

Normally agents solubilize tissue samples especially if combined with ultrasonic agitation. Colour is removed completely or removed sufficiently for instrument colour correction to provide good results. However, if either should be a problem, combustion of the tissue sample to H_2O and CO_2 is the only alternative. Packard supplies an apparatus to carry out this procedure which is relatively straightforward. Packard's recommendations and protocol should be followed carefully.

10.2.4 Precipitates on filters

A variety of processes produce solids on filters: extraction, gradient centrifugation, paper, column and thin-layer chromatography, ligand binding, electrophoresis and filtration. Glass fibre and cellulose nitrate filters are used most frequently. Three possible counting conditions occur upon presentation of the sample to a cocktail: total removal, partial removal or no removal of the sample from the filter. Cocktails formulated with an appropriate combination of emulsifiers can solubilize certain sample types completely, for example samples from extractions, centrifugation and column chromatography. Cocktail manufacturer's literature should be consulted.

In these cases sample counting is normal and quench monitors provide accurate

Cocktail

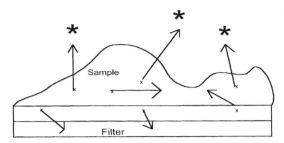

Figure 28 Beta particle absorption in heterogeneous samples.

d.p.m. values. Two samples similarly prepared with different counting efficiencies can be compared. If partial or no solution occurs, accurate d.p.m. results are not possible and sample comparison is dangerous. Consider *Figure 28*, a representation of a solid sample on a filter. Any beta track leading into the bulk of the sample or into the filter does not produce a scintillation. An LSC reports only those events reaching the cocktail. This problem is greatest with ^{3}H because of the low penetration of its radiation. The amount of a sample distributed on the filter paper is a contributing factor. Count rate is lower than that obtained if the sample is completely solubilized. Additional count time is required to obtain the same count precision. Cellulose nitrate filters have been favoured by some writers because they dissolve in some cocktails. Glass fibre filters do not. However, the important point is that the sample forms a homogeneous solution, not that the filter dissolves.

10.2.5 Samples for solid scintillators

Solid scintillators in the form of ytterium silicate doped with Ce serve as coatings on either porous or non-porous filter materials. They are in the form of 25 mm filters, filter sheets and caps designed for 200 µl samples. These formats can be where uncoated filters traditionally have been used, used with automated harvesters and for direct counting applications without filtration. Recalling that the counting efficiency advantage depends on the reduction of absorption quench, sample drying is critical, especially for low energy beta emitters like ^{3}H. Heat lamps, hot-air blow dryers, microwave ovens and vacuum centrifuges have been used to decrease drying time compared with air drying. Dry [^{3}H]5-fluorouracil has been reported to count with five-fold improvement in efficiency compared with the same wet sample.

Types of sample that have been used with this approach are whole cells, membrane receptor assays, transport studies, enzyme assays, radioimmunoassays, thymidine incorporation and tumour response assays. External quench monitors cannot be used, but internal monitors can. Volatile samples and non-volatile solvents should not be used because drying must remove solvent not sample. Buffer molarity should be kept below 1 M.

10.2.6 Čerenkov counting

Čerenkov counting requires no cocktail. No chemical quench is present and a sample can be recovered. In the context of usual biological samples, only ^{32}P is available for this treatment. If appropriate, then the only requirement is that the sample be an aqueous solution.

10.3 Vials

Four characteristics of sample vials should be considered: their photon transparency, size, composition and contribution to background count rate. Standard LSC vials are made from common glass, polyethylene, or borosilicate glass. The latter two are preferred because of their transparency and low contribution to the background count rate. Special Teflon vials are available for counting extremely low-level activity should such demand arise. Standard volumes are 18–20 and 5–6 ml ('mini-vial'), depending on the manufacturer. Smaller vials require appropriate sample/cocktail loading. They provide advantages of lower cost for the vial, smaller cocktail volume and improved waste disposal. On occasion, the use of low sample volumes may be unavoidable because of sample scarcity. A 100 or 200 μl sample plus an equivalent volume of a 50% sample loading cocktail is feasible. Such samples can be counted using centrifuge tubes placed inside standard mini-vials as carriers.

At one time plastic vials were a concern because of their propensity to enhance static electricity discharges. Also, because solvent can migrate into the vial wall, external standard measurements of quench could vary with time. Both concerns have been addressed by modern LSCs.

11 Gamma counters

Gamma counters monitor gamma rays and X-rays. Detection depends on the photoelectric effect. Detectors are inorganic crystals that have been doped or activated by an 'impurity'. Electrons bound at lattice sites in insulators or semi-conductor materials occupy the valence band. If valence band electrons receive sufficient energy, then they transfer to the conductance band. The solid arrow on the left of *Figure 29* depicts this situation. The return of an electron to the valence band has low probability for a pure crystal, as noted by the crossed return arrow.

Addition of a second metal ion, an activator, in low concentration (about 10^{-3} mole fraction) introduces an alternative energy transfer path for an electron in the conduction band to lose energy and return to the valence band. With a properly chosen host and activator, the photon emitted as the electron passes from the activator's excited state to its ground state will have a visible wavelength. Thallium-activated NaI (NaI(Tl)) is an example of such a detector and finds general use in gamma counters.

The peak of the emission spectrum for NaI(Tl) is 415 nm, which is transmitted through the crystal very efficiently. About 38 000 photons are produced for each

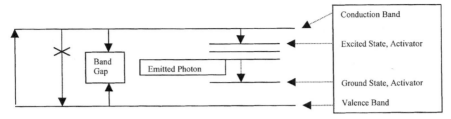

Figure 29 Electron band structure in a doped crystal.

MeV absorbed by NaI(Tl). This excellent light output, combined with the ability to machine NaI(Tl) into a variety of shapes and sizes, has made it the standard detector for routine use. NaI(Tl) can be damaged by mechanical or thermal shock and is hydroscopic. Crystals are enclosed in aluminium to protect against moisture. Inside surfaces are coated with an oxide that reflects photons emitted by the crystal except for one side designed to transmit photons to the PMT. The detector is cylindrical with diameters of 1, 2 or 3 inches (25, 51 or 76 mm). Counting efficiencies for higher-energy emitters improve as the diameter of the detector increases. Energetic gamma rays can transmit and escape small crystals without being detected. For example, the counting efficiency of ^{137}Cs increases from 30% to 42% if the detector diameter increases from 51 to 76 mm. For ^{125}I, the increase is only 3%, from 72% to 75%. *Table 10* lists other examples together with their emission peaks. *Figure 30* illustrates the detector design with a sample in place.

Both the NaI(Tl) and the PMT are surrounded by $\geq$5 cm of lead to reduce the background count arising from cosmic sources and from samples adjacent to the one being counted.

11.1 Gamma counter operational parameters

With any type of commercial gamma instrument, specific operating details should be obtained from the manufacturer's manual. Knowing about general

Table 10 Nuclides, emission energy and counting efficiency

Nuclide	Emission peaks (keV)[a]	Counting efficiency	
		51 mm	76 mm
^{137}Cs	660 (32)	32	41
^{51}Cr	320 (4.9)	3	6
^{57}Co	122 (6.4)	82	85
^{60}Co	1170, 1330	40	58
^{59}Fe	1100, 1290		36
^{125}I	55, 27.5	72	77
^{131}I	364, 640		55
^{22}Na	1280, 1020, 510	67	82
^{65}Zn	1100		19

[a]X-ray energy in parenthesis.

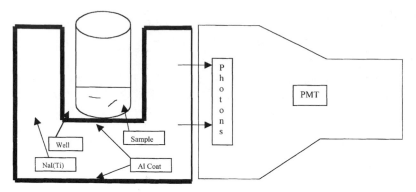

Figure 30 Schematic of a NaI(Tl) crystal machined for a sample.

operational parameters is helpful, however. Consider four concepts: calibration, resolution, stability and efficiency.

11.1.1 Calibration

Calibration refers to the process of setting a given instrument's count window (or windows if more than one nuclide is used) so that the peak of a given nuclide is window 'centred'. No sample counts will be lost. Setting gain or voltage controls may be required first. Standard sources are available from metrology laboratories and from a variety of commercial suppliers of radionuclides. A major advantage of gamma counting is that sample cocktails are not used. Quench does not occur. Consequently standard sources may be used to set an instrument's window to count samples prepared in the local laboratory. All manufacturers provide modules that automatically set count window(s) correctly for a given nuclide so that the nuclide's entire spectrum is observed by the detector. A general protocol includes the following steps.

Protocol 6

Setting counting windows

Equipment and reagents

- Gamma counter
- Nuclide standard
- A background count sample (i.e. a 'blank')

Method

1   Follow the manufacturer's procedure to set the gain and voltage to the PMT correctly.

2   Place a nuclide standard in the sample counting chamber and select by keyboard or otherwise the automatic calibration for the desired nuclide.

3   If manual calibration is necessary, then systematically change the upper and lower

Protocol 6 continued

levels of count window discriminators (or count window controls) until a maximum count (or count rate) is obtained.

4   If dual-labelled counting is needed, two count windows are required. Change each upper and lower discriminator (or count window control) systematically while monitoring both nuclides to ensure maximum count rate of each nuclide in its window and with minimum overlap of each into the other window.

5   After the count window has been set correctly, count a 'blank' to determine the background of the counter in the defined window region. Subtract this background from each sample's total count rate to obtain count rates produced by the sample only.

11.1.2 Resolution

Resolution is the ability to measure accurately two energy peaks that lie close to each other. Narrow bands improve energy resolution. A formal definition of resolution, R, is

$$R\% = 100(\text{FWHM})/E_0$$

where FWHM is the full width of the band at half its maximum height and E_0 is the energy of the peak position. This ratio is expressed as a percentage. Resolution of gamma ray detectors depends on their size and typically should be in the 8–15% range. Good practice requires measurement of resolution. Poor resolution frequently means that a detector has been damaged. A general protocol proceeds as follows using a ^{137}Cs standard with peak position at 662 keV. If the gamma counter has an MCA, a graph of the ^{137}Cs band can be obtained automatically.

Protocol 7

Determination of a gamma counter's resolution

Equipment and reagents

- Gamma counter
- ^{137}Cs standard

Method

1   Place a ^{137}Cs standard in the sample chamber.

2   Set the count window width to 40 keV with the lower edge at 580 and the upper edge at 620 keV.

3   Count the standard for 1 min.

4   Move the lower edge up 20 keV, maintaining a 40 keV window width, and count for 1 min.

5   Repeat this process eight times. The window position for the last count is 760–800 keV.

Protocol 7 continued

6   Plot the count values against the centre position for each value's window.

7   Determine the width of the peak at half maximum and the peak position in keV.

8   Compute R%.

11.1.3 Stability

Comparison of samples within a set is required frequently. If the time needed to count a set is long, then stability of the gamma counter is essential. Stability refers to maintaining the peak position of a standard gamma ray band at the same window setting or at some acceptable per cent change for a specified time. A general protocol involves setting the desired window to position the peak of a ^{137}Cs standard at a known position. Repetitively count the ^{137}Cs standard for the time period required for sample comparison, say 6 h. At the end of that period, either determine the position of the peak relative to the initial position or adjust the count window to restore the counts observed at the beginning of the process. The peak position should change by <3%.

11.1.4 Efficiency

As noted previously, the counting efficiency of ^{137}Cs should be >30% if a 2 inch (51 mm) NaI(Tl) crystal is used and >40% if a 3 inch (76 mm) one is used. If the efficiency falls below these values, then the NaI(Tl) crystal or the PMT may need replacement. After establishing the gain and correct window settings for counting ^{137}Cs, count a standard for $\geq$1 min (c.p.m. std) to obtain the desired counting statistics, $2\sigma\% = 1$. A value for the background (c.p.m. bkg) is also needed. Half-life corrections must be made for decay of the ^{137}Cs from its standardization date to the date of counting to obtain d.p.m.std. Then compute

$$\% \text{Efficiency} = 100(\text{c.p.m.}_{\text{std}} - \text{c.p.m.}_{\text{bkg}})/\text{d.p.m.}_{\text{std}}$$

12 Flow counters

Radiolabelled chemical mixtures separated by high performance liquid chromatography (HPLC) have been fraction-collected and serially counted on an LSC instrument. Attaching a flow counter to an HPLC system as a detector is more straightforward and saves time compared with collecting fractions and counting on an LSC. In addition, flow detectors are designed to operate fraction collectors automatically. That permits recovery of the original sample, frequently an attractive feature. Flow detectors operate on the same principles discussed for liquid scintillation counters. A flow cell rather than a scintillation vial is placed between two PMTs. (Some flow counters use only one PMT.) The electronics following the PMTs are essentially the same as those for the scintillation counter. Specially formulated flow cocktails accommodate low and high salt conditions, high sample loads and stability over an extremely wide pH range. Some systems use a solid scintillant as the primary detector. The solid can be packed into a

Table 11 Flow detector counting efficiencies

Nuclide	Counting efficiency (%)
^{3}H	35–45
^{14}C	75–85
^{125}I	40–50
^{32}P	20–30

counting cell or into a cell external to the sample tubing. Emitted particles or rays must pass through the sample tubing into the detector.

Regardless of which form of primary detector is used, the basic flow cell design must consider the detector-active volume, sample flow rate and specific activity. These with counting efficiency determine counting statistics. A primary effect of small active volumes and short count times is a lower counting efficiency compared with a static count of the same sample. Emitted particles with long path lengths may pass through a solid scintillant without being detected. Examples of counting efficiencies shown in *Table 11* are for systems with two PMTs and liquid scintillators.

In addition to those nuclides listed in *Table 11*, many others can be counted by this technique: ^{35}S, ^{45}Ca, ^{22}Na, ^{36}Cl, ^{129}I, ^{51}Cr, ^{65}Zn, ^{57}Co, ^{59}Fe, ^{99m}Tc, etc. Typical applications include monitoring HPLC effluents and receptor-binding assays. Flow detectors are frequently used in conjunction with other HPLC detectors such as UV and fluorescence. Dual-label counting is also available on some flow detectors.

13 Multi-detector liquid scintillation and gamma counters

Sample formats and requirements for high sample throughput have led to the development of multi-detector liquid scintillation and gamma counters. Such systems depend on the same basic concepts as single detector counters.

Gamma counters are available with two, five or ten detectors and capability of processing 1000 samples. LSCs are available with two, three, six or 12 detectors, each with two PMTs.

Some LSC instruments count directly from microtitre plates, filters or tubes and with sample capacity of 3072 samples using 96-well plates. Direct access to 24-well culture plates has been developed. Packard Instruments provide systems that count 12 samples at a time taken from 24-, 96- and 384-well microtitre plate formats. An external stacker provides for 15 000 samples.

14 Storage phosphor technology

Storage phosphor technology is an imaging technique that replaces X-ray film in autoradiography, or in any experiment where a radiolabelled sample has been traditionally visualized by X-ray film. (This also includes digital radiography, a system for obtaining radiographs for any portion of the body.) Discovered in the

1970s, developed in the 1980s, with major improvements in the 1990s, storage phosphor technology is faster, more accurate and has a greater dynamic range than X-ray film (5). A dark room and its associated chemicals are not required. ^{14}C, ^3H, ^{125}I, ^{131}I, ^{32}P, ^{33}P, ^{35}S, ^{59}Fe and other radionuclides have been monitored. The PhosphorImager system (Molecular Dynamics) and the Cyclone (Packard Instruments) are examples of this technology. (Fuji, Agfa and Kodak have developed systems used specifically for digital radiography.)

Such a system consists of a cassette to hold the sample, a phosphor screen to collect the sample's image, a scanning unit, software to analyse the image and a means to erase the image on the phosphor screen allowing its re-use.

14.1 Phosphor screens

Phosphor screens as used in this technology are typically 20 × 25 cm or about 35 × 45 cm. The active phosphor layer consists of $BaFBr:Eu^{2+}$ crystals bonded with polyurethane and given a thin protective coat of cellulose acetate. Alpha, beta, gamma radiation and light of wavelengths ≤380 nm activate the phosphor. Eu^{2+} is oxidized and the electron is trapped in an empty site in the BaFBr lattice known as an F centre. The transferred electron can be represented with the notation $BaFBr^{1-}$. In summary,

$$\beta + Eu^{2+} \rightarrow Eu^{+3} + e^-$$

$$e^- + BaFBr \rightarrow BaFBr^{1-} \text{ (electron trapped in an F centre)}$$

Although some loss of the oxidized and reduced species begins immediately, >50% of the stored energy remains ≤24 h after the ionizing radiation has been removed. If a previously prepared gel sample were being run, then the gel would be positioned on the phosphor screen. Particles emitted from all radionuclides on the gel are detected by the phosphor screen as described.

14.2 Scanning unit

The phosphor screen with its latent image is placed on a stage that is scanned by a 10 mW HeNe laser focused to an 88 μm diameter spot. This translation stage holding the phosphor screen moves past the 633 nm laser beam. A galvanometer-controlled mirror sweeps the laser beam back and forth over the phosphor screen. The laser stimulates release of the trapped electron which in turn produces a blue photon in the range 350–450 nm peaking at 385 nm as shown by *Figure 31*. Eu^{3+} accepts the free electron and is reduced to Eu^{2+}. Only those areas of the phosphor screen exposed to radionuclide emissions produce such photons. Laser scanning releases ~80% of the stored energy. Blue emitted photons are collected by a fibre optic bundle and directed on to the face of a blue-sensitive PMT. Signal output from the PMT is amplified, digitized and stored on a computer.

Image analysis software converts the stored image into a usable and quantifiable form. Most importantly, image analysis provides better results than autoradiography film. Spatial resolution depends on the storage phosphor reader, imaging plates and the nature of autoradiography itself. However, the contribution from the reader is minimal because the pixel size and the laser beam radius

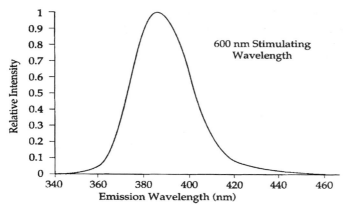

Figure 31 Emission spectrum from the phosphor screen.

are the same. Image diffusion is greater as nuclide energy increases. An image from a ^{32}P source is more diffuse than that from ^{14}C. Diffusion of the image also increases with the source–plate distance.

14.3 Phosphor screen erasure

All signals remaining on the screen must be erased prior to running another sample. Exposure to visible light erases the screen. Alternatively, conventional laboratory light boxes can be used. Systems are supplied with a convenient and

---

Protocol 8

General procedure for obtaining phosphor image

Equipment and reagents

- Experimental sample (e.g. a gel)
- Phosphor screen and holder
- Laser scanning system

- Phosphor screen eraser (500 W photoflood tungsten bulb), e.g. from Molecular Dynamics

Method

1 Prepare a sample, a gel for example, following the normal protocol.

2 Erase the phosphor screen by illumination with the 500 W photoflood tungsten bulb; exposure time < 1 min.

3 Place the sample in a holder and position the phosphor screen on top for exposure.

4 Expose the film. Exposure time is about one-tenth that of an X-ray film exposure time for the same sample.

5 Laser scan the sample and store the digitized data. This requires 1–12 min depending on the sample.

6 Perform software analysis of the stored data. This is an automated process and requires variable times dependent on the sample and the nature of the analysis.

rapid eraser in the form of a 500 W photoflood tungsten bulb. Erasure by this lamp leaves a residual signal $\sim 10^{-5}$ that of the original.

14.4 Quantitative analysis

Reliable quantitative analysis depends on care exercised in the preparation of standards. Standards should be prepared in the same manner as the sample. For example, gel standards should be prepared as gel dots or strips duplicating the concentration of the gel used for the sample. Consistent standard preparation can be monitored if four or five standards with different activities are prepared, run and analysed. Counting efficiency of such standards should not change over the five orders of linear range for the system. The counting efficiency for ^{35}S, for example, is almost constant over five orders of magnitude of activity. Tritium can be analysed. Its small energy of emission requires a special screen made with a layer of phosphor crystals thicker than normal.

Dual-label samples have been quantified using storage phosphor imaging (6). Three requirements must be met to make this successful: linear quantification, a large dynamic range and differential attenuation of the two nuclides. Sample and standards are exposed under two conditions: first, normal conditions without a filter; second, by placing an appropriate thickness of copper foil, about 35 μm, between the sample and the storage phosphor screen. The foil blocks particles emitted by the less energetic of the two nuclides. Analysis of the data is similar to that discussed for the dual-labelled liquid scintillation case. Errors are typically <15% unless one nuclide is present as a small fraction of the other.

Quantitative analysis has been performed on alpha particles. By appropriate use of filters, alpha particles can be distinguished from beta particles or gamma rays with storage phosphor technology. A summary of properties is shown in Table 12.

Table 12 Summary of primary properties

Property	General characteristic
Time savings	exposure times are about one-tenth that of conventional autoradiography
Sensitivity	
^{14}C and ^{35}S	25 disintegrations/mm^2 or 1 d.p.m./mm^2 with 25 min exposure
^{32}P	1 disintegration/mm^2 or 0.04 d.p.m./mm^2 with 25 min exposure
^{59}Fe	has been detected at 6 fg/mm^2
Dynamic range	five orders of magnitude

References

1. Fox, B.W. (1977). Techniques of Sample Preparation for Liquid Scintillation Counting. North-Holland Publishing Co., Amsterdam
2. Peng, C.T. (1977). Sample Preparation in Liquid Scintillation Counting. Amersham International, Amersham.
3. L'Annunziata, M.F. (ed.) (1998). Handbook of Radioactivity Analysis. Harcourt Brace & Co., San Diego, USA.

4. Dodson, C. (1991). In Ross, H., Noakes, J.E. and Spaulding, J.D. (eds), Liquid Scintillation Counting and Organic Scintillators, p. 335. Lewis Publishers, Inc., Michigan, USA.
5. Johnston, R.F., Pickett, S.C. and Barker, D.L. (1990). Electrophoresis 11, 355–360.
6. Johnston, R.F., Pickett, S.C. and Barker, D.L. (1991). Methods: A Companion to Methods in Enzymology 3, 128–134.

Further reading

Cheng, Y.T., Soodprasert, T. and Hutchinson, J.M.R. (1996). Radioactivity measurements using storage phosphor technology. Int. J. Appl. Radiat. Isot. 47, 1023.

Cook, G.T., Harkness, D.D., Mackenzie, A.B., Miller, B.F. and Scott, E.M. (eds) (1994). Radiocarbon. Advances in Liquid Scintillation Spectrometry, University of Arizona, Tucson, USA.

Horrocks, D.L. (1974). Applications of Liquid Scintillation Counting. Academic Press, London.

Knoll, G.F. (1989). Radiation Detection and Measurement, 2nd edn. John Wiley & Sons, New York.

Noakes, J.E., Schönhofer, F. and Polach, H.A. (eds). (1992). Radiocarbon. Advances in Liquid Scintillation Spectrometry, University of Arizona, Tucson, USA.

Wunderly, S.W. (1989). Solid scintillation counting: a new technique for measuring radiolabeled compounds. Int. J. Appl. Radiat. Isot. 40, 569–573.

Chapter 6

In vitro labelling of nucleic acids and proteins

M. W. CUNNINGHAM, A. PATEL, A. C. SIMMONDS and D. WILLIAMS

Amersham Pharmacia Biotech, Amersham Laboratories, White Lion Road, Amersham, Buckinghamshire HP7 9LL, UK

1 Introduction

Radioactive tracers are uniquely suited for the labelling of biological molecules because their presence generally has no effect on the properties of those molecules. Indeed, most of the radioisotopes used can be incorporated into biomolecules in place of the equivalent non-radioactive element. This is particularly important when the biomolecule is itself being used as an enzyme substrate since enzymes tend to display a strong selectivity for their specific substrates.

Despite the growth over recent years of non-radioactive methods for labelling and detecting biological molecules, these methods often rely on relatively bulky organic groups which themselves can influence the behaviour of the biomolecule. Examples of such labels are enzymes and fluorophores, the latter tending to be small enough to allow their use whilst retaining the native activity of the biological molecule, though careful choice of labelling strategy is required to limit any deleterious effect.

1.1 Choice of radiolabel

A number of elements which are normally found in biological molecules are available in radioisotopic forms, e.g. phosphorus in nucleic acids (available in radioisotopic form as ^{32}P or ^{33}P), sulfur in proteins (available as ^{35}S), carbon (available as ^{14}C) and hydrogen (available as ^{3}H) in all organic compounds. Other elements can be incorporated which produce a limited perturbation of the structure, for example sulfur in place of oxygen in nucleic acids, or iodine (available as ^{125}I and ^{131}I) in proteins. The choice of radioisotope for *in vitro* labelling of biological molecules can depend on a number of parameters: the ability to incorporate the radiolabel without disrupting the properties or behaviour of the biological molecule; the sensitivity required; resolution of detection; the means of detection; safety considerations; and regulatory requirements.

Sensitivity, resolution and means of detection are to a large extent inter-dependent. Sensitivity is determined broadly by the type of emission, the energy of the emission and the means of detection. The isotopes generally used for labelling purposes in the biological sciences tend to be either beta emitters or gamma emitters (see Appendix). Of the isotopes that can be relatively easily incorporated into biological molecules, ^{35}S has a relatively weak emission and for nucleic acids has been largely superseded by ^{33}P, although for proteins it is still widely used, especially for metabolic labelling applications. ^{32}P is very widely used for nucleic acid blotting and other applications where its relative ease and sensitivity of detection are most useful. It is available at high specific activity (110 TBq/mmol, 3000 Ci/mmol) in a variety of nucleotides. Southern and northern blotting are techniques involving a number of steps at which material can be lost, and material fixed to a filter may be relatively inaccessible to a probe. Sensitive detection methods can provide a useful safety net to ensure the successful completion of a long experiment. 'In-gel' detection of nucleic acids such as used in footprinting techniques requires an energetic emitter such as ^{32}P to avoid internal absorption by the sample matrix. ^{33}P has become widely used for applications such as *in situ* hybridization and manual sequencing where its energy of emission provides better resolution than ^{32}P but greater speed of detection than ^{35}S. For some *in situ* hybridization experiments where resolution is of primary importance, ^{35}S or even ^{14}C is used at the expense of longer auto-radiography exposure times (see Chapter 4). *Figure 1* gives a visual indication, in a

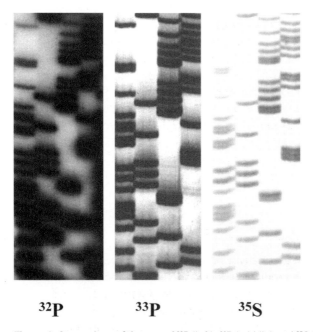

^{32}**P** ^{33}**P** ^{35}**S**

Figure 1 Comparison of the use of ^{32}P (left), ^{33}P (middle) and ^{35}S (right) for labelling the products of DNA sequencing reactions. Equivalent amounts of radiolabelled products were loaded in each lane and, following electrophoresis, the dried gel was exposed overnight to film (Hyperfilm β max, Amersham Pharmacia Biotech).

standard DNA sequencing gel, of the relative intensities and resolution of ^{32}P, ^{33}P and ^{35}S under equivalent conditions.

The low energy of ^{3}H and ^{14}C tends to limit their use to specific applications such as receptor and metabolic studies where other elements obtainable as radio-isotopes are either not normally present or would cause deleterious effects to the performance of the assay. Complex organic chemistry may also be required to label a molecule with ^{14}C or tritium. The maximum specific activities of ^{14}C and tritium are 2.29 GBq/mmol (62.4 mCi/mmol) and 1.05 TBq/mmol (28.8 Ci/mmol) respectively. Several atoms of ^{14}C or tritium can be substituted in a molecule, but the specific activity obtained is still very much lower than that of isotopes such as ^{125}I. However, as there is an inverse relationship between specific activity and half-life, tracers containing these isotopes can remain useful for years, as compared with a few months for some of the other radioisotopes.

^{125}I is primarily used for the labelling of peptides and proteins for radio-immunoassays and receptor studies. The isotope decays by electron capture followed by X-ray emission, which can be counted directly in a gamma counter. ^{125}I is used *in vivo* for imaging due to its non-particulate emission, which reduces radiation damage to the biological material. An advantage in using ^{125}I over ^{14}C or tritium is the higher specific activity of the former. It has a maximum theoretical specific activity of 79.75 TBq/mmol (2175 Ci/mmol) and is usually obtainable at ~73.3 TBq/mmol (2000 Ci/mmol) so that very small amounts of radio-iodinated material can be used whilst maintaining sensitivity. In practice, the count rate obtained from ^{125}I can be 100 times greater than that from tritium and 35 000 times greater than that from ^{14}C. It should, however, be remembered that the advantage of a high specific activity is countered by the disadvantage of a shorter half-life, a factor that should be taken into account when using ^{125}I and ^{32}P in particular, although a shorter half-life is advantageous for waste disposal. In most cases iodine is a foreign label, that is it does not normally occur in the molecule. The replacement of a proton with a large iodine atom can have a considerable effect on the properties of the protein although this can usually be overcome if the label is some distance from the site of biological activity. This can be a particular problem in receptor studies as even a small change in structure, such as oxidation of one amino acid in an iodination, can completely block binding to the receptor.

2 Applications of nucleic acid labelling

In vitro, radiolabelled nucleic acids are generally used as probes for hybridization, or in sequencing, footprinting and gel mobility shift assays. Hybridization probes can be used in a number of different formats. Most commonly hybridization is carried out against purified target material immobilized on a solid support such as a blotting membrane or the well of a microtitre plate. However, other formats include hybridization to target present in the solution phase, and *in situ* hybridization against nucleic acids present in cells or tissue sections. The information obtained from such processes can vary. It may simply be the presence or absence of target when hybridizing to a dot blot on a membrane such as in a high density array or to a colony/plaque lift. Alternatively it may be target size when hybridizing to a Southern or northern blot where the target has previously been sized by gel electrophoresis. It may also be chromosomal or tissue localization in the case of *in situ* hybridization. RNase protection and S1 nuclease digestion assays are also widely used in expression analysis and transcript quantification. In RNase protection assays, labelled RNA probes are hybridized in solution to total or messenger RNA samples followed by RNase A and/or T1 digestion. In S1 nuclease assays, labelled DNA probes are hybridized to RNA and, after S1 nuclease digestion of single-stranded DNA probe, the amount of DNA probe remaining hybridized to RNA is determined. In both these assays, reaction products are usually run on a gel, but selective precipitation of labelled nucleic acid can also be used.

Although sequencing of nucleic acids has become dominated by high throughput techniques based on fluorescence, a significant amount is still carried out manually using radioisotopes. The use of manual cycle sequencing will be described in more detail in a later section. Footprinting and gel mobility shift assays examine interactions between proteins and nucleic acids. These techniques have been particularly important in identifying and studying transcription factors. Interactions between nucleic acid and protein can easily be disrupted by modifications to the binding site of either of the partners and so non-radioactive labels are less appropriate for these applications. These are techniques where the unique ability of radioisotopes to replace elements within the nucleic acid structure itself is most useful.

3 Labelling methods for hybridization probes

All of the methods to be described in this section are standard enzyme-catalysed reactions. However, it is still useful to mention some general points that may help towards obtaining good, reproducible labelling. The reactions should be set up on ice, except when the buffer contains spermidine, which may cause precipitation of nucleic acid. Reagents should be added in the order described in the protocols with the final additions being radiolabel followed by enzyme. Radiolabel is added next to last to reduce handling of radioactive solution. Successful results require thorough mixing of all reagents before addition of enzyme; this can usually be achieved by a brief (2 sec) spin in a microcentrifuge. Enzyme is then added and mixed by gently pipetting two or three times. This ensures good mixing without denaturation of the enzyme, which is frequently added in a very small volume and in a viscous glycerol-containing solution.

Poor labelling can occur for a variety of reasons. A common cause is impure DNA, and it is therefore often useful to compare the efficiency of labelling with that obtained with a pure, control DNA. Commercially available plasmids or phage DNA are useful in this case. If impure substrate is the cause of poor labelling, a further ethanol precipitation, phenol extraction or dialysis followed by ethanol precipitation can be of benefit. Frequently, it is possible to improve labelling efficiency, even with impure DNA, by the use of more enzyme or by extending the reaction time. It is also necessary to ensure that efficient labelling is theoretically possible by determining both the amount of label present and the maximum amount of label that can be incorporated. This point is explored further below but if, for example, in an end-labelling reaction, there is ten times more label present than there are ends available for labelling, then a maximum incorporation of only 10% is possible.

When establishing a labelling method or using a new preparation of nucleic acid as substrate, it may be valuable to determine the efficiency of labelling. A variety of methods are available, including precipitation by trichloroacetic acid (TCA) (1), binding to DE81 paper or nitrocellulose, and thin layer chromatography. The following protocol describes the use of DE81 paper.

Protocol 1

DE81 chromatography

Equipment and reagents
- Whatman DE81 paper discs (2.4 cm diameter)
- 0.5 M Na_2HPO_4
- 95% ethanol
- infra-red lamp
- liquid scintillation counter

Method

1 Spot 2–10 μl (10^3–10^5 c.p.m.) of each sample on to four 2.4 cm discs of Whatman DE81 paper. Designate two filters A and two B. The A filters will be used to give the total number of counts in the sample, A. For most labelling reactions an initial dilution of 1 μl reaction mixture into 100 μl H_2O or 0.2 M EDTA will be necessary to give counts in the required range.

2 Wash the B filters six times, 5 min per wash, in 0.5 M Na_2HPO_4. Then wash twice in H_2O (1 min per wash) and in 95% ethanol (1 min per wash). The B filters give the incorporated counts, B.

3 Dry A and B filters using an infra-red lamp and count by liquid scintillation. The percentage incorporation is $(B/A \times 100)$%.

$$\text{specific activity} = \frac{\text{total incorporated counts}}{\text{mass of hybridizable nucleic acid}}$$

For ^{32}P-labelled probes it is possible to quantify the radioactivity by Čerenkov counting without scintillation fluid. In this case the same filter can be counted initially before washing for total counts, and after washing for incorporated counts.

If required, unincorporated label can be removed by ethanol precipitation or using a Sephadex or Sephacryl spin column (1,2) or one of a wide variety of commercially available columns (for example AP Biotech Microspin columns, Qiagen QIAquick columns). In our experience, this is not necessary for most filter hybridization applications if incorporation is $\geq 60\%$, but we would advise that the user verifies this under their experimental conditions before adoption.

If a labelling method is performed routinely and unincorporated nucleotides are usually removed, then it is often sufficient simply to monitor the incorporated label in the probe either quantitatively by scintillation counting or, in the case ^{32}P, semi-quantitatively with a Geiger counter. It is advisable to monitor the efficiency of a labelling reaction; a method for calculating this is given for each labelling method.

Finally, virtually all the enzymes used for labelling require a divalent metal cation as cofactor, so the chelating agent EDTA can be added to terminate the reaction effectively.

3.1 Random prime labelling

This method uses oligonucleotides of random sequence to prime synthesis along the length of a single-stranded DNA template (3,4). They anneal to short stretches of complementary DNA on the template and, in the presence of appropriate DNA polymerase enzyme and deoxynucleoside triphosphates (dNTPs), a new complementary DNA strand is synthesized. If a radioactive nucleotide is present, this will be incorporated into the new strand.

The most commonly used enzyme is the Klenow fragment of *Escherichia coli* DNA polymerase I (5), which synthesizes the new strand in a 5'–3' direction. Other polymerases are sometimes used, including bacteriophage T7 DNA polymerase and an exonuclease-free derivative of the Klenow fragment which also lacks the 3'–5' exonuclease activity of DNA polymerase I and has been observed to give somewhat higher levels of incorporation. Primers of length ranging from 6-mer to 14-mer are used, with 6- and 9-mers perhaps the most common.

The method has a number of useful properties. It can be used to label small quantities of DNA, as little as 25 ng, and high probe specific activities are achievable, up to 5×10^9 d.p.m./μg using high specific activity [α-^{32}P]dNTPs (110–220 TBq/mmol, 3000–6000 Ci/mmol). The reaction can be carried out very rapidly, in 5–10 min at 37°C, or over a longer period, for example overnight, at room temperature. The ability to label small amounts of DNA is useful when labelling insert DNA which can only readily be purified in microgram quantities. Impure DNA, for example in low melting point agarose, can also be labelled efficiently.

Careful choice of appropriate nucleotide concentrations and ratio of primer to template concentrations is important in order to synthesize probe of an optimal length for hybridization. The protocol below should give acceptable results over a range of template concentrations. Commercially available systems containing pre-tested reagents to ensure optimal labelling have also proved popular (for example, Megaprime System, from Amersham Pharmacia Biotech). Dried pre-aliquoted reactions requiring only the addition of label and denatured template (for example, Ready To Go Beads, from Amersham Pharmacia Biotech) additionally have significant advantages in terms of throughput, convenience and reproducibility.

Protocol 2

Random prime labelling

Equipment and reagents

- DNA to be labelled
- 10× labelling buffer (600 mM Tris–HCl, pH 7.8, 100 mM Mg Cl$_2$, 100 mM 2-mercaptoethanol)
- TE buffer, pH 8.0 (10 mM Tris–HCl, pH 8.0, 1 mM EDTA)
- 300 μM dATP in TE buffer
- 300 μM dCTP in TE buffer
- 300 μM dGTP in TE buffer
- 300 μM TTP in TE buffer

- [α-^{32}P]dNTP at 110 TBq/mmol, 3000 Ci/mmol, 10 mCi/ml
- DNA polymerase I (Klenow fragment)
- Random nonamer primers dissolved at a concentration of 30 absorbance units/ml in TE buffer containing nuclease-free bovine serum albumin (BSA) at 4 mg/ml
- 0.2 M EDTA, pH 8.0
- boiling water bath

Method

1 Prepare an appropriate nucleotide mix from the 300 μM stocks by mixing 3.3 μl of each of three dNTP stocks, excluding that to be used as label. Alternatively, if the same labelled nucleotide is to be used frequently, a stock 5× nucleotide mix can be prepared containing each dNTP at 100 μM, omitting that to be used as label. For example, if [α-^{32}P]dCTP is to be used, the 5× nucleotide mix contains 100 μM dATP, 100 μM dGTP and 100 μM TTP.

2 Dissolve the DNA to be labelled[a] in either distilled water or TE buffer[b] to a concentration of 2–25 μg/ml. Place all solutions, except for the enzyme, at room temperature. Leave the enzyme in a freezer until required, and return immediately after use.

3 Place 25 ng[c] of the DNA to be labelled into a microcentrifuge tube and to it add 5 μl of primers and the appropriate volume of water to give a final volume of 50 μl in the labelling reaction. Denature by heating to 95–100°C for 5 min in a boiling water bath. Spin the tube briefly in a microcentrifuge to bring the contents to the bottom of the tube.

4 Keeping the tube at room temperature, add the following in the order given:

10× labelling buffer	5 μl
5× nucleotide mix	10 μl
[α-^{32}P]dNTP	5 μl
Klenow enzyme	2 units

5 Mix gently by pipetting up and down, and cap the tube. Spin for a few seconds in a microcentrifuge. Avoid vigorous mixing as this can cause loss of enzyme activity.

6 Incubate at 37°C for 10 min.[d]

7 Terminate the reaction by the addition of 5 μl of 0.2 M EDTA.

8 Remove a small aliquot (1–2 μl) for determination of incorporation.[e]

Protocol 2 continued

9 For use in hybridization, denature the DNA by heating to 95–100°C for 5 min, then chill it on ice.

Calculation of probe's specific activity

During random primer labelling there is net synthesis of DNA while the initial substrate remains unlabelled. Both newly synthesized DNA and template can participate in the subsequent hybridization, so:

$$\text{probe yield} = \text{ng template DNA} + \text{ng DNA synthesized.}$$

The average molecular weight of a nucleotide in DNA is ~350, so for a labelled nucleotide of specific activity $X \times 10^3$ Ci/mmol,

$$\text{ng DNA synthesized} = (\mu\text{Ci incorporated} \times 0.35 \times 4)/X$$

Note that a multiplication factor of 4 is included, as there are four nucleotides, only one of which is labelled. This assumes equal abundance of each nucleotide.

Once the probe yield has been calculated, the specific activity can be determined:

$$\text{specific activity (d.p.m./}\mu\text{g)} = \frac{\text{total activity incorporated (d.p.m.)}}{\text{probe yield (}\mu\text{g)}}$$

Specific example

Assume 70% incorporation of labelled nucleotide at 3000 Ci/mmol with 25 ng template DNA

amount of labelled nucleotide incorporated = 50 μCi $\times$ 0.7 = 35 μCi
amount of DNA synthesized = (35 $\times$ 0.35 $\times$ 4)/3 = 16.3 ng
total DNA = 25 ng + 16.3 ng = 41.3 ng = 0.041 μg

As 1 μCi = 2.2×10^6 d.p.m., total activity incorporated = $35 \times 2.2 \times 10^6$ d.p.m. = 7.7×10^7 d.p.m. So probe specific activity = $(7.7 \times 10^7)/0.041 = 1.9 \times 10^9$ d.p.m./μg.

[a] When labelling DNA in low melting point agarose, first place the tube containing the stock DNA in a boiling water bath for 30 sec to melt the agarose before removing the required volume. The volume of low melting point agarose DNA should not exceed 25 μl in a 50 μl reaction.

[b] DNA in most restriction enzyme buffers can also be used.

[c] More than 25 ng may be labelled in a 50 μl reaction but the highest specific activity is obtained if the whole reaction is scaled up appropriately.

[d] Purified DNA can be labelled to high specific activity in 10 min at 37°C but, if desired, can be labelled for up to 1 h at this temperature. When labelling DNA in low melting point agarose, longer incubations of 15–30 min at 37°C are required for optimum labelling. Longer incubations of up to 1 h are required when nucleotide analogues such as [^{35}S]dNTPαS are used. Reactions can also be routinely left to proceed overnight at room temperature with any radioisotope.

[e] With purified DNA template, the reaction should give >60% incorporation of label, equivalent to a specific activity of $>1.7 \times 10^9$ d.p.m./μg.

3.2 Nick translation

The nick translation reaction (6) was developed several years before the random prime approach and has to a large extent been superseded by it. It is, however, particularly suitable for labelling large (microgram) quantities of DNA. The reaction uses two enzymes, deoxyribonuclease I (DNase I) and DNA polymerase I, both from *E. coli*. Single-strand nicks are introduced into a double-stranded DNA template by the action of DNase I. Starting from the nicks, the 5'-3' exonuclease activity of DNA polymerase I then progressively removes nucleotides from the exposed 5'-end, while the polymerase activity replaces them in the same direction from the exposed 3'-end, incorporating radioactive nucleotides present in the reaction mix. The position of the nick therefore moves along the DNA in a 5'-3' direction.

Nick translation avoids an initial denaturation step for double-stranded template and the probe concentration for hybridization is readily determined as there is no net synthesis of DNA during the reaction. When labelling microgram amounts of DNA, specific activities approximately ten-fold lower than with random primer labelling are generally obtained. However, it is possible to adapt the reaction to label as little as 25–50 ng of DNA, giving specific activities of $>10^9$ d.p.m./µg within 30 min. Because of the presence of DNase I and the 5'-3' exonuclease activity of DNA polymerase I, it is necessary to control the reaction time and temperature carefully to avoid removal of incorporated nucleotides. This is generally achieved by incubating the reaction at 15–16 °C for a maximum of 2–3 h, or less when using less template. The DNase I:template ratio largely determines probe size and, again, commercially available kits are useful as a source of pre-optimized reagents (for example, Nick Translation Kit from Amersham Pharmacia Biotech).

Protocol 3

Nick translation

Reagents

- DNA to be labelled
- 10× labelling buffer (600 mM Tris–HCl, pH 7.8, 100 mM MgCl$_2$, 100 mM 2-mercaptoethanol)
- TE buffer, pH 8.0 (10 mM Tris–HCl, pH 8.0, 1 mM EDTA)
- 300 µM dATP in TE buffer
- 300 µM dCTP in TE buffer
- 300 µM dGTP in TE buffer
- 300 µM TTP in TE buffer
- [α-^{32}P]dNTP at 110 TBq/mmol, 3000 Ci/mmol, 10 mCi/ml
- Enzyme mix (0.006 units/ml DNase I and 500 units/ml DNA polymerase I)
- 0.2 M EDTA, pH 8.0

Method

1   Prepare an appropriate nucleotide mix from the 300 µM stocks by mixing 3.3 µl of each of three dNTP stocks, excluding that to be used as label. Alternatively a stock 5× nucleotide mix can be prepared containing each dNTP at 100 µM, omitting that to be used as label. For example, if [α-^{32}P]dCTP is to be used, the mix should contain 100 µM dATP, 100 µM dGTP and 100 µM TTP.

Protocol 1 continued

2   Dissolve the DNA to be labelled in distilled water or TE to a concentration of 5–50 μg/ml.[a] Place all solutions, except the enzyme mix, at room temperature. Leave the enzyme mix in a freezer until required and return immediately after use.

3   Add the following in the order given to a microcentrifuge tube on ice:

DNA	50–500 ng[b]
5× nucleotide mix (dATP, dGTP, TTP)	10 μl
10× labelling buffer	5 μl
[α-^{32}P]dNTP	10 μl
Enzyme mix	5 μl
Water to a final reaction volume of 50 μl	

4   Mix gently by pipetting up and down, and cap the tube. Spin for a few seconds in a microcentrifuge. Avoid vigorous mixing as this can cause loss of enzyme activity.

5   Incubate at 15°C for 60 min.[c]

6   Terminate the reaction by the addition of 5 μl of 0.2 M EDTA.

7   Remove a small aliquot (1–2 μl) for determination of incorporation.[d]

8   For use in hybridization, denature the DNA probe by heating to 95–100°C for 5 min, then chill it on ice.

Calculation of probe's specific activity

During nick translation, nucleotides are excised and replaced and there is usually no net synthesis of DNA. Thus the probe's specific activity is calculated simply as:

$$\text{specific activity (d.p.m./μg)} = \frac{\text{total activity incorporated (d.p.m.)}}{\text{amount of template DNA added (μg)}}$$

Thus, 50% incorporation in the above reaction would give a specific activity of 2×10^8 d.p.m./μg with 500 ng template DNA and 2×10^9 d.p.m./μg with 50 ng (as 1 μCi = 2.2×10^6 d.p.m./μg).

[a] Any double-stranded DNA can be used as template for the nick translation reaction. DNA solutions too dilute to be used should be concentrated by ethanol precipitation and redissolved in an appropriate volume of water or TE buffer.

[b] This protocol may be used to label up to 2 μg of DNA but the volume of [α-^{32}P]dNTP should be increased to 20 μl (66 pmol). For lower amounts of DNA (50–100 ng), 5 μl labelled nucleotide will be adequate. Nucleotides with lower specific activity (for example, 400 Ci/mmol) can be used, particularly with the higher amounts of DNA, but the rate of reaction will be lower.

[c] Shorter reaction times (30 min) can be used for smaller amounts of DNA (50–100 ng), while longer times (≤3 h) can be used with larger amounts (250–500 ng) or if using nucleotide analogues such as [^{35}S]dATPαS. Careful control of temperature is necessary to avoid the generation of 'snap-back' regions in the labelled probe.

[d] With purified DNA template, the reaction should give >50% incorporation of label.

3.3 Transcription labelling

By cloning a DNA sequence downstream of an RNA polymerase promoter in a suitable vector, it is possible to use the polymerase to synthesize large amounts of transcript from the insert in the presence of the ribonucleoside triphosphates ATP, CTP, GTP and UTP (7,8). By replacing one of the nucleotides with a radio-labelled equivalent, it is possible to synthesize smaller amounts of highly labelled transcript. The method has the advantage that strand-specific probes free of vector sequence can be readily prepared.

The most frequently used polymerases are from the bacteriophages SP6, T7 and T3, and a variety of vectors are available incorporating one or more of their promoters next to multiple cloning sites for a variety of restriction enzymes. Transcription of vector sequence can be avoided by cutting the vector with a restriction enzyme just downstream of the insert, so that a run-off transcript is produced. Frequently two different promoters in opposite orientation flank the cloning site so that, by choice of polymerase, it is possible to produce a transcript of either strand of the insert. This can provide a valuable negative control, for example for *in situ* hybridization.

A relatively large amount of template (1–2 μg) is used in standard labelling reactions but, as this is intact vector rather than purified insert, there is usually little difficulty in obtaining adequate quantities. Transcript, and hence probe, size will vary with the size of the insert, usually between 100 bases and several kilobases. To avoid a high proportion of prematurely terminated transcripts, the chemical concentration of the radioactively labelled nucleotide should be at least equal to that of the K_m of the enzyme for that nucleotide, ~12 μM for most nucleotides. UTP and CTP are most frequently used as the labelling nucleotide. The non-radioactively labelled nucleotides are, as with other uniform labelling reactions, present in excess. Under these conditions, it is possible to obtain <80% incorporation of labelled nucleotide within 1 h at 37–40°C, generating probes of ~1–2 × 10^9 d.p.m./μg with ^{32}P-labelled nucleotide and 5 × 10^8 d.p.m./μg with ^{35}S-labelled nucleotide.

RNA probes have a number of properties that distinguish them from DNA probes. They are single-stranded and hence do not require denaturation before hybridization, and will not re-anneal to themselves in solution during hybridization. In the presence of formamide, RNA–RNA and RNA–DNA hybrids are more stable than DNA–DNA hybrids, and it is often found that the optimum ratio of signal to background for RNA probe hybridizations is obtained at a higher temperature than with an equivalent DNA probe. Non-specifically bound probe can also be removed by treatment with RNase A, which is highly specific for single-stranded RNA.

In general, RNA probes are less frequently used than DNA probes for membrane hybridizations, but are often used in particular applications such as RNase mapping and *in situ* hybridization. For the latter application the method is most commonly used with ^{35}S-labelled and ^{33}P-labelled nucleotides. *Protocol 4* describes the incorporation of [^{35}S]UTPαS. Probe can be used directly in hybridization following the transcription stage. Lower backgrounds may result if the initial DNA template is removed. For *in situ* hybridization, optimal probe size is generally 200–400 bases; therefore, if transcripts are longer than this, it is advisable to carry out an alkaline hydrolysis before hybridization (*Protocol 4*). Ideally, optimum probe size should be determined empirically for each application.

Protocol 4

Transcription labelling

Reagents

- TE buffer (10 mM Tris–HCl, pH 8.0, 1 mM EDTA).
- Transcription buffer (200 mM Tris–HCl, pH 7.5, 30 mM $MgCl_2$, 10 mM spermidine, 0.05% (w/v) BSA)
- 20 mM ATP
- 20 mM GTP
- 20 mM CTP
- Human placental ribonuclease inhibitor (HPRI) (e.g. Amersham Pharmacia Biotech catalogue no. E2310Y)
- 0.2 M dithiothreitol (DTT) (30.8 mg DTT in 1 ml sterile water), freshly prepared
- Linearized DNA template containing an SP6, T7 or T3 RNA polymerase promoter upstream of the sequence to be transcribed
- [^{35}S]UTPαS, >37 TBq/mmol, 1000 Ci/mmol, 20 mCi/ml (Amersham Pharmacia Biotech catalogue no. SJ603)[a]
- 0.4 M $NaHCO_3$ (3.66 g in 100 ml water), sterilized by autoclaving

- RNA polymerase as appropriate for promoter and probe sequence required (SP6, Amersham Pharmacia Biotech catalogue no. E2520; T7, catalogue no. E70001; T3, catalogue no. E70051)
- 0.2 M EDTA, pH 8.0
- RNase-free DNase I (e.g. Amersham Pharmacia Biotech catalogue no. E2210) freshly diluted to 10 units/80 μl in sterile water
- 0.6 M Na_2CO_3 (6.36 g in 100 ml water), sterilized by autoclaving
- Sterile water
- Glacial acetic acid
- 3 M sodium acetate (24.6 g anhydrous sodium acetate dissolved in 90 ml water; pH adjusted to 5.2 by the addition of glacial acetic acid; made up to 100 ml and sterilized by autoclaving)
- 10 mg/ml yeast tRNA (e.g. Sigma catalogue no. R8759)
- Ethanol

Method

1 Dissolve the DNA in either distilled water or TE buffer at a concentration of ~500 μg/ml. Place all solutions, except the enzyme, on ice to thaw. Leave the enzyme in a freezer until required and return immediately after use.

2 Add the following, in the order given, to a microcentrifuge tube at room temperature.[b]

Transcription buffer	4 μl
DTT solution	1 μl
HPRI	20 units
ATP	0.5 μl

CTP	0.5 μl
GTP	0.5 μl
Linearized template DNA[c]	1 μg
[^{35}S]UTPαS	10 μl
RNA polymerase (SP6, T7 or T3)	20 units

3 Mix gently by pipetting up and down, and cap the tube. Spin for a few seconds in a microcentrifuge to bring the contents to the bottom of the tube. Avoid vigorous mixing as this can cause loss of enzyme activity.

4 Incubate for ≥1 h.[d] For SP6 polymerase incubate at 40°C and for T7 and T3 incubate at 37°C.

5 If the reaction is to be used in a hybridization with no further processing, it may be terminated by the addition of 2 μl 0.2M EDTA and a small aliquot (1–2 μl) can be removed for determination of incorporation[e] (*Protocol 1*). Denaturation is not required for use in hybridization as the probe is single stranded.

Template removal

6 If the DNA template is to be removed before hybridization, proceed from step **4** above. Do not add EDTA as it will inhibit the action of DNase I.

7 Add 10 units of RNase-free DNase I to the labelling mix from step **4** above.[f]

8 Incubate at 37°C for 10 min.

9 If the reaction is to be used in a hybridization at this stage, it may be terminated by the addition of 2 μl 0.2 M EDTA and a small aliquot (1–2 μl) can be removed for determination of incorporation (*Protocol 1*). Denaturation is not required for use in hybridization as the probe is single stranded.

Alkaline hydrolysis

10 To the reaction mix from step **8** above, add 20 μl 0.4 M $NaHCO_3$, 20 μl 0.6 M Na_2CO_3 and 60 μl sterile water. Mix gently.

11 Incubate at 60°C for a time based on the transcript length and the probe size required. This can be determined from

$$t = \frac{L_o - L_f}{k\,L_o\,L_f}$$

where t is incubation time (in minutes), L_o is the primary transcript length (in kilobases), L_f is the average probe length required (in kilobases) and the rate constant k is 0.11 cuts/kbase/min.

12 Add 1.3 μl glacial acetic acid, 20 μl 3 M sodium acetate solution, 2 μl yeast tRNA solution and 500 μl ethanol. Precipitate the RNA at −20°C for ≥2 h.

13 Pellet the RNA in a microcentrifuge at 13 000 r.p.m. for 15 min. Remove supernatant.

14 Rinse the pellet in 70% (v/v) ethanol at $-20\,^{\circ}$C. Centrifuge for 5 min at 13 000 r.p.m. Remove the supernatant.

15 Redissolve the RNA probe in sterile water (or sterile 0.1 M DTT for ^{35}S-labelled probes) at ten to 20 times the concentration required in the hybridization procedure. Store the RNA probe at $-20\,^{\circ}$C.[g] Denaturation is not required for use in hybridization as the probe is single stranded.

Calculation of probe's specific activity

As unlabelled vector DNA does not participate in the hybridization, the specific activity of the newly synthesized probe depends solely on the specific activity of the labelled nucleotide, while probe yield can be calculated from the percentage incorporation of label. The specific activity of the probe is independent of labelling efficiency.

Specific example

Assume 70% incorporation of 200 μCi of [^{35}S]UTPαS at >37 TBq/mmol, >1000 Ci/mmol.

(i) *Calculation of probe yield*

The amount of labelled nucleotide incorporated is $200 \times 0.7 = 140$ μCi. This is equivalent to 0.14 nmol (1000 Ci = 1 mmol), or $0.14 \times 350 = 49$ ng of incorporated label (molecular weight of a nucleoside monophosphate in DNA is ~350). With a single labelled nucleotide species this is equivalent to $49 \times 4 = 196$ ng of probe. Note that the amount of probe is significantly less than the amount of whole vector present in the reaction.

(ii) *Calculation of specific activity*

1 mCi = 2.2×10^9 d.p.m., so the specific activity of the [^{35}S]UTPαS can be expressed as

$$(1000 \times 2.2 \times 10^9)/350 = 6.3 \times 10^9 \text{ d.p.m./μg}$$

Therefore, the specific activity of the probe = $6.3 \times 10^9/4 = 1.6 \times 10^9$ d.p.m./μg. If unlabelled UTP is added to the reaction, the specific activity of the probe will be reduced, although the yield may be improved.

[a] There are several formulations of [^{35}S]UTPαS available which are appropriate for use with phage polymerase labelling reactions. However, with some formulations of high specific activity, it is difficult to achieve the required nucleotide concentration of 12.5 μM, based on the apparent K_m of the enzyme for UTP, to ensure a predominance of full-length transcripts. As transcripts may therefore be of variable length, it is then more difficult to predict optimum conditions for alkali digestion. This concentration is more readily achieved with nucleotides of lower specific activity (for example, 15 or 30 TBq/mmol, 400 or 800 Ci/mmol), but the specific activity of the probe is correspondingly lower. The formulation suggested (1000 Ci/mmol, 20 mCi/ml) is at a sufficiently high concentration to allow 12.5 μM to be achieved in the reaction whilst also maximizing specific activity. Higher specific activities for *in situ* hybridization are achievable with [α-^{33}P]UTP, which is available at 37–110 TBq/mmol, 1000–3000 Ci/mmol, 20 mCi/ml (Amersham Pharmacia Biotech catalogue no. BF1002), but the transcript length is likely to be reduced due to low nucleotide concentration using 10 μl per 20 μl reaction. As there are currently no alternative formulations, the hydrolysis time must be re-optimized when using this nucleotide. As a general rule, the hydrolysis time can be calculated for the theoretical full-

length transcript and then halved. Several formulations of $[\alpha\text{-}^{32}\text{P}]$UTP are also available. Again, it is not possible to achieve a concentration of 12.5 μM in the labelling reaction with the highest specific activities (110 TBq/mmol, 3000 Ci/mmol, 10 mCi/ml; Amersham Pharmacia Biotech catalogue no. PB10203). Alternatives with lower specific activity (e.g. 30 TBq/mmol, 800 Ci/mmol, 20 mCi/ml; Amersham Pharmacia Biotech catalogue no. PB20383) produce probes of adequate specific activity (1.3×10^9 d.p.m./μg) for most *in situ* and membrane hybridizations.

[b] A similar reaction can be set up with the same concentration of labelled CTP with unlabelled ATP, GTP and UTP. ATP and GTP are not generally recommended as labels because GTP is involved in the transcription initiation step and ATP has a higher K_m with some enzymes.

[c] Plasmids containing an insert to be used as the template for RNA probe production should be linearized with an appropriate restriction enzyme. The restriction site should ideally be as close as possible to the end of the insert sequence, to avoid production of labelled plasmid sequences which may cross-hybridize with target sequences.

[d] Nucleotide analogues such as $[^{35}\text{S}]$UTPαS are incorporated less efficiently than normal nucleotides. With $[\alpha\text{-}^{32}\text{P}]$UTP and $[\alpha\text{-}^{33}\text{P}]$UTP label it is possible to use 4–10 units of polymerase. Under these conditions, a reaction time of 1 h is adequate for all nucleotides.

[e] With purified DNA template, the reaction should give >70% incorporation of label. RNA probe at this stage should not be denatured before hybridization as this will also render the unlabelled template single stranded. This is generally present in excess over labelled probe and will also hybridize with target.

[f] If a DNase is used that is not known to be RNase-free, then an additional 20 units of HPRI should be added.

[g] For longer-term storage, RNA probes can be stored at -70°C.

3.4 End-labelling of oligonucleotides

Methods are available for labelling either the 3′- or the 5′-ends of oligonucleotides or longer nucleic acid molecules.

3.4.1 3′-End-labelling with terminal transferase

In the presence of dNTPs, the enzyme terminal deoxynucleotidyl transferase (TdT) will introduce a series of nucleotides at the 3′-end of DNA molecules in a manner that is not dependent on the presence of a template strand (9). By including only a single nucleotide species, a 3′-homopolymer tail can be synthesized, the length of which can be controlled by a number of factors such as reaction time, divalent cation species and the molar ratio of nucleotides to 3′-ends available for labelling. If a radiolabelled dNTP is used, the addition of multiple residues will increase the specific activity of the probe and, hence, maximize the sensitivity of detection. However, in some cases, the presence of too many additional nucleotide residues can reduce the sequence specificity of the probe during hybridization. The use of a 2′,3′-dideoxynucleotide, available as $[\alpha\text{-}^{32}\text{P}]$ddATP (Amersham Pharmacia Biotech catalogue nos PB10233 and PB10235), restricts the addition to a single residue as there is no 3′-hydroxyl group available for formation of a phosphodiester bond to a further nucleotide.

For preparation of oligonucleotide probes for *in situ* hybridization it is possible to use either [α-^{33}P]dATP (Amersham Pharmacia Biotech catalogue no. BF1001) or [^{35}S]dATPαS which is available in a formulation specifically designed for use in 3′-end-labelling reactions (Amersham Pharmacia Biotech catalogue no. SJ1334). *Protocol 5* details the use of [^{35}S]dATPαS for labelling an oligonucleotide. The reaction uses 50 pmol labelled nucleotide and 10 pmol oligonucleotide, so that only a short chain of labelled residues will be introduced. The 3′-end-labelling reaction will also label double-stranded DNA, with more efficient incorporation reported for molecules with protruding 3′-ends.

Protocol 5

3′-End-labelling

Reagents

- Cacodylate buffer (1.4 M sodium cacodylate, pH 7.2, 10 mM cobalt (II) chloride, 1 mM DTT). **Note**: Cacodylate is an arsenic-containing compound which is highly toxic by contact with skin or if swallowed. It may be a carcinogen and with danger of cumulative effects. Cobalt (II) chloride is also harmful if swallowed, inhaled or absorbed through the skin and may cause eye and skin irritation. Manufacturers' safety data sheets should be consulted for these compounds and appropriate safe handling procedures followed.

- Oligonucleotide to be labelled[a]
- [^{35}S]dATPαS, 37 TBq/mmol, 1000 Ci/mmol, 10 mCi/ml (Amersham Pharmacia Biotech, catalogue no. SJ1334)
- TdT
- 0.2 M EDTA, pH 8.0

Method

1   Place all solutions except the enzyme on ice to thaw. Leave the enzyme in a freezer until required and return immediately after use.

2   Add the following in the order given to a microcentrifuge tube on ice:

cacodylate buffer	5 μl
oligonucleotide	10 pmol[b]
water to a final reaction volume of 50 μl	
[^{35}S]dATPαS[c]	5 μl
TdT	10 units

3   Mix gently by pipetting up and down, and cap the tube. Spin for a few seconds in a microcentrifuge to bring the contents to the bottom of the tube. Avoid vigorous mixing as this can cause loss of enzyme activity.

4   Incubate for 1–2 h at 37 °C.

5   Terminate the reaction by adding 5 μl of 0.2 M EDTA.

6 Remove a small aliquot (1–2 μl) for determination of incorporation[d] as described in *Protocol 1*.

7 If a single-stranded oligonucleotide is labelled then the probe can be used directly in hybridization without denaturation. If the substrate is double stranded, then denature the probe by heating to 95–100°C for 5 min then chill on ice. For longer-term storage keep the probe at −20°C.

Calculation of specific activity

$$\text{Specific activity (d.p.m./μg)} = \frac{\text{total activity incorporated (d.p.m.) (from } \textit{Protocol 1}\text{)}}{\text{amount of substrate added (μg)}}$$

An incorporation of 50% in the above reaction would give the following specific activity:

total activity incorporated = $0.5 \times 50 \times 2.2 \times 10^6 = 5.5 \times 10^7$ d.p.m.

amount of substrate added = 50 ng = 0.05 μg (for a 15-mer oligonucleotide)

specific activity = 1.1×10^9 d.p.m./μg

[a] It is recommended that the probe to be labelled is dissolved or diluted in sterile distilled water.

[b] Calculation of oligonucleotide concentration from absorbance (A) varies according to the base composition. The molar extinction coefficient (ε) at 260 nm (pH 8.0) for a given oligonucleotide can be obtained by summing the contribution of each nucleotide: A = 15 200, C = 7050, G = 12 010 and T = 8400. Concentration (mol/l) = A_{260}/ε. The weight of oligonucleotide corresponding to 10 pmol is dependent on the length of the sequence. As an approximate estimate, for each additional residue the amount required to give 10 pmol increases by 3.3 ng. Thus, for example, 50 ng of a 15-mer oligonucleotide is equivalent to 10 pmol, while 166 ng of a 50-mer is required.

[c] The reaction is illustrated for use with a formulation of [^{35}S]dATPαS that has been optimized for efficient incorporation in 3′-end-labelling reactions. Other formulations may contain levels of DTT that inhibit the reaction. ^{32}P- and ^{33}P-labelled nucleotides do not contain DTT and may be used successfully in this reaction.

[d] A typical incorporation obtainable using an oligonucleotide in the reaction detailed above would be ~50% labelled nucleotide.

3.4.2 3′-End-repair catalysed by Klenow polymerase

In the presence of suitable deoxynucleotides, the 5′–3′ polymerase activity of the Klenow fragment can be used to fill in from a recessed 3′-end, produced by annealed oligonucleotides or restriction endonuclease cleavage, using the corresponding 5′-overhang as template. The nucleotides chosen for labelling will depend on the sequence of the 5′-overhang. It is often possible to fill in and label with only one nucleotide or with several, depending on the length of the overhang, its sequence and the labelled nucleotides supplied. Thus, some variation in specific activity is possible. By careful choice of nucleotide it is also possible to label selectively one species of end in a mixture of fragments cut with different restriction enzymes. Single fragments produced by a double cut can also be selectively labelled at one end, thereby producing strand-specific probes. Klenow polymerase can also introduce label at blunt ends; the 3'–5' activity of the enzyme is adequate for the removal of terminal 3' nucleotides, allowing subsequent replacement by a labelled equivalent. The enzyme can also add an additional non-specific nucleotide to a blunt end.

When used for end-labelling, it should be remembered that Klenow polymerase will continue to carry out pyrophosphate exchange (removal of terminal nucleotide by 3'–5' exonuclease activity followed by repolymerization) when all residues have been filled in. At low nucleotide concentrations, this can cause conversion of all free dNTP corresponding to the terminal nucleotide to dNMP, resulting in loss of the terminal nucleotide. For this reason, short reaction times (10–15 min at room temperature) are recommended. Termination by methods other than heat inactivation is also advisable, as raised temperature increases the rate of the exchange reaction.

The major advantages of Klenow-catalysed end-labelling are that it can generally be used after restriction digestion with no intermediate purification, relatively high specific activities can be achieved and selective labelling is possible, facilitating production of strand-specific probes. The major disadvantage is that it cannot be used efficiently for 3'-overhangs. A method suitable for most situations is given in *Protocol 6*. As with other end-labelling methods, the relative amounts of available ends and picomoles of label can be altered depending on requirements. A variety of radionucleotides may be used but, because label density is low, ^{32}P is often the label of choice.

Protocol 6

End-repair using Klenow polymerase

Reagents

- DNA to be labelled[a]
- 10× end-repair buffer (500 mM Tris–HCl, pH 7.5, 100 mM $MgCl_2$, 10 mM DTT)
- 5× nucleotide mix (containing each nucleotide at 100 μM, except those to be used as label) in 10 mM Tris–HCl, pH 8.0, 1 mM EDTA

- [α-^{32}P] dNTP(s) at 110 TBq/mmol, 3000 Ci/mmol, 10 mCi/ml[b]
- Klenow polymerase
- 0.2 M EDTA
- Chase solution (1 mM each of dATP, dCTP, dGTP and dTTP)

Method

1   Set up the following reaction in a microcentrifuge tube on ice:

    DNA to be labelled (equivalent to 100 pmol ends for a 30 bp oligo)   1 μg

    H_2O to a final volume of 20 μl

    10× end-repair buffer   2 μl

    5× nucleotide mix   2 μl

$[\alpha\text{-}^{32}P]dNTP(s)$ (10 μCi, 3 pmol)	1 μl
Klenow polymerase	2 units

2 Incubate at room temperature for 15 min.

3 Add 2 μl of chase solution and incubate for a further 5 min at room temperature before addition of EDTA.[c]

4 Terminate reaction by addition of 5 μl 0.2 M EDTA. Unincorporated nucleotides may be removed by ethanol precipitation or chromatography on Sephadex G-50 (1,2).

5 Denature double-stranded probe at 95–100 °C for 5 min before hybridization.

Calculation of probe's specific activity
Specific activity can be calculated as in Section 2.4.1.

[a] It is also possible to label DNA in most restriction enzyme buffers without prior purification, as the primary requirements for Klenow polymerase are Mg^{2+} ions and a roughly neutral pH.

[b] The use of high specific activity nucleotides is shown above, in order to provide high labelling efficiency. However, it is also possible to use lower specific activity label (15 or 30 TBq/mmol, 400 or 800 Ci/mmol) or a higher amount of high specific activity label. As discussed for other end-labelling techniques, a variety of DNA:label ratios are possible, depending on whether the probe's specific activity or efficient incorporation of label is the chief priority.

[c] If it is important that all end-labelled molecules are of the same length, for example when using 3'-recessed DNA fragments in a cloning step, it is advisable to carry out a cold chase step after labelling.

3.4.3 5'-End-labelling

An advantage of 5'-end-labelling is that oligonucleotides can, unlike 3'-end-labelled oligonucleotides, be used as primers, for example in sequencing, PCR and probe capture experiments. Polynucleotide kinase (PNK) does not catalyse the incorporation of a nucleotide but the transfer of a phosphate group from the terminal γ-position of a ribonucleoside triphosphate (most commonly ATP) to the 5'-end of the terminal nucleotide of a nucleic acid molecule (10). The nucleic acid may be DNA or RNA, including chemically synthesized oligonucleotide. Labelling is achieved in one of two ways. First, if the 5'-end contains a hydroxyl group (as is the case with chemically synthesized oligonucleotides or with DNA that has been dephosphorylated with alkaline phosphatase), a direct transfer of phosphate can take place (the forward reaction). With $[\gamma\text{-}^{32}P]ATP$ or $[\gamma\text{-}^{33}P]ATP$ as donor, probe can be labelled with a high degree of efficiency but transfer of a phosphorothioate group from $[^{35}S]ATP\gamma S$ occurs with lower efficiency and it is necessary to use a higher enzyme concentration and longer reaction times for effective labelling. Second, if the 5'-end contains a phosphate group, then PNK can be used to catalyse an exchange reaction (11) in the presence of an excess of ADP. This occurs with a lower efficiency than the forward reaction. During the reaction, the 5'-phosphate of the nucleic acid is transferred to ADP, converting it to ATP, then the 5′-end is rephosphorylated using a nucleoside triphosphate donor (usually ATP) as in the forward reaction.

Protocol 7 can be employed for labelling oligonucleotides, DNA or RNA using the forward reaction. It includes details of 5′-dephosphorylation with calf intestinal alkaline phosphatase. If an oligonucleotide is to be labelled, this can be omitted and the protocol started from step **6**.

Protocol 7

5′-End-labelling

Reagents

- DNA or RNA to be labelled
- Calf intestinal alkaline phosphatase (CIAP)
- 10× CIAP buffer (0.5 M Tris–HCl, pH 9.0, 10 mM $MgCl_2$, 10 mM $ZnCl_2$, 10 mM spermidine)
- 10 mM Tris–HCl, pH 8.0
- Phenol saturated with TE buffer
- 24:1 (v/v) chloroform:isoamyl alcohol, or ether
- 5 M NaCl
- Ice-cold absolute ethanol

- TE buffer (10 mM Tris–HCl, pH 8.0, 1 mM EDTA)
- 10× kinase buffer (0.5 M Tris–HCl, pH 7.6, 0.1 M $MgCl_2$, 50 mM DTT, 1 mM spermidine)
- T4 polynucleotide kinase
- [γ-^{32}P]ATP (110 TBq/mmol, 3000 Ci/mmol, 10 mCi/ml, Amersham Pharmacia Biotech catalogue no. PB10168)
- 0.2 M EDTA, pH 8.0

Method

1 Dissolve the nucleic acid at a concentration of 10 pmol 5′-ends in 5–35 μl 10 mM Tris–HCl, pH 8.0. Place all solutions, except the enzymes, on ice to thaw. Leave the enzymes in a freezer until required, and return immediately after use.

2 Add the following in the order given to a microcentrifuge tube on ice:

DNA/RNA/oligonucleotide	10 pmol ends[a]
10× CIAP buffer	5 μl
water to final reaction volume of 50 μl	
CIAP	0.05 units

3 Mix gently by pipetting up and down, and cap the tube. Spin for a few seconds in a microcentrifuge to bring the contents to the bottom of the tube. Avoid vigorous mixing as this can cause loss of enzyme activity.

4 Incubate at 37 °C for 30 min for duplex DNA or at 55 °C for 30 min for RNA.

5 Briefly centrifuge as in step **3** above, then add an equal volume of buffer-saturated phenol. Extract twice with phenol and twice with chloroform–isoamyl alcohol or ether. Add 0.1 vol of 0.5 M NaCl followed by 2 vols of cold absolute ethanol for DNA or 3 vols for RNA. Precipitate the nucleic acid at −20 °C overnight or at −80 °C for 30 min.

Protocol 7 continued

6 Resuspend the pellet in 10 μl TE buffer and add the following in the order given to a microcentrifuge tube on ice:[b]

DNA/RNA	10 μl
10× PNK buffer	5 μl
water to a final reaction volume of 50 μl	
[γ-^{32}P]ATP	20 μl[c]
T4 polynucleotide kinase	10 units

7 Mix gently by pipetting up and down, and cap the tube. Spin for a few seconds in a microcentrifuge to bring the contents to the bottom of the tube. Avoid vigorous mixing as this can cause loss of enzyme activity.

8 Incubate at 37°C for 30–60 min.

9 Terminate the reaction by adding 5 μl 0.2 M EDTA.

10 Remove a small aliquot (1–2 μl) for determination of incorporation (*Protocol 1*).

11 If a single-stranded oligonucleotide is labelled then the probe can be used directly in hybridization without denaturation. If the substrate is double stranded, then denature the probe by heating to 95–100°C for 5 min then chill on ice. For longer-term storage keep the probe at −20°C.

Calculation of probe's specific activity

Specific activity can be calculated as for 3'-end-labelled probes described in *Protocol 5*.

[a] Calculation of oligonucleotide concentration is as described for 3'-end-labelled probes.

[b] The forward and exchange reactions are most efficient with single-stranded molecules or double-stranded molecules with protruding 5'-ends. If the DNA molecules have blunt or recessed 5'-ends, then a short incubation at 70°C followed by rapid chilling on ice prior to labelling (i.e. between steps **5** and **6**) may improve efficiency.

[c] The reaction contains 10 pmol 5'-ends and 67 pmol [γ-^{32}P]ATP in order to maximize the probe's specific activity. As the label is in molar excess, the percentage incorporation will be low and removal of unincorporated label is advisable. This can be achieved by precipitation or gel filtration chromatography or, for oligonucleotides, by thin-layer chromatography or polyacrylamide gel electrophoresis in the presence of 7 M urea (12). This has the additional benefit of separating labelled and unlabelled oligonucleotide based on the different mobilities of 5'-phosphorylated and 5'-hydroxyl oligonucleotides, although it will be found that, for most hybridization applications, this further purification will not be necessary. A more economical use of labelled nucleotide can be achieved by adding a smaller amount of nucleotide (20–50 μCi), although a smaller proportion of labelled oligonucleotides may result.

3.5 Choosing a radioactive labelling system for nucleic acid probes

Table 1 summarizes the main properties of the labelling methods described above and indicates some of the factors involved in choosing a labelling approach. While in many cases it will be possible to choose from more than one approach,

Table 1 Properties of nucleic acid labelling reactions

Method	Labelling density	Template	Amount of template (typical)	Reaction time	Incorporation efficiency	Nature of probe	Amount of probe	Specific activity of probe (d.p.m./μg)
Random prime	uniform	ssDNA (and denatured dsDNA)	25 ng	5 min–3 h	~75%	DNA	40–50 ng	5×10^9
Nick translation	uniform	dsDNA	0.5–1 μg	~2 h	~60%	DNA	0.5–1 μg	5×10^8
Transcription labelling	uniform	dsDNA	0.5–1 μg	1 h	~75%	RNA	~250 ng	5×10^9
3'-End-labelling	end	oligo or DNA fragment	~10 pmol ends	30–60 min	variable	oligo or DNA	10 pmol	5×10^6
End-repair	end	oligo or dsDNA fragment	~100 pmol ends	15 min	variable	oligo or DNA	100 pmol	5×10^5
5'-End-labelling	end	oligo or DNA fragment[a]	~10 pmol ends	~1 h	variable	oligo or DNA	10 pmol	5×10^6

[a] Dephosphorylated for forward reaction.

each reaction has specific features which may make it more appropriate for a particular experimental situation. The following is a brief summary for each of the above reactions which, it is hoped, will be of help in guiding decisions.

Random prime labelling is the most frequently used method for uniform labelling of DNA to provide probes with high label density, particularly ^{32}P-labelled probes for blotting applications. Incorporation is efficient and it is effective with small amounts of template and radiolabel. To facilitate higher throughput or convenience, alternative formulations are commercially available. For example, Ready-To-Go DNA Labelling Beads (Amersham Pharmacia Biotech) use a technology that converts pre-dispensed reaction mixes into a small bead which can be rapidly reconstituted. Each bead is capable of labelling between 10 ng and 1 μg of DNA and requires the addition of radiolabelled dCTP. Specific activities of $>10^9$ d.p.m./μg can be achieved within 5 min. Similar specific activities are achievable within 5–10 min using radiolabelled dCTPs with the Rediprime system (Amersham Pharmacia Biotech) in which individual pre-aliquoted reaction mixes are dried in separate tubes. Although the nick translation reaction can also be adapted to use small amounts of template with acceptable incorporation efficiency, it is generally used as a means of incorporating label at high density into relatively large (microgram) amounts of template. A common application for this technique is to introduce ^{33}P or ^{35}S into probes for *in situ* hybridization.

Transcription labelling produces radiolabelled RNA probes from inserts situated downstream of an appropriate polymerase promoter. If the template sequence has been cloned between two paired promoters, then it is possible to produce asymmetric probes consisting of a sense and an anti-sense sequence. These are of particular use in *in situ* hybridization to provide positive results and negative controls and, again, ^{33}P and ^{35}S are most frequently used for this application. Increasingly, sequences are transcribed from promoters not in vectors but in sequences that have been amplified by PCR (13,14) and the RNA products can be used in a coupled transcription–translation reaction.

End-labelling methods are the primary approach for labelling oligonucleotides. A major application of radioactive 3′-end-labelling (and Klenow in-filling for double-stranded fragments) is in the production of ^{32}P-labelled probes for less sensitive screening applications involving colonies, plaques or PCR clones. [33]P or ^{35}S end-labelled probes are also used for *in situ* hybridization. 5′-End-labelled oligonucleotides can be used in similar applications but, additionally, can be subjected to further polymerase-based reactions such as PCR. 5′-End-labelled primers can be used for increasing the sensitivity or quantification of PCR reactions. They act primarily as a tracer, as the final specific activity of the PCR product is generally too low to be used as a probe except on the lowest sensitivity applications. End-repair reactions are often used with annealed oligonucleotides with sticky ends for examining the binding of double-stranded DNA to binding proteins.

PCR reaction products can also be labelled either during or after PCR. As there is a relatively high concentration of nucleotide present during the PCR reaction, radiolabel is greatly diluted unless present in very high quantities, so that the final specific activity of the probe is again low. If PCR products are to be used as probes, it is most efficient to use a technique such as random prime labelling with an appropriate aliquot of the reaction mix. If the dilution is sufficient, it may be possible to label to high specific activity directly, but it is generally advisable first to remove unincorporated nucleotides and primer either by column purification or by selective precipitation (1).

4 Cycle sequencing with ^{33}P-labelled terminators

DNA sequencing for some years has been largely based on the use of 2′,3′-dideoxynucleoside triphosphate (ddNTP) terminators which, when incorporated at a low level by a polymerase into growing DNA products, result in the formation of a ladder of terminated fragments; the lack of a 3′-hydroxy group prevents any further extension of the chain (15). This sequencing technique also requires that extension fragments be labelled to allow their detection following gel separation. While high and medium throughput sequencing is invariably carried out non-radioactively, short sequences are commonly determined manually with radioactivity. In radioactive sequencing, the method of labelling extension fragments has traditionally been either through the use of labelled primers or, more commonly, through the use of labelled dNTPs.

Recent advances have resulted in a most convenient and accurate way of manual sequencing based on temperature cycling and the use of [α-^{33}P]ddNTPs as terminators. These advances have been based firstly on the fact that an efficient way of manufacturing [α-^{33}P]ddNTPs has been found which has resulted in the commercial availability of all four labelled ddNTPs. Secondly, a thermostable sequencing enzyme has been developed which will incorporate dideoxy-nucleotides as efficiently as deoxynucleotides (16).

Labelled ddNTPs have been used in fluorescence sequencing for a number of years. They ensure that only correctly terminated DNA strands are detected. Methods that rely on detection of fragments through the use of labelled primers or dNTPs suffer from the appearance of background bands that have not been properly terminated by a ddNTP. In extreme cases, heavy background bands are found across all four lanes at the same position, presumably resulting from secondary structure within the template. By using a labelled ddNTP terminator, only correctly terminated fragments are labelled and a much clearer result is obtained. Hence, the development of radiolabelled terminators has made a significant improvement to manual sequencing. In addition to this, ^{33}P provides significant advantages over ^{35}S, which had previously been used for radioactive sequencing. In order to obtain the resolution required to achieve long read lengths with ^{35}S, direct autoradiography was used, which typically requires several days to obtain a result. However, the use of ^{33}P gives acceptable resolution and sensitivity in an overnight exposure to film. The presence of sulfur in the α-phosphate position of nucleoside triphosphates also renders the triphosphate a less efficient substrate for DNA polymerases.

Protocol 8

Cycle sequencing with ^{33}P-labelled terminators

Reagents

- DNA to be sequenced
- Sequencing primer
- ThermoSequenase DNA polymerase (Amersham Pharmacia Biotech)
- Reaction buffer (260 mM Tris–HCl, pH 9.5, 65 mM MgCl$_2$)

- Termination master mix (15 μM each of dATP, dCTP, dGTP and dTTP)
- Set of all four [^{33}P]ddNTPs (radiolabelled terminators), 55.5 TBq/mmol, 1500 Ci/mmol (450 μCi/ml) (Amersham Pharmacia Biotech)

Method

1 Prepare reaction mix as follows:

Reaction buffer	2 μl
DNA	50–500 ng or 2.5–250 fmol
Primer	0.5–2.5 pmol
H$_2$O	to 20 μl total volume
ThermoSequenase polymerase	2 μl (added last)

2 Prepare four tubes, each labelled A, C, G or T, and transfer 2 μl of the termination master mix to each. Then add 0.5–1 μl of the appropriate [α-^{33}P]ddNTP (G, A, T, C) to each tube.

3 Transfer 4.5 μl of reaction mix (from step **1** above) to each tube. Cap tubes, mix gently and centrifuge briefly. If the thermocycler does not have a heated lid, add an overlay of 10–20 μl of mineral oil. Place tubes in a thermal cycler.

4 Heat at 95 °C for 30 sec, then at 60 °C for 30 sec and then at 72 °C for 60–120 sec. Repeat, typically for 30 cycles (taking 2–3 h). Fewer cycles may produce better results when using 250–500 fmol DNA.

5 Gel electrophoresis: after the cycling is complete, add 4 μl of standard sequencing stop solution (1) to each of the termination reactions, separate the oil by centrifugation (if required) and load 3–5 μl on to a denaturing acrylamide gel (1), after a 2 min denaturation at 70 °C.

^{a} Cycling conditions may vary if nucleotide analogues are used to avoid compression artefacts. For example, if dITP is used, the extension temperature should be reduced to 60 °C, the extension time increased to 4–10 min and the annealing temperature reduced to 55 °C.

With regard to terminator incorporation, whereas ddNTPs are good substrates for T7 DNA polymerase and commercially available versions of this such as Sequenase, they are poor substrates for other enzymes such as *Taq* DNA polymerase. The reason for this is a single amino acid difference in the active site. A phenylalanine to tyrosine mutation in the active site of *Taq* DNA polymerase allows the mutant enzyme to incorporate ddNTPs on average 3000 times more efficiently than an equivalent wild-type enzyme (17), with the result that there is little difference in incorporation efficiency between the two substrates. This has made it possible to use this mutant enzyme (ThermoSequenase, Amersham

Pharmacia Biotech) in cycle sequencing, a form of sequencing which enables the use of very small amounts of template.

Other nucleotide analogues, such as dITP, can be used in place of dGTP to avoid compression artefacts seen on sequencing gels, due to the presence of secondary structure in the DNA. The procedure for sequencing is given in *Protocol 8*.

5 Applications of labelled proteins

Radioactive peptides and proteins are used extensively in many areas of biochemistry, pharmacology and medicine. For example, they are used in radio-immunoassay and related techniques, receptor studies, protein blotting, photo-affinity and chemical cross-linking, membrane studies, specific enzyme inhibition studies, etc. Several different radioisotopes can be employed for labelling protein, such as ^{14}C, ^{3}H, ^{125}I and ^{131}I. However, in practice, most are labelled with ^{125}I due to its radiochemical characteristics, ease of use and lower costs.

6 Labelling of proteins with ^{125}I

6.1 General considerations

Radio-iodination is the process of chemically modifying a molecule to contain one or more atoms of radioactive iodine. In general the procedures for radio-iodinating proteins can be grouped into two categories which can be optimized to produce a satisfactory ligand (18):

(1) direct incorporation of ^{125}I into tyrosines (and/or) histidine residues of proteins, usually in the presence of an oxidizing agent;

(2) conjugation methods, in which the radioactive moiety is conjugated with lysine or terminal amino residues via amide bonds.

The choice of iodination method depends on a number of factors. Direct methods are generally rapid and convenient, giving high yields and specific activities. They involve the oxidation of $^{125}I^-$ to '$[^{125}I^+]$' (iodonium), followed by electrophilic substitution of $^{125}I^+$. Several reagents, such as chloramine-T, Iodo-Gen™, lactoperoxidase/hydrogen peroxide, may effect this oxidation. Other less commonly used methods involve the use of iodine monochloride, chlorine, sodium hypochlorite and *N*-bromosuccinimide (19–23).

One of the problems associated with direct methods is the unwanted oxidation of susceptible groups such as methionine and tryptophan residues. Methionine residues can be oxidized to sulfoxides and sulfones, which can often render the protein biologically inactive (23). Procedures that use strong oxidizing agents, such as chloramine-T, are more likely to lead to oxidation problems, whereas the milder procedures are less likely to oxidize the protein.

6.2 Direct labelling

6.2.1 Radioiodination using chloramine-T

The procedure described in *Protocol 9* was introduced by Hunter and Greenwood (24) and is probably the most widely used method for iodination of peptides and proteins to give products of high specific activity. The advantages of the chloramine-T method are its reproducibility, economy, rapidity and efficiency even with less than half a nanomole (1 mCi) of ^{125}I. Reaction at neutral pH in aqueous, isotonic solvents helps protect against denaturing the proteins.

Chloramine-T (the sodium salt of the *N*-monochloro derivative of *p*-toluene sulfonamide) breaks down slowly in aqueous solution, producing hypochlorous acid. This oxidizes sodium [^{125}I]iodide to '[^{125}I$^+$]', which is then incorporated into aromatic rings. The reaction is terminated usually with an agent such as sodium metabisulfite, which reduces excess chloramine-T and free iodine. During purification, carrier potassium iodide or a protein-containing buffer is often added to prevent losses of labelled material. The efficiency of incorporation depends on concentration as well as the relative amounts of reacting components. At optimum conditions, determined by experimentation for any particular protein, this method normally yields higher levels of incorporation than other direct methods of radio-iodination.

Oxidation damage, especially to methionine residues, can be minimized by using only a small excess of chloramine-T. The optimum pH for the iodination of tyrosine is pH 7.2–7.4, with reduced incorporation being obtained at pH <6.5 or >8.5. It is therefore necessary to buffer the reaction to pH 7.2, although to iodinate the imidazole ring of histidine a pH of 8.1 is required. Reagents in the chloramine-T iodination are added in rapid succession into a polypropylene or Sarstedt microcentrifuge tube, whilst continuously agitating to ensure thorough mixing of reagents. Poor mixing is probably the commonest cause of low yield of labelled protein by this method. Glass should be avoided if the protein is known to adhere non-specifically—this can be minimized by silanization. Chloramine-T, cysteine and sodium metabisulfite solutions should be freshly prepared prior to use.

Protocol 9

Radio-iodination using chloramine-T

Equipment and reagents

- Protein to be labelled (0.5 mg/ml) in sodium phosphate buffer (0.2 M, pH 7.4)
- Fresh chloramine-T (1.0 mg/ml) in sodium phosphate buffer (0.2 M, pH 7.4)
- Fresh sodium metabisulfite (1.0 mg/ml) in sodium phosphate buffer (0.2 M, pH 7.4).
- Sodium [^{125}I]iodide (>69.5 GBq/mmol, >1875 Ci/mmol)
- Pre-packed column of Sephadex G-25 (Amersham Pharmacia Biotech, PD-10, catalogue number 17-0851-01; see Section 6.4)

Method

1   Add chloramine-T (50 μl, 50 μg) to a microcentrifuge tube containing protein (50 μl, 25 μg).

2   Add sodium [^{125}I]iodide (1.0 mCi, 10 μl, 0.5 nmol) and mix thoroughly for 30 sec.

3 Terminate reaction by addition of sodium metabisulfite (50 μl, 50 μg) and mix thoroughly.[a]

4 Separate the radioactive labelled peptide from the remaining radioactive iodide by gel filtration using a pre-packed column of Sephadex G-25.

[a] The reaction can also be terminated by addition of cysteine solution (20 μl, 10 mM) although its use is not advisable in peptides and proteins that contain disulfide bridges. Tyrosine solution (0.2 mg/ml in phosphate buffer (0.2 M, pH 7.4)) can be used as an iodine scavenger. However, in cases when the specific activity is to be determined before termination of the reaction, a small sample of the reaction mixture should be terminated in sodium metabisulfite solution.

6.2.2 Radio-iodination using Iodo-Gen™

Iodo-Gen™ is an iodination reagent (1,3,4,6-tetrachloro-3α,6α-diphenylglycoluril) first described by Fraker and Speck (25) as an effective solid phase oxidation reagent for iodination of proteins and cell membranes. It is dissolved in a suitable organic solvent and then used to coat the walls of the reaction vessel by evaporating the solvent in a stream of dry nitrogen. The coated vials can be stored in dry conditions to allow batch iodinations. Once the vessel is coated with Iodo-Gen™, it may be stored at −20 °C, in the dark, for up to a month. Buffer, sodium [125I]iodide and protein are added to start the reaction. Due to the heterogeneous nature of the reaction, incubation times are typically in the 10–20 min range. The reaction is terminated by removal from the reaction vessel and loading straight on to the gel filtration column or by addition of a reducing agent. Although Iodo-Gen™ is almost completely insoluble in water, the iodination does not stop immediately after removal of solution from the Iodo-Gen™ vial. Therefore a small amount of reducing agent or iodine scavenger such as tyrosine should be added after removal. The coating of the reaction vial requires careful attention, otherwise small parts of the coated Iodo-Gen™ will become dislodged and contaminate the reaction mixture. Pre-coated Iodo-Gen™ iodination tubes are available from Pierce Chemical Co. (catalogue no. 28601V). An indirect method of labelling with Iodo-Gen™ known as the Chizzonite method can also be used to reduce contact of protein with oxidant (26). In this method the sodium [125I]iodide is initially activated by adding to an Iodo-Gen™-coated vial and then removed and added to the protein-containing vial. The activated iodide is allowed to react for 6–9 min before quenching with scavenging buffer. The procedure is described in *Protocol 10*.

Protocol 10

Direct radio-iodination using Iodo-Gen™

Equipment and reagents

- Protein to be labelled (1 mg/ml in sodium phosphate buffer)
- Iodo-Gen™ (Pierce Chemical Co.)
- Sodium phosphate buffer (0.2 M, pH 7.4)
- A source of nitrogen gas
- Dichloromethane
- Sodium [125I]iodide (>69.5G Bq/mmol, >1875 Ci/mmol)
- Tyrosine solution (50 μg/ml in sodium phosphate buffer)

Protocol 10 continued

Method

1 Dissolve 0.5 mg Iodo-Gen™ in dichloromethane (1.0 ml).

2 Dispense Iodo-Gen™ solution (10 μl) into the reaction tube and evaporate the solution to dryness by blowing a gentle stream of nitrogen on to the surface.

3 Wash the Iodo-Gen™ reaction vial by dispensing 50 μl sodium phosphate buffer into it and withdraw the solution to remove any non-adherent flakes of dry Iodo-Gen™.

4 Add 50 μl sodium phosphate buffer, 10–20 μl protein and 10 μl sodium [^{125}I]iodide (1.0 mCi) to the Iodo-Gen™ reaction vial.

5 Allow to react at room temperature for 10–15 min.

6 Terminate the reaction by transferring to another vial containing 250 μl tyrosine solution (50 μg/ml) in sodium phosphate buffer (0.2 M, pH 7.4).

7 Separate the labelled protein from the remaining radioactive iodide by gel filtration (Section 6.4).

6.2.3 Radio-iodination using Iodo-Beads™

Markwell (27) introduced a new oxidizing reagent, *N*-chloro-benzene sulfonamide, coupled covalently to non-porous polystyrene spheres (3.175 mm), referred to as Iodo-Beads™, to facilitate reproducible iodinations. The procedure is described in *Protocol 11*. Iodo-Beads™ are commercially available from Pierce Chemical Co. The oxidative capacity is about 0.55 μmol/bead and is limited to the outer surface of the bead which allows reaction times to be extended to ~15 min rather than 30 sec as with chloramine-T. The rate of reaction may be changed depending on the number of beads and the concentration of sodium [^{125}I]iodide. The reagent is useful over a broad range of pH (pH 6.5 is best) as well as temperature conditions. Removing the Iodo-Beads™ from the reaction, which avoids the use of reducing agents, terminates the reaction. Azide, detergents, urea or high salt concentration do not inhibit the performance of Iodo-Beads™. They are, however, inactivated by reducing agents, moisture on storage and organic solvents that dissolve the surface of the beads.

Protocol 11

Radio-iodination using Iodo-Beads™

Reagents

• Protein to be labelled (0.25 mg/ml in sodium phosphate buffer)

• Sodium phosphate buffer (0.2 M, pH 7.4)

• Iodo-Beads™ (Pierce Chemical Co.)

• Sodium [^{125}I]iodide (>69.5 GBq/mmol, >1875 Ci/mmol)

Method

1 Wash three Iodo-Beads™ with sodium phosphate buffer (500 μl; 0.2 M; pH 7.4).

Protocol 11 continued

2 Add the beads to 400 μl protein solution and 10 μl sodium [^{125}I]iodide (1.0 mCi, 0.5 nmol).

3 Allow to react at room temperature for 2–15 min.

4 Terminate the reaction by removing the solution from the beads. This can be done by pipetting the solution away from the beads or by physically removing the beads. Wash the beads with the iodination buffer to ensure complete recovery of protein.

5 Separate the radioactive labelled protein from the remaining radioactive iodide by gel filtration (see Section 6.4).

6.2.4 Radio-iodination using lactoperoxidase

Marchalonis first described this method for radio-iodination of immunoglobulin (28). Lactoperoxidase is used to catalyse the oxidation of iodide in the presence of a very small amount of hydrogen peroxide (29). The resulting iodine may react with tyrosine and histidine residues within the protein. The reaction is terminated by dilution with buffer or quenching the enzyme action with cysteine. Reaction times are longer (~20 min) than other direct radioiodinations but, as no strong oxidizing or reducing agents are used, immunological damage is kept to a minimum. In general, this method gives low yields and a product of low specific activitity (200–400 Ci/mmol) . The optimum pH for iodination is dependent on the nature of the protein, hence a number of experiments are often required to establish the best protocol.

Buffers incorporating sodium azide as a preservative must be avoided, as this is a potent inhibitor of lactoperoxidase. During the iodination, lactoperoxidase is self-iodinated, thus increasing the iodide loss and complicating the separation of labelled protein from labelled enzyme and determination of specific activity. Separation can be achieved typically by reversed phase HPLC; Sephadex G-25 is generally unsuitable as the self-iodinated [^{125}I]lactoperoxidase will co-elute in the void volume with the protein. The procedure is described in *Protocol 12*.

Protocol 12

Radio-iodination using lactoperoxidase

Reagents

- Protein to be labelled (1.0 mg/ml in sodium phosphate buffer, 0.2 M, pH 7.4)
- Lactoperoxidase (25 units/ml; Calbiochem)
- Sodium phosphate buffer (0.2 M, pH 7.4)
- Sodium [^{125}I]iodide (>69.5 GBq/mmol, >1875 Ci/mmol)
- Hydrogen peroxide (0.3%, 88 mM)

- Sodium azide (0.1% (w/v) in sodium phosphate buffer). Sodium azide is highly toxic. Manufacturer's safety data sheets should be consulted for this compound and appropriate safe handling procedures followed.

Protocol 12 continued

Method

1   Immediately before use, dilute the hydrogen peroxide (0.3%) to 0.003%.

2   Add 10 μl hydrogen peroxide (0.003%, 8.8 nmol) to the reaction vial containing 200 μl sodium phosphate buffer (0.2 M, pH 7.4), 25 μg protein and 10 μl lactoperoxidase.

3   Add 10 μl sodium [^{125}I]iodide (1.0 mCi, 0.5 nmol), mix and allow the reaction to proceed for 20 min at room temperature.

4   Stop the reaction by adding 800 μl 0.1% sodium azide in sodium phosphate buffer (0.2 M, pH 7.4) and separate the radioactive labelled peptide from the remaining radioactive iodide by gel filtration (Section 6.4) or HPLC.

6.3 Indirect methods

6.3.1 Radio-iodination using Bolton–Hunter reagent

The most widely used indirect method of radiolabelling is with Bolton–Hunter reagent (*N*-succinimidyl-3(4-hydroxy-5-[^{125}I]iodophenyl)propionate), which is an acylating reagent (30,31). The procedure is described in *Protocol 13*. The reagent itself is prepared by chloramine-T iodination. Conjugation occurs mainly with ε-amino groups of lysine residues and *N*-terminal amino groups. Rapid conjugation is essential as the acylating agent in aqueous medium is quickly degraded by hydrolysis to 3-(4-hydroxy)-5-[^{125}I]iodophenylpropionic acid, hence the reaction is concentration dependent. The conjugation takes place under mildly alkaline conditions (pH 8–8.5). This reaction overcomes problems of contact with oxidizing or reducing agents. It can be used when tyrosine residues occur in the biologically active regions of the protein or if they are not readily accessible for iodination unless some amount of unfolding of the protein molecule occurs.

The reaction needs to be carried out as rapidly as possible to minimize hydrolysis. The conjugation involves three steps: evaporation of the benzene, addition of the protein and purification. This operation *must* be performed in a well-ventilated fume cupboard or similar facility. Additional safety precautions may be taken by using a charcoal trap to absorb volatile ^{125}I; this is usually available on request from the manufacturer of the reagent. The method is technically quite complex requiring significant manipulation of radioactive material, although this can be reduced by using commercial sources of the reagent. This reagent is available commercially from a number of suppliers, e.g. Amersham Pharmacia Biotech (catalogue no. IM.5861), NEN and ICN Biomedicals. It is supplied in dry benzene at 74 TBq/mmol, 2000 Ci/mmol and has been purified by HPLC. The activity of the product can be increased by using the di-iodo [^{125}I]Bolton–Hunter reagent (148 TBq/mmol, 4000 Ci/mmol), but this is at the expense of reduced stability.

The Bolton–Hunter reagent can be used to conjugate to amine-containing proteins. The derivative can then be iodinated using one of the above methods. This is useful if the Bolton–Hunter derivative can be stored under stable conditions until required for iodination.

Protocol 13

Conjugation with [^{125}I]Bolton–Hunter reagent

Equipment and reagents

- Protein to be labelled (0.5 mg/ml in sodium borate buffer)
- [^{125}I]Bolton–Hunter reagent in benzene containing 0.2% dimethylformamide (74 TBq/mmol, 2000 Ci/mmol, 185 MBq/ml, 5mCi/ml)
- Sodium borate buffer (0.1 M, pH 8.5)
- 0.2 M glycine in sodium borate buffer
- nitrogen source
- two hypodermic needles
- charcoal trap (prepared by forming a sandwich of cotton wool/charcoal/cotton wool within the body of a syringe, then attaching the needle)

Method

1 Evaporate to dryness 0.2 ml [^{125}I]Bolton & Hunter reagent by blowing a gentle stream of dry nitrogen on to the surface of the solvent. This is achieved by inserting two hypodermic needles through the seal in the top of the vial and attaching a tube leading from the nitrogen source to one of the needles. The other needle acts as an outlet to which a charcoal trap is attached for the gaseous solvent.

2 Add the 10 μl protein to the dry reagent, mix and incubate the vial for 15–20 min at room temperature.

3 Add 50 μl glycine solution to quench the conjugation and incubate the vial for an additional 5 min at room temperature.

4 Separate the radioactive labelled peptide from the remaining radioactive iodide by gel filtration[a] (Section 6.4).

[a] Low molecular weight products may bind to serum albumins such as BSA, therefore buffers containing it should be avoided during the separation stage. Avoid thiol reagents, sodium azide and buffers containing free amines as these will react with the Bolton–Hunter reagent.

6.4 Purification of radio-iodinated proteins

Most procedures that employ radio-iodinated proteins require isolation of the labelled protein from the iodination mixture, which contains low molecular weight products (e.g. unreacted iodide, oxidizing and reducing agents). Gel filtration chromatography is the simplest and most common method used for the purification. It is usually carried out on a pre-packed column of Sephadex G-25 (Amersham Pharmacia Biotech, PD-10, catalogue no. 17-0851-01) which has been equilibrated with sodium phosphate buffer (0.2 M, pH 7.4, 10 ml) containing bovine serum albumin (0.2%) or gelatin (0.2%). The carrier protein is added to reduce non-specific binding, which may result in significant losses, especially when working with very small amounts of starting material. The benefit of not using a carrier protein is that the radiolabelled protein can be quantified by a protein assay. The tracer recovered will be contaminated with damaged protein (i.e. multi-iodinated protein, aggregates and oxidized protein), thus requiring additional purification. The methods used for additional purification will depend upon the individual cases. Powerful separation techniques such as affinity chromatography, ion exchange and reversed phase HPLC are commonly used in most laboratories.

Affinity chromatography is particularly useful when labelling immunologically active materials and receptor ligands. The conditions required to remove the labelled ligand from the affinity material are usually fairly harsh and can damage the products. Affinity chromatography does not remove the unlabelled material. Ion exchange chromatography, which separates according to charge, can be used to remove di-iodinated products and unlabelled material as well as lactoperoxidase, but will not separate various multi-iodinated species or oxidized material. Reversed phase HPLC can be used to separate various mono-iodinated materials from all the other reaction products and yield a pure product at maximum specific activity. The latter technique is ideally suited to small peptides.

6.5 Determination of specific activity

Specific activity is defined as the unit of radioactivity per mole, which in the case of iodine is usually quoted as TBq/mmol or Ci/mmol. The introduction of a single atom of ^{125}I into a protein molecule causes the least alteration to its structure and thus keeps to a minimum any substitution effects. In general, further incorporation accelerates radiolytic decomposition. For receptor studies, specific activity must be known accurately as it is used to quantify the density of a ligand and its affinity for receptors. In order to calculate the specific activity, the amounts of protein and radioactive iodide used need to be known, as well as the yield of the product. The radioactivity should be measured in a calibrated ionization chamber and in the same type of vial to ensure the same geometry on each occasion.

There are several ways of determining specific activity. If gel filtration is used as a method of purification, unreacted [^{125}I]iodide and labelled protein can both be measured in a calibrated instrument. [^{125}I]Iodide is not adsorbed on to the surfaces of plastic or gel filtration columns. Thus, any losses are normally due to the adsorption of [^{125}I]protein on to the gel filtration column. Measurement of radioactivity originally taken and that of the salt peak (free iodide) allows the calculation of the amount of radioactivity in the protein. The method is suitable for most methods of iodination. It assumes that the absorption of the unlabelled protein and labelled protein remain the same. However, with lactoperoxidase methods, self-iodination of the enzyme gives rise to complications, and with insoluble oxidizing media such as Iodo-Gen™, a certain proportion of the [^{125}I]iodide remains attached to the plastic in the absence of protein.

Ascending thin-layer chromatography can also be used to determine specific activity using a suitable medium and solvent. In such cases, using the appropriate mobile phase, the labelled protein remains at the origin, whereas the [^{125}I]iodide moves up the plate. By determining the percentage incorporation and knowing the amount of activity and protein used, the specific activity can be determined. It should be stressed that, in the case of proteins, the radioligand is, in reality, a heterogeneous mixture of labelled and unlabelled protein.

Trichloroacetic acid (TCA) precipitation can also be used to determine specific activity. A small sample of reaction mixture is taken, diluted and precipitated in 10% TCA. After 30 min on ice, the precipitate is pelleted by centrifugation for 5 min at 10 000*g*. An aliquot of the supernatant is counted and compared with the counts taken before precipitation to determine the level of incorporation in the reaction.

7 Labelling of proteins with ^{3}H and ^{14}C

7.1 Tritium labelling

Tritiated proteins can be prepared by a method similar to the one using Bolton–Hunter reagent. *N*-Succinimidyl [2,3-^{3}H]propionate ([^{3}H]NSP) is a general purpose labelling reagent which labels free amino groups. It reacts in a similar manner to [^{125}I]Bolton–Hunter reagent and, although offering a lower specific activity, the propionyl group introduced is small and hence there is less alteration to the protein structure. The relative rate of reaction of [^{3}H]NSP with the ε-amino lysine and other groups such as the α-amino groups of N-terminal amino acids and the thiol group of cysteine have been compared to establish the general conditions for use of this reagent (32). [^{3}H]NSP also offers the convenience of a one-step reaction, described in *Protocol 14*.

Protocol 14

^{3}H-labelling with *N*-succinimidyl [2,3-^{3}H]propionate

Equipment and reagents

- Sodium phosphate buffer (25 mM, pH 8.0)
- *N*-Succinimidyl [2,3-^{3}H]propionate (NSP; 2.8–4.1 TBq/mmol, 75–110 Ci/mmol, 37 MBq/ml, 1 mCi/ml)
- Protein to be labelled (0.5 mg/ml in sodium phosphate buffer)
- nitrogen source
- two hypodermic needles

Method

1 Evaporate the toluene containing [^{3}H]NSP by blowing a gentle stream of dry nitrogen on to the surface of the solvent. This is achieved by inserting two hypodermic needles through the seal in the top of the vial and attaching a tube leading from the nitrogen source to one of the needles. The other needle acts as an outlet for the gaseous solvent. **Note:** This operation must be carried out in a well-ventilated fume cupboard or similar facility.

2 Add the protein to the vial in sodium phosphate buffer to give molar equivalence of protein to [^{3}H]NSP.^{a}

3 Incubate at room temperature for 2–4 h.

---

4   Purify by the method of choice as discussed previously.

^{a} Generally reactions are carried out in the range 10–50 mCi/mg protein. The specific activity of the product varies with protein concentration and the age/quantity of reagent but is generally in the range 1–10 mCi/mg.

7.2 Carbon-14 labelling

[^{14}C]Formaldehyde reacts with amino groups to form a Schiff's base, stabilized by reducing with borohydride. Potassium borohydride is a milder reducing agent than sodium borohydride. It has the advantage of being stable under neutral aqueous solution for short periods, unlike sodium borohydride, which quickly hydrolyses in the absence of base. Sodium cyanoborohydride has fewer side reactions (such as hydrolysis) and can be used at a neutral pH. In the range pH 6–7, it is completely inert to ketone and aldehyde groups (33). This procedure is described in *Protocol 15*.

Protocol 15

^{14}C-Labelling by reductive methylation

Equipment and reagents

- Protein to be labelled (1 mg/ml in potassium phosphate buffer)
- Potassium phosphate buffer (40 mM, pH 7.0)
- [^{14}C]Formaldehyde (10 μl; 30 mCi/mmol, 1.11 GBq/mmol)
- Sodium cyanoborohydride, 6 mg/ml in phosphate buffer (freshly prepared)
- Dialysis tubing or cassette

Method

1   Add 10 μl [^{14}C]formaldehyde to 25 μl protein solution.

2   Immediately add freshly prepared sodium cyanoborohydride solution and incubate at 25 °C for 1 h with occasional shaking.

3   Increase the volume of the reaction mixture to 250 μl by addition of potassium phosphate buffer.

4   Purify the labelled protein by extensive dialysis (16 h) at 4 °C against potassium phosphate buffer.

8 Protein kinase assays

Protein kinases are involved in a wide variety of cellular responses, including cell growth, cell differentiation and inflammation, and are classified according to their functional properties and their location within cells (34). These cellular mechanisms are important sites for therapeutic intervention (35). Phosphorylation events within cells are triggered by extracellular signals; they lead to the enzymatic amplification of the initial signal via signalling cascade mechanisms involving many different kinases. This can be through second messenger systems, such as cAMP, or by a variety of kinases which phosphorylate specific substrates. Kinases can also undergo autophosphorylation.

Protein kinases transfer the γ-phosphate group from ATP to a hydroxyl group of an acceptor amino acid, which can be serine, tyrosine or threonine, with the subsequent release of ADP. They can be studied *in vitro* by the use of radio-labelled ATP (either [γ-^{32}P]- or [γ-^{33}P]-ATP). This means the transfer of the labelled γ-phosphate group can be monitored; therefore detection and activity measurements, inhibitor studies and kinetics experiments can be undertaken. ^{32}P was formerly the label of choice but it has been superseded by ^{33}P, which has a lower emission energy and therefore lower radiation hazard.

A simple approach to studying kinases is to use specific substrates made from consensus sequences recognized by the enzyme or proteins containing that sequence, for example myelin basic protein. Alternatively, generic or specific peptides can be used. Chemical modifications can also be applied to the substrate to assist in the detection of that substrate. Once phosphorylated, the substrate can be captured using filter binding papers. In the method described in *Protocol 16*, the reactants are spotted on to the filter paper (usually phosphocellulose) after incubation. Excess label is removed by various washing strategies, the filters are dried and placed into glass vials, scintillant is added and the radioactivity in the vials is counted. Any substrate that is labelled with ^{32}P or ^{33}P will produce a signal. The substrate is captured on to the membrane through weak electrostatic forces. The efficiency of the system can be improved by using biotinylated substrate and streptavidin-coated filter papers.

Protocol 16

Filter method for kinase assay

Equipment and reagents

- Assay buffer (dependent on the particular kinase being assayed; for example, some kinases have an absolute requirement for Mg^{2+} or Mn^{2+} ions); a typical buffer is 75 mM Hepes, 270 mM sodium orthovanadate, pH 7.4
- [γ-^{32}P]ATP (~0.11 TBq/mmol, ~3.0 Ci/mmol) or [γ-^{33}P]ATP (>92.5 TBq/mmol, >2500 Ci/mmol)(magnesium salt), diluted to 200 μCi/ml using 1.2 mM ATP, 80 mM MgCl$_2$, 33 mM Hepes, pH 7.4
- 'Stop' solution: 150 mM orthophosphoric acid

- Substrate, diluted (if necessary) using assay buffer[a]
- Kinase: recombinant or derived from lysed cells/tissues (if necessary dilute using assay buffer)[a,b]
- scintillation vials
- scintillant
- scintillation counter
- Binding paper, Whatman 881 cellulose phosphate paper squares numbered in pencil

Method

A. Assay

1 Pipette 10 μl assay buffer, inhibitor or inhibitor solvent as a control into an appropriate well or tube.

2 Pipette 10 μl substrate into each well/tube.

3 Pipette 5 μl recombinant enzyme or cell lysate into each appropriate well/tube (concentration dependent on the specific activity of the kinase).

4 Start the reaction by adding 5 μl magnesium [γ-^{32}P]ATP solution.

5 Incubate for 1 h at 30 °C. During the incubation period dilute some stop solution to give 75 mM orthophosphoric acid to wash the papers (for step **8** below).

6 Terminate the reaction by adding 10 μl 'stop' reagent and separate phosphorylated substrate as described below.

B. Separation

7 Aliquot ≤30 μl of the terminated reaction mixture on to an appropriately sized numbered square of binding paper. Allow the solution to completely soak into the paper.

8 Place the binding papers in 75 mM orthophosphoric acid. Use at least 10 ml of this wash reagent per paper. Leave for 10 min with intermittent gentle mixing.

9 Decant the wash solution and dispose of liquid waste by an appropriate disposal route according to your Local Rules. Add a similar volume of wash reagent and leave the papers for a further 10 min with gentle mixing.

10 Wash the papers twice using distilled water then air dry.

11 Once dry, place the papers into individual scintillation vials. Add 10ml scintillant and count using a scintillation counter.

C. Calculation of results

The binding papers quantify ^{33}P or ^{32}P incorporated into the peptide. In the presence of enzyme, the [^{33}P]/[^{32}P] counted on the papers is the sum of non-specific [^{33}P]/[^{32}P]ATP binding, specific binding of phosphorylated peptide and binding of phosphorylated proteins in the cellular extract (A).

In the absence of enzyme, the [^{33}P]/[^{32}P] counted on the papers is the non-specific binding of [^{33}P]/[^{32}P] ATP or its radiolytic decomposition products (B). Kinase activity is therefore obtained from (A − B)

(i) Calculation of specific activity (R) of 1.2 mM magnesium [^{33}P]/[^{32}P]ATP:
   5 μl of 1.2 mM Mg[^{33}P]/[^{32}P]ATP contains 6 × 10^{-9} mol ATP

$$R = \frac{(\text{c.p.m. per 5 μl Mg[}^{33}\text{P]/[}^{32}\text{P]ATP c.p.m./nmol)}}{6}$$

(ii) Calculation of total phosphate (T) transferred to peptide and endogenous proteins:
   30 μl spotted on to binding paper and total terminated volume 40 μl
   $T = (A - B) \times 40/30$

(iii) Calculation of pmol phosphate (P) transferred per minute:

$$P = \frac{T \times 1000 \text{ pmol/min}}{IR}$$

where I is incubation time (in minutes).

167

Protocol 16 continued

(iv) Calculation of % inhibition:

$$\% \text{ inhibition} = \frac{\text{non-inhibited control c.p.m.} - \text{sample c.p.m.}}{\text{non-inhibited control c.p.m.}} \times 100$$

[a]Concentrations of the substrate or kinase will either have to be determined experimentally or suggested values obtained from the literature.

[b]Cell samples from tissue culture should be lysed and homogenized in a buffer containing protease and phosphatase inhibitors. Cells may be lysed in 10 mM Tris, 150 mM NaCl, 2 mM EGTA, 2 mM DTT, 1 mM orthovanadate, 1 mM PMSF, 10 μg/ml leupeptin, 10 μg/ml aprotinin, pH 7.4 measured at 4°C. Cellular debris should be precipitated at 25 000g for 20 min and the supernatant retained.

Alternatively the substrate can be captured using scintillation proximity assay (SPA) beads (36). SPA (Amersham Pharmacia Biotech) involves the use of scintillant beads with an acceptor molecule, in this case streptavidin, to which binding occurs in a selective manner. This is a homogeneous assay whereby the substrate, which is biotinylated, is captured by the beads through the interaction of biotin and streptavidin. Any substrate that contains a labelled phosphate group in close proximity will cause the bead to emit light, which can be detected by a scintillation counter. The advantage of this system is that, because it is homogeneous, it is amenable to automation and thus useful for high throughput screening.

9 *In vitro* translation with radiolabelled amino acids

Radiolabelled amino acids can be incorporated into proteins by *in vitro* translation through the action of cellular extracts on added mRNA species (*Protocol 17*). The amino acids most widely used in this type of reaction are ^{35}S-labelled methionine or cysteine or [3]H- or ^{14}C-labelled leucine. The source of the extract is most commonly rabbit reticulocyte lysate (37) though wheatgerm extracts (38) and occasionally other sources are also used. The advantages of the rabbit reticulocyte system are that it has a high translational activity and supports a range of post-translational activities such as acetylation, myristoylation, isoprenylation, phosphorylation and proteolysis. Other post-translational modifications such as signal peptide cleavage and core glycosylation can be obtained by the addition of canine microsomal membranes.

Endogenous mRNAs are usually removed from the lysate by digestion with micrococcal nuclease before use. This is calcium dependent and so EGTA is subsequently added to avoid digestion of added mRNA. For optimal activity of the reticulocyte lysate, extra tRNAs, haemin (to prevent inactivation of an initiation factor) and creatine phosphokinase with creatine phosphate (as an energy-producing system) are added. Potassium and magnesium concentrations may also need to be optimized for efficient translation of any particular mRNA species.

Linked transcription and translation reactions can be carried out by the inclusion of purified RNA polymerases or viral core particles that act as a source of RNA polymerase. This requires a DNA template with a promoter upstream of the region to be transcribed. The process also appears to increase the amount of protein produced in the reaction.

Protocol 17

In vitro translation

Reagents

- Rabbit reticulocyte lysate (e.g. Amersham Pharmacia Biotech, catalogue no. RPN3151)
- Potassium acetate, 2.5 mM
- Potassium chloride 2.5 mM
- Magnesium acetate, 25 mM
- RNase-free water
- RNA sample diluted in RNase-free water

- 12.5× translation mix minus particular amino acid used for radiolabel (methionine, leucine or cysteine), containing 25 mM DTT, 250 mM Hepes, pH 7.6, 100 mM creatine phosphate and 19 amino acids at 312.5 µM each (also available commercially)

Method

1 For a standard 50 µl translation reaction mix,[a] prepare:

12.5× translation mix (minus appropriate amino acid)	4 µl
2.5 mM potassium acetate	2 µl
25 mM magnesium acetate	1 µl
radioactive amino acid (e.g. [^{35}S]methionine, >1000 Ci/mmol, 15 mCi/ml)	4 µl
RNA sample in solution	0.5–2.0 µg
rabbit reticulocyte lysate	20 µl

   Make up to 50 µl with RNase-free water.

2 Incubate at 30 °C for 60–90 min.

3 Place reactions on ice prior to analysis.

[a] Always include a blank reaction without added RNA to determine the small amount of background due to translation of residual endogenous mRNA in the lysate.

The products of an *in vitro* translation reaction are commonly analysed by denaturing polyacrylamide gel electrophoresis and autoradiography or fluorography. Incorporation of labelled amino acids can be checked by TCA precipitation followed by collection of the precipitate and washing on glass fibre or cellulose nitrate filters and subsequent scintillation counting.

Acknowledgements

Thanks are due to Garth Brown who has given valuable help with the sections relating to protein labelling and to Cliff Smith who has provided help with and data for the DNA sequencing section.

References

1. Sambrook, J., Fritsch, E.F. and Maniatis, T. (eds) (1989). *Molecular Cloning: a Laboratory Manual* (2nd edn). Cold Spring Harbor Laboratory Press, Cold Spring Harbor, NY.
2. Mundy, C.R., Cunningham, M.W. and Read, C.A. (1991). In Brown, T. A. (ed.), *Essential Molecular Biology: A Practical Approach*, Vol. 2, p. 57. Oxford University Press, Oxford.
3. Feinberg, A.P. and Vogelstein, B. (1983). *Anal. Biochem.* **132**, 6.
4. Feinberg, A.P. and Vogelstein, B. (1984). *Anal. Biochem.* **137**, 266.
5. Lehman, I. R. (1981). In Boyer, P.D. (ed.), *The Enzymes*, Vol. XIV, p. 16. Academic Press, London.
6. Rigby, P.W.J., Dieckmann, M., Rhodes, C. and Berg, P. (1977). *J. Mol. Biol.* **113**, 237.
7. Melton, D., Krieg, P.A., Rebagliati, M.R., Maniatis, T., Zinn, K. and Green, M.R. (1984). *Nucleic Acids Res.* **12**, 7035.
8. Durrant, I. and Cunningham, M.W. (1995) In Hames, B. D. and Higgins, S. (eds), *Gene Probes: A Practical Approach*, Vol. 1, p. 189. Oxford University Press, Oxford.
9. Bollum, F.J. (1974). In Boyer, P.D. (ed.), *The Enzymes*, Vol. X, p. 145. Academic Press, London.
10. Richardson, C.C. (1981). In Boyer, P.D. (ed.), *The Enzymes*, Vol. XIV, p. 299. Academic Press, London.
11. Berkner, K.L. and Folk, W.R. (1977). *J. Biol. Chem.* **252**, 3176.
12. Thein, S.L., Ehsani, A. and Wallace, R.B. (1993). In Davies, K.E. (ed.), *Human Genetic Disease Analysis: A Practical Approach*, p. 21. Oxford University Press, Oxford.
13. Bales, K.R., Hannon, K., Smith II, C.K. and Santerre, R.F. (1993). *Mol. Cell. Probes* **7**, 269.
14. Sitzmann, J.H. and Le Motte, P.K. (1993). Rapid and efficient generation of PCR-derived riboprobe templates for *in situ* hybridization histochemistry. *J. Histochem. Cytochem.* **41**, 773.
15. Sanger, F., Nicklen, S. and Coulson, A.R. (1977). *Proc. Natl Acad. Sci. USA* **74**, 5463.
16. Fan, J., Ranu, R.S., Smith, C., Ruan, C. and Fuller, C.W. (1996). *Biotechniques* **21**, 1132.
17. Tabor, S. and Richardson, C.C. (1995). *Proc. Natl Acad. Sci. USA* **92**, 6339.
18. Anon. (1993). *Guide to Radioiodination Techniques*. Amersham International plc, Little Chalfont, UK.
19. McFarlane, A.S. (1958). *Nature (London)* **182**, 53.
20. Butt, W.R. (1972). *J. Endocrinol.* **55**, 453.
21. Redshaw, M.R. and Lynch, S.S. (1974). *J. Endocrinol.* **60**, 527.
22. Reay, P. (1982). *Ann. Clin. Biochem.*, **19**, 129.
23. Houghten, R.A. and Li, C.H. (1979). *Anal. Biochem.* **98**, 36.
24. Hunter, W.M. and Greenwood, F.C. (1962). *Nature* **194**, 495.
25. Fraker, P.J. and Speck, J.C. (1978). *Biochem. Biophys. Res. Commun.* **80**, 849.
26. Chen, T., Repetto, B., Chizzonite, R., Pullar, C., Burghardt, C., Dharm, E., Zhao, Z., Carroll, R., Nunes, P., Basu, M., Danho, W., Visnick, M., Kochan, J., Waugh, D. and Gilfillan, A.M. (1996) *J. Biol. Chem.* **271**, 25308.
27. Markwell, M.A.K. (1982). *Anal. Biochem.* **125**, 427.
28. Marchalonis, J.J. (1969). *Biochem. J.* **113**, 299.
29. Huber, R.E., Edwards, L. A. and Carne, T. J. (1989). *J. Biol. Chem.* **264**, 1381.
30. Bolton, A.E. and Hunter, W.M. (1973). *Biochem. J.* **133**, 529.
31. Wilbur, D.S. (1992). *Bioconj. Chem.* **3**, 433.
32. Tang, Y.S., Davis, A.M. and Kitcher, J.P. (1983). *J. Lab. Comp. Radiopharm.* **20**, 277.
33. Dottavio-Martin, D. and Ravel, J.M. (1978). *Anal. Biochem.* **87**, 562.
34. Krebs, E.G. (1994). *Trends Biochem. Sci.* **19**, 439.
35. Levitzki, A. and Gazit, A. (1995). *Science* **267**, 1782.
36. Bosworth, N. and Towers, P. (1989). *Nature* **341**, 6238.
37. Pelham, H.R.B. and Jackson, R.J. (1976) *Eur. J. Biochem.* **67**, 247.
38. Anderson, C.W., Straus, J.W. and Dudock, B.S. (1983). In Wu, R., Grossman, L. and Moldave, K. (eds), *Methods in Enzymology*, Vol. 101, p. 635. Academic Press, London.

Chapter 7
Subcellular localization of biological molecules

RICHARD CUMMING, IAN DURRANT and
RACHEL FALLON
Amersham Pharmacia Biotech, Amersham Laboratories, White Lion Road,
Amersham, Buckinghamshire HP7 9LL, UK

1 Introduction

1.1 Compartmentalization within the cell

The complexity of higher organisms is due in part to the wide variety of tissues and cells which have become specialized for different functions. Cells can often be differentiated from each other on the basis of size, shape, fine structural differences and phenotypic molecular markers. Techniques such as tissue culture can help to elucidate the function of individual cell types by growing and manipulating them under controlled conditions. A further degree of complexity comes from the organization of structures within the cell; at the simplest level we can divide the cell into nucleus and cytoplasm, while sophisticated techniques allow distinction between many different structures such as mitochondria, lyzosomes, ribosomes, etc. It is now widely appreciated that these structures perform varied cellular functions and that different molecules are localized within different subcellular compartments (e.g. neuronal compartmentalization of cytoskeletal proteins) (1). In addition to the well-known movement of nucleic acids between the nucleus and the cytoplasm, it is now known, for example, that translocation of specific molecules from one region of the cell to another appears to represent the mechanism of action of molecules including steroids, cyclic nucleotides and receptors.

Organelles including mitochondria and nuclei, and cytoskeletal structures such as microtubules and intermediate filaments can also be biochemically separated by fractionation. However, certain subcellular structures, for example endosomes, cannot be routinely observed in electron micrographs and require labelling with markers for identification. Furthermore, biochemical fractions such as microsomes and synaptosomes represent preparations that are closely related but not identical to subcellular structures as they occur *in vivo*.

A commonly used technique is cell fractionation based on tissue homogenization followed by differential centrifugation or density gradient centrifugation.

Differential centrifugation exploits the difference in size and density of different organelles; these have different sedimentation coefficients and can be separated with an ultracentrifuge (for a review of practical centrifugation, see ref. 2).

Care must be taken to ensure that subcellular fractions are as pure as possible and the use of enzyme markers, for example acid phosphatase for lyzosomes, or *N*-acetylglucosamine galactosyl transferase for the Golgi complex, together with confirmatory electron microscopy is essential. Recent developments in biochemical subcellular fractionation have included the use of free flow electrophoresis, flow cytometry for sorting organelles and the use of more selective biochemical/immunochemical markers of subcellular compartments. Diffusion between subcellular compartments is always a concern but will naturally vary depending on the organelle and the molecule being identified.

A general scheme that has been used for separation of different subcellular fractions from brain is shown in *Protocol 1*.

Protocol 1

General scheme for separation of subcellular fractions from brain using density gradient centrifugation

Equipment and reagents

- Homogenizer
- Sterile plastic centrifuge tubes
- Brain tissue in ~10 vols 0.32 M sucrose

- 1.2 M sucrose
- 0.8 M sucrose

Method

1 Homogenize brain tissue in sucrose using a glass homogenizer.[a]

2 Centrifuge at 1000g for 10 min; remove the supernatant and centrifuge this at 17 000 g for 60 min. Resuspend the pellet in 0.32 M sucrose (~2–3 ml per gram of starting tissue).

3 Carefully layer over a density gradient (made up at least 1 h before use) consisting of layers of 10 ml of 1.2 M sucrose and 10 ml of 0.8 M sucrose.[b] Spin for 2 h at 53 000g. The material separates into three layers in the tube, with small myelin fragments at the top, synaptosomes in the middle and mitochondria at the bottom. These may then be separated and analysed or purified further.

[a] All solutions and manipulations must be maintained at 4 °C.

[b] Layer the solutions of different sucrose concentrations by slowly pipetting down the side of the centrifuge tube.

2 Subcellular localization of proteins

2.1 Introduction

A multitude of techniques exists for identifying molecules within biochemical fractions, but many of them lack specificity and sensitivity. For this reason we have chosen to focus on procedures using antibodies and nucleic acid probes which are now available for detecting biological molecules with high specificity and high sensitivity. In order to determine the presence of a particular molecule in a cell or tissue homogenate or subcellular fraction, the preparation is commonly separated on a gel (polyacrylamide for proteins, agarose for nucleic acids), transferred to a membrane and then detected using the probe and detection system. Although techniques such as radioimmunoassay and dot blotting enable the presence and quantity of a molecule to be determined, they do not allow the molecular size to be determined or the relationship to other separated molecules to be ascertained. Membrane transfer is frequently used as it avoids the problems of spurious reactions within gel matrices. Two examples are given here: (a) western blotting, for identification of specific protein(s) in a mixture described below, and (b) northern blotting, for identification of specific mRNAs in a mixture described in Section 3.9.1(i).

An alternative technique, known as immunoprecipitation, which does not require membrane transfer, is discussed in Section 2.3.

2.2 Western blotting

For further details on gel separation and blotting methods, see refs 3–6.

2.2.1 Protein separation

In order to identify a particular protein in a complex mixture using western (electro-) blotting, the first step is polypeptide separation using gel electrophoresis under denaturing conditions using sodium dodecyl sulfate–polyacrylamide gel electrophoresis (SDS–PAGE). 2D gel separation by isoelectric focusing can also be used (7) but is beyond the scope of this chapter.

Protocol 2

Protein separation using polyacrylamide gels

Equipment and reagents

- Slab gel apparatus suitable for PAGE
- Sample buffer: 2.5% SDS, 2% 2-mercaptoethanol, 4 mM EDTA, 20% sucrose, 0.25 M Tris–HCl, pH 6.8
- Stacking gel: 3.5% acrylamide, 0.09% bis-acrylamide, 0.1% SDS, 2 mM EDTA, 0.125 M Tris–HCl, pH 6.8, ammonium persulfate 0.04% (w/v), N,N,N',N'-tetramethylethylene diamine to 0.08% (v/v)
- Total protein stain: 0.025% (w/v) Coomassie Blue R in 50% (v/v) methanol, 5% (v/v) acetic acid

- Separating gel: select from 5.5% to 12.5% acrylamide in 0.1% SDS (w/v), 2 mM EDTA, 0.375 M Tris–HCl, pH 8.9, 0.04% (w/v) ammonium persulfate, 0.08% (v/v) N,N,N',N'-tetramethylethylene diamine
- Molecular weight markers (e.g. Rainbow™ markers, Amersham Pharmacia Biotech)
- Electrode buffer: 6 g Tris, 28.8 g glycine, 1 g SDS in 1 litre of distilled water, pH 8.3
- Destaining reagent: 7.5% (v/v) acetic acid and 5% (v/v) methanol

---

Protocol 2 continued

Method

1 Solubilize samples (protein concentration ~2 mg/ml) by mixing 1:1 with sample buffer. Boil samples immediately to prevent proteolysis of susceptible polypeptides.

2 Set up slab gels (160 × 160 × 2 mm) consisting of a stacking gel overlaid on a separating gel. The latter can be made to an acrylamide concentration ranging between 5.5% and 12.5% depending on the molecular weight of the polypeptides of interest to be studied. The ratio of acrylamide:*bis*-acrylamide in the separating gel is 37:1. Polymerize the gels by adding solid ammonium persulfate to 0.04% (w/v) followed by *N,N,N′,N′*-tetramethylethylene diamine to 0.08% (v/v) immediately before pouring.

3 Apply samples to wells in the stacking gel (formed by using a plastic 'comb') and overlay with electrode buffer. Electrophorese at 40 mA until the refractive front reaches the bottom of the gel. Coomassie Blue dye may be included to check the process of separation. Alternatively, add coloured molecular weight markers as an additional separate sample.

4 Use the gels directly for electroblotting or stain for total protein overnight at room temperature with 0.025% (w/v) Coomassie Blue R in 50% (v/v) methanol, 5% (v/v) acetic acid. Destain in 7.5% (v/v) acetic acid and 5% (v/v) methanol at room temperature. For higher sensitivity, silver staining methods can be used.

2.2.2 Electroblotting

A standard procedure for electroblotting is described in *Protocol 3*.

Protocol 3

Electroblotting

Equipment and reagents

- Tank for electroblotting (e.g. Bio-Rad Transblot cell, Amersham Pharmacia Biotech gel destaining tank)
- Nitrocellulose or polyvinylidene difluoride (PVDF) membrane, domestic nylon scouring pads, blotting paper, disposable gloves
- Protein stain: 0.2% Ponceau S (Sigma) in 1× PBS

- Blotting buffer: 15.15 g Tris, 72.1 g glycine, 0.1% SDS (w/v), 1000 ml of methanol in 5 litres of distilled water
- Phosphate-buffered saline (PBS): 10 mM sodium phosphate, pH 7.2, 130 mM sodium chloride (make a 10× solution)

Method

1 Prepare to make a sandwich with the gel and nitrocellulose or PVDF membrane between two pieces of blotting paper and two layers of domestic nylon scouring pads, and plastic racks. Use gloves when handling the gel and membrane and also for assembly.

Protocol 3 continued

2   Make up the blotting buffer for SDS–PAGE transfer (adding 0.1% SDS (w/v) to the buffer will help in the transfer of high molecular weight proteins).

3   Place the components of the sandwich and the gel itself into buffer for a few minutes prior to assembly. The gel and membrane must be carefully opposed to ensure that air bubbles are not trapped; also take care when handling the fragile gels, especially those with a low percentage of acrylamide.

4   Cut the membrane slightly larger than the gel and mark the top left-hand corner of the gel and the membrane using, for example, a scissors cut. Submerge the sandwich completely and allow it to rest on its long axis on the base of the tank. Place weights on top if necessary to ensure it does not tip over. Put the loose-fitting lid on the tank and connect the electrodes to a stabilized power supply. For blotting of SDS–polyacrylamide gels the membrane *must* be at the anode (+); electroblot at 30 V for ~16 h.

5   After transfer, disconnect the electrodes and carefully separate the components of the sandwich. Cut the membrane to exactly the size of the gel before separation, remembering to mark the top left-hand corner again. Remove the gel and stain with Coomassie Blue (see *Protocol 2*) to detect any residual polypeptides on the gel that have failed to transfer.

6   Wash the membrane thoroughly in 1× PBS and process as a sheet (for comparing different protein samples with the same antibody) or cut into tracks to compare different antibodies. This is conveniently carried out by lining up the membrane with the gel 'comb'. When handling nitrocellulose strips, flat-bladed forceps should be used to prevent marking the surface.[a]

7   To determine the fidelity of polypeptide transfer, cut out selected strips and stain with Ponceau S for ~5 min. After staining, wash out the dye (if required) with 1× PBS and then probe the blots with antibody. If necessary bands can be marked with pencil lines for re-identification after immunostaining.

[a] The membrane must be kept moist at all stages after electroblotting.

2.2.3 Immunostaining of nitrocellulose or PVDF replicas

Procedures commonly used for the detection of individual proteins include non-radioactive techniques (eg enhanced chemiluminescence (ECL) (8) and ECL Plus (9) systems, Amersham Pharmacia Biotech), isotopically labelled reagents (generally ^{125}I) or biotin–streptavidin techniques, for example using [^{35}S]streptavidin. These approaches confirm that a specific molecule is present in the mixture or subcellular fraction. The precise size of a protein is determined and the presence of related molecules (e.g. proteolytic fragments that may show immunological cross-reactivity to the protein under investigation) can be detected. A method based on the use of ^{125}I detection reagents is described in *Protocol 4*.

Protocol 4

Procedure for staining membrane replicas

Equipment and reagents

- Clean plastic trays
- Saran Wrap™ (Dow)
- Tris-buffered saline (TBS): 50 mM Tris–HCl, pH7.4, 130 mM sodium chloride (make a 10× stock solution)
- Primary antibody diluted in 0.05% (w/v) Tween in 1× TBS
- Blocking agent: 1% (w/v) non-fat dried milk or 1% (w/v) bovine serum albumin (BSA) in 1× TBS
- Secondary antibody: ^{125}I-labelled anti-immunoglobulins or [125I]protein A/protein G (final concentration 2.4 MBq/ml), diluted in 0.05% (w/v) Tween in 1× TBS

Method

1 Block the nitrocellulose or PVDF membrane (sheet or strips) in blocking agent for 1 h at room temperature.

2 Incubate primary antibodies at selected dilutions in 0.05% (w/v) Tween in TBS for several hours at room temperature. Take care to ensure that cross-contamination does not occur between strips of membrane incubated with different reagents; suitable vessels include 10 ml plastic tubes on a rotating support, shallow plastic dishes and silicone rubber moulds.

3 After thorough washing of the membrane in TBS, detect the antibody using ^{125}I-labelled anti-immunoglobulins or [125I]protein A/protein G (final concentration 2.4 MBq/ml), diluted in 0.05% (w/v) Tween/TBS, for 1 h at room temperature. Wash thoroughly (until counts in the buffer read background levels), allow blots to air-dry, then seal with Saran Wrap (Dow) to prevent adhesion of the blots to the autoradiography film. Run controls including the omission of primary antibodies.

4 After appropriate exposure using X-ray film (see Chapter 4) develop the film according to the manufacturer's instructions.

2.3 Immunoprecipitation

In this technique the antigen is isolated from a radiolabelled mixture by specific precipitation with antibody and analysed by polyacrylamide gel electrophoresis followed by autoradiography. This allows detection of the antigen, characterization of its molecular weight (especially useful in biosynthetic studies) and identification of other proteins closely associated with it. In addition, the method of radiolabelling the initial mixture may be varied and information obtained on the structure and orientation of the antigen. Generally, the proteins in a mixture are radiolabelled with iodine (see Chapter 6) or newly synthesized proteins are biosynthetically labelled with ^{35}S-labelled amino acids.

Immunoprecipitation is usually applied to cell extracts and is carried out in the presence of detergents. The most common means of precipitating the antibody involves binding it to *Staphylococcus aureus* bacteria or protein A–Sepharose. Unfortunately, not all classes or subclasses of immunoglobulin bind tightly to protein A; for those that do not, a second antibody must be added to induce precipitation. 'Pre-clearing' the labelled extract decreases non-specific binding. Non-immune IgG, immune complexes or, more simply, *S. aureus* bacteria can be used to pre-clear the sample according to manufacturer's instructions. A control precipitation with pre-immune sera should always be carried out.

The procedure described in *Protocol 5* is given as an example of an immuno-precipitation technique that has been successfully used for the identification of receptors, cytoskeletal proteins and enzymes.

Protocol 5

Protocol for immunoprecipitation

Equipment and reagents

- Sterile plastic ware
- Microcentrifuge tubes
- Rotary mixer
- Slab gel apparatus suitable for PAGE
- Solubilization buffer: 50 mM Tris–HCl pH 7.4, 150 mM sodium chloride, 5 mM EDTA, 1 mM sodium fluoride, 2 mM phenylmethylsulfonyl fluoride (PMSF), 1% (v/v) Triton X-100
- Cell growth medium: methionine-free Dulbecco's modified Eagle's medium (DMEM; Flow Laboratories)

- $[^{35}S]$methionine (3.7 MBq, 100 μCi per ml of cell growth medium)
- PBS: 10 mM sodium phosphate, pH 7.2, 130 mM sodium chloride (make a 10× solution)
- Pansorbin (insoluble protein A; Calbiochem)
- Tween 20
- Sample buffer: (125 mM Tris–HCl, pH 6.75, 4% (w/v) SDS, 20% (v/v) glycerol, 0.1 M 2-mercaptoethanol, plus 2 mg of Pyronin Y per 100 ml)
- Slab dryer

Method

1 Suspend cells in 5 ml of methionine-free DMEM containing $[^{35}S]$methionine. Incubate for 4 h at 37°C.

2 Wash cells in 1× PBS, centrifuge at 2000g for 5 min, then re-suspend in 1 ml of ice-cold solubilization buffer and incubate for 30 min while vortex mixing at 5 min intervals.

3 Centrifuge lysate at 12 500g for 15 min and collect the supernatant. Store the super-natant on ice.

4 Dispense convenient volumes of lysate into microcentrifuge tubes. Incubate with monoclonal antibody for 2–4 h at room temperature or overnight at 4°C (ideally the antibody should be in excess; the precise quantity required can be established in a pilot titration experiment).

5 Separate the immune complexes by adding Pansorbin to give a final 1 in 10 dilution, and incubate for 1 h at room temperature with constant agitation.

6   Centrifuge at 12 500g for 10 min. Resuspend the pellet in 1× PBS, 0.1% Tween 20 then spin again as before. Discard the wash solution, resuspend the pellet in PBS, and spin at 12 500g for 10 min.

7   Discard the supernatant and resuspend the pellet in 50–100 μl of sample buffer. Boil the resuspended pellet for 2 min and centrifuge at 12 500g for 5 min to pellet the immunoabsorbent. The supernatant can be used directly or stored at −20 °C.

8   Load samples of the supernatant on to SDS–polyacrylamide gels (see *Protocol 2*) and run until the tracking dye has traversed the length of the gel.

9   Dry the gel on a suitable slab dryer then expose the dried gel to autoradiography film (see Chapter 4).[a]

[a] If high levels of non-specific protein interactions are observed, the lysate may be pre-incubated with Pansorbin (Calbiochem) followed by centrifugation between steps 2 and 3.

2.4 Histological techniques

Subcellular fractionation techniques cannot always isolate particular parts of a cell. Furthermore, they have the disadvantage that they will not allow subtle differences in molecular location to be observed. The way in which this information may be obtained is to use techniques to visualize the location of molecules within microscopic structures *in situ*. While light microscopy, with a maximum resolution of 0.2 μm, will enable distinction of some subcellular organelles, electron microscopic methods are needed to resolve all subcellular structures with high resolution.

For reasons of brevity we will not discuss autoradiography using *in vivo* administered labelled compounds such as [³H]thymidine or ³H-labelled amino acids (for nucleic acid synthesis and protein synthesis respectively). Aspects of this technique are discussed in Chapter 4.

In this chapter we will discuss immunocytochemistry (ICC) which utilizes highly specific antibody probes and *in situ* hybridization (ISH) which uses nucleic acid probes. ICC is a powerful technique that can be used to identify the position of closely related molecules at different cellular and subcellular locations (see, for example, ref. 12). One of the many applications of ISH is the demonstration of precise sites of protein synthesis within the cell by the use of nucleic acid probes for the visualization of specific mRNA transcripts. The principles underlying these two techniques are similar.

2.4.1 Immunocytochemistry

(i) Cell/tissue preparation

In order to locate a molecule of interest precisely, it is essential that the morphology of the cells is well preserved. Fixation is used to maintain structural integrity whilst also preventing diffusion of molecules. The choice of fixation method is dependent upon the antigen of interest and the technique employed (11). For optimal results in animals, perfusion fixation is the method of choice. Rapid freezing of tissue may be used to 'fix' cells and chemical fixatives may be applied in the liquid and vapour phases on freeze-dried tissue. In order to fix cells it is important that a chemical fixative penetrates rapidly to all cells in a tissue and all subcellular sites in order to prevent molecular diffusion. In order for the antibody to gain access to subcellular compartments in a cell, the cell membrane can be made permeable using detergents and/or enzyme treatment. Subcellular preparations may also be fixed and processed for molecular localization (12).

(ii) Selection of antibody

Polyclonal and monoclonal antibodies (as well as fragments) may be used for ICC, monoclonals being preferred for high specificity due to recognition of a single epitope. It should not be assumed that an antibody that has been shown to be specific by criteria such as radioimmunoassay is suitable for ICC (10). Techniques such as western blotting may also be used to confirm specificity but again may not be directly applicable to ICC since blotting techniques separate molecules prior to detection and may present them in a different configuration from that on a tissue section.

2.4.2 Radioimmunocytochemistry

Although ICC is predominantly a non-radioactive technique, one of the main advantages of utilizing radiolabelled antibodies is that quantification is made simpler by the ability to count silver grains in an emulsion. Quantification is more complex for non-isotopic techniques, particularly with the light microscope. ^{125}I-Labelled antibodies have been widely used in many areas of immunology. The longer path length of the radiation means that they have poor resolution compared with weak beta emitters such as ^{3}H and therefore are not widely suitable for subcellular localization studies, particularly at the electron microscope level of resolution. Labelled antibodies have certain advantages over labelled ligands; for example, radiolabelled antibodies to receptors can detect occupied and unoccupied receptor. Several papers have been published on the use of isotopically labelled antibodies for use in ICC either using labelled polyclonal antibodies, labelled monoclonal antibodies or using radiolabelled biotin for detection. The latter technique will be described in detail here as it is the most generally applicable and offers good resolution since it uses tritium (^{3}H).

The advantages of the technique are the ability to perform semi-quantification by grain counting with light microscopy and electron microscopy, although this requires statistical analysis. Disadvantages include the fact that sensitivity does not appear to be as high as with non-radioactive methods, probably because of the low penetration of beta particles from the ^{3}H source. Resolution is still not as good in electron microscopy as for non-radioactive markers such as colloidal gold. The development time is naturally longer for the radioimmunocytochemical method.

Although the isotopic technique has a number of disadvantages it is likely that, with reports of simple *in vitro* labelling of monoclonal antibodies using ^{3}H, the technique may become more widely used (11). A procedure is provided in *Protocol 6*.

Protocol 6

Light microscopic radioimmunocytochemistry using biotinylated protein A and (strept)avidin–[^{3}H]biotin complex (11)

Equipment and reagents

- Microtome
- Rotary mixer
- Coplin jars
- Fixative: 4% paraformaldehyde solution in 0.1 M sodium phosphate buffer, pH 7.4, containing lysine (3.4 g/l) and sodium periodate (0.55 g/l)
- Wash buffer: 0.1 M sodium phosphate buffer, pH 7.4, 30% (w/v) sucrose
- Biotinylated protein A

- 10 µg of avidin or streptavidin and 370 kBq (10 µCi) of [^{3}H]biotin in 1 ml 0.1 M Tris buffer, pH 7.4
- PBS: 10 mM sodium phosphate, pH 7.2, 130 mM sodium chloride (make a 10× solution)
- Incubation buffer: 1× PBS with 1% (w/v) BSA and 0.3% (v/v) Triton X-100 (for all antibody incubations)

Method

1 Fix tissue in fixative at room temperature for an appropriate length of time determined by the size of the tissue. Remove the tissue and post-fix in fresh fixative for 2–4 h.

2 Wash the tissue overnight in wash buffer.

3 Freeze the tissue and section at 30 µm using a microtome.

4 Incubate tissue sections in an optimal dilution (to obtain maximum specific signal: background ratio) of primary antibody in incubation buffer at 4 °C with continuous agitation (also use pre-adsorbed or non-immune antibodies in control experiments).

5 Wash in incubation buffer for 30 min.

6 Incubate in 1 µg biotinylated protein A/ml of incubation mixture for 1 h at 37 °C.

7 Wash in incubation buffer for 30 min.

8 Incubate for 1 h at 37°C in a solution containing 10 µg of avidin or streptavidin and 370 kBq (10 µCi) of [^{3}H]biotin in 1 ml 0.1 M Tris buffer pH 7.4 pre-mixed for 20 min beforehand.

9 Wash in incubation buffer and prepare for autoradiography using nuclear track emulsion.

3 Subcellular localization of nucleic acids

In 1871 Miescher isolated cell nuclei from white blood cells using a combination of dilute acid and proteases. He was able to demonstrate acid-precipitable 'nuclein' within the isolated nuclei. Since this discovery, experiments detailing subcellular localization of DNA and RNA molecules have been, and continue to be, invaluable in the elucidation of function and mechanisms of action.

Early work included the use of histological markers or stains to demonstrate total DNA or RNA within whole cells. Such staining procedures remain vitally important today, for example in the analysis of chromosomes, cell sorting and screening (13).

In 1958, Meselson, Stahl and Vinograd developed the important technique of density gradient centrifugation in caesium chloride solutions for separating nucleic acids. This technique is still important in recombinant DNA technology; for example, to purify bacterial plasmid DNA from chromosomal DNA or RNA (14).

One of the most powerful early techniques for localizing nucleic acids in cells involved the use of radioisotopes that can be detected using autoradiography and microscopy (15); for example, autoradiography of cells incubated with the radioactive DNA precursor [^{3}H]thymidine demonstrates localization of DNA within the eukaryotic cell nucleus. With the advent of recombinant DNA technology in the early 1970s, radioisotopes have played, and continue to play, a pivotal role as detection systems for nucleic acid hybridization and analysis in solution, on solid supports and *in situ*.

3.1 Identification of nucleic acid molecules isolated from subcellular fractions

A number of *in vitro* techniques for nucleic acid detection are appropriate to the identification of nucleic acids isolated from subcellular fractions. Their use is summarized below; detailed protocols can be found in previous volumes in this series (16,17).

3.1.1 Southern blotting

DNA is purified from a cell homogenate and digested with appropriate restriction enzymes. The DNA fragments are separated according to size using agarose gel electrophoresis. The DNA fragments are then transferred to a suitable membrane support, usually by capillary action. The membrane is probed subsequently using a sequence-specific radiolabelled probe. This technique produces information with respect to the location, size and quantity of a given target sequence.

3.1.2 Northern blotting

RNA is purified from a cell homogenate. The RNA molecules are separated according to size using denaturing agarose gel electrophoresis. The RNA molecules are then transferred to a suitable membrane support and probed in a similar way to the Southern blot described in Section 3.1.2. This technique produces information about the size, quantity and integrity of target RNA. A detailed protocol for this technique is given in Section 3.9.1 of this chapter.

3.1.3 Dot blotting

In this procedure, purified DNA or RNA is immobilized on a solid support by direct application. The membrane is then probed with a sequence-specific radio-labelled probe. This technique is quantitative and useful for rapid screening of multiple samples but does not give information about size; cross-hybridization of the probe with other nucleic acids in the sample can give false results.

3.1.4 S1 nuclease mapping and primer extension

These techniques are used to identify the termini or splice sites of RNA molecules.

RNA is hybridized to an end-labelled DNA probe (usually a sequenced restriction fragment of a genomic clone) that is complementary to the RNA over part of its length and overlaps a site of interest. After hybridization, S1 nuclease can be used to digest away unhybridized single-stranded DNA and RNA. The size of the labelled DNA strand in the protected hybrid is then determined using denaturing gel electrophoresis. This length then corresponds to the distance between the labelled end of the DNA restriction fragment and the RNA terminus or splice site. In addition, if the DNA probe is used in excess, the hybridization goes to completion such that quantification of the RNA species is possible.

Primer extension is often used to confirm S1 mapping data and vice versa. Similar information is generated with respect to the start of transcription in a particular DNA sequence. Reverse transcriptase is used to extend the DNA probe to the precise 5′ terminus of the RNA molecule. The extension product is sized using denaturing gel electrophoresis. In addition, sequence data can be generated for the site of interest by incorporating dideoxynucleotides using the Klenow fragment of DNA polymerase I.

3.1.5 Study of nucleic acids using the electron microscope

The electron microscope has greatly increased our understanding of the properties and localization of nucleic acids. A range of protocols are available (17) covering the application of electron microscopy to the study of nucleic acid hybridization, the interaction of proteins with nucleic acids, and the visualization of transcriptional complexes and chromatin. The use of radioisotopes has played a significant role in electron microscopy work to date, although non-radioactive markers such as colloidal gold are increasingly used for improved resolution.

3.2 Localization of nucleic acids using ISH

ISH involves the detection of nucleic acid sequences within single cells or on chromosomes using labelled sequence-specific probes. This extremely powerful subcellular technique complements the current range of nucleic acid technologies in terms of the information generated. This method will be discussed in detail.

ISH supplies spatial information about the nucleic acid sequence of interest, e.g. whether a particular cell type is expressing a transcript of interest, or the precise location of a gene on a chromosome. Furthermore, extreme sensitivity is possible using ISH since detection of target molecules is confined to a single cell and visualized by virtue of the microscope. Extraction of nucleic acids followed by detection in solution or on a solid support can result in a dilution of the target sequence when only a few cells from a mixed population contain the sequence of interest. This results in loss of sensitivity following detection.

ISH is carried out using the following basic steps:

(1) Preparation and fixation of the cell or tissue sample to retain cellular morphology and target nucleic acid in a form compatible with hybridization.

(2) Pre-hybridization of the sample. This step serves to establish denaturing conditions, block non-specific binding of probe and facilitate probe entry where necessary.

(3) Hybridization of the nucleic acid probe with target nucleic acid in the tissue sample under the correct conditions of salt, temperature, probe concentration and blocking agents.

(4) Removal of labelled probe that has not specifically hybridized to target nucleic acid.

(5) Detection of the labelled probe *in situ*.

(6) Visualization of the labelled probe by light or electron microscopy.

Since this technology was first described by Gall and Pardue in 1969 (18), radioactive labelling and detection of the probe continue to be the most widely used approach (*Figures 1* and *2*). Radioactive detection currently has advantages

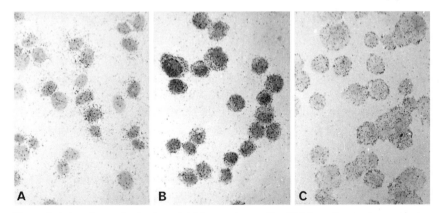

Figure 1 Autoradiograph of cultured chicken embryo cells which have been cytospun on to glass slides and hybridized with a chicken α-actin RNA probe. The autoradiographs demonstrate the degree of resolution produced by different radioactive labels. The cells are shown in bright-field illumination. Cells hybridized with (A) ^{3}H-labelled RNA probe (specific activity 8.6×10^8 d.p.m./μg) exposed for 14 days; (B) ^{35}S-labelled RNA probe (specific activity 6.6×10^8 d.p.m./μg) exposed for 6 days; (C) ^{32}P-labelled RNA probe (specific activity 1.3×10^9 d.p.m./μg) exposed for 4 days.

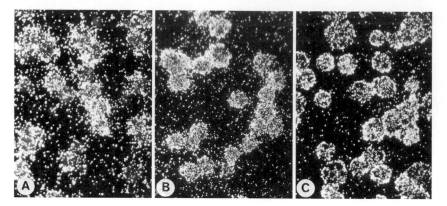

Figure 2 Autoradiographs identical to those detailed in *Figure 1* but shown in dark-field illumination.

over non-radioactive detection in terms of high sensitivity, e.g. detection of unique gene sequences and the ability to quantify the signal by grain-counting an autoradiograph (particularly with reference to radioactive samples of defined activity). The future of this technology, however, lies with non-radioactive detection, which yields results more quickly (e.g. in 24–48 h rather than the 4–6 weeks required when using ^{3}H-labelled probes), has greater resolution resulting in better spatial information (19), and does not involve using hazardous reagents. Currently the non-radioactive procedures are generally perceived as being less sensitive and less amenable to quantification than radioactive counterparts.

3.2.1 Applications of ISH
The applications of ISH fall into four main areas:

(i) Gene assignment
ISH of nucleic acid probes to the DNA of condensed metaphase chromosomes is an important application of this technology. Many genes have now been assigned to particular chromosomes and the position of the gene mapped according to histological staining procedures. In addition the importance of translocations in disease states emphasizes the potential of this approach for diagnosis (19). Detailed protocols and references covering ISH to metaphase chromosomes are available (20–22).

(ii) Detection of gene sequences and transcripts in interphase cells
Gene sequences have been detected in interphase cells showing a highly ordered arrangement of chromosomes in interphase nuclei with implications for the phenotype of particular organisms (23). In addition, specific transcripts have been localized within certain areas of the cell cytoplasm (24). This has required the high resolution afforded by non-radioactive markers (19).

(iii) Detection of RNA transcripts in whole cells or tissue sections

ISH has been used to study expression at the cellular level, providing data which are not available from other techniques. This has included the expression of a particular gene with respect to cell type during development (25) or establishing the site of gene expression, which may be different from the location of protein molecules, e.g. in the central nervous system (26).

(iv) Detection of intracellular viral nucleic acids

Cells infected with virus contain both viral genomic nucleic acid molecules and viral RNA transcripts. It is possible to detect either or both of these types of molecule using ISH. There are now many publications detailing the detection of viral sequences by ISH (27–29). In addition it is possible, using strand-specific probes, to identify viral nucleic acid replicative intermediates (30).

The application of ISH to detection of virally infected cells is likely to be important for diagnosis where there is a latent viral infection, and when both the morphology of the tissue specimen is required in conjunction with the diagnosis of viral infection, for example detection of human papillomavirus in cervical biopsies (31).

Unfortunately, there is a lack of consensus in the literature with respect to ISH methods and the worker wishing to use the technique is faced with a plethora of protocols. We attempt to provide a reasoned approach to the protocols for detecting nucleic acids within whole cells or tissue sections.

3.3 Safety considerations when using ISH (32)

This section highlights some precautions to be taken when carrying out ISH. It is not meant to be a comprehensive guide to laboratory safety. Clearly, before planning any ISH experiment, set laboratory procedures should already be in place for current radioactive licensing requirements. Appropriate tissue-handling facilities and Genetic Manipulation Advisory Group (GMAG) approval for genetic manipulation work are also required.

3.3.1 Use of radioisotopes for ISH

A range of isotopes is useful for ISH (33), including ^{32}P, ^{33}P, ^{35}S, ^{3}H and ^{125}I. It is important to have an understanding of the different properties and handling procedures for each prior to use (see Appendix 1).

It is recommended that ISH experiments are carried out in a contained area, close to a sink designated for radioactive liquid waste disposal. This is particularly useful when disposing of the stringency wash solutions. Solid waste, such as contaminated pipette tips, cover-slips, etc., should be carefully stored in a watertight container prior to bulk disposal. The contained area should comprise a flat, non-absorbent tray containing several layers of readily disposable absorbent paper (keep tissues and detergent close by to mop up any spills).

The hybridization and washing chambers should be constructed of material, such as glass or plastic, that can be readily decontaminated by soaking or wiping with strong detergent.

Contamination of the hybridization chamber can be minimized by carefully drying around each slide and tissue section so that 'hot' hybridization buffer does not travel by capillary action to the underside of the slide and then to contact points within the chamber.

Finger dosimeters are recommended for this type of work: personal dose received from ^{32}P or ^{125}I through handling slides should be minimized by remote handling using forceps.

^{32}P, ^{33}P and ^{125}I can be detected at the bench using hand-held Geiger and gamma counters respectively. These monitors should be available at all times during the experiment when these isotopes are used. The Geiger counter is extremely inefficient at detecting ^{35}S and does not detect ^{3}H (see Appendix 2).

The amounts of solid and liquid waste should be carefully recorded. After each experiment, monitor the area with a hand-held monitor if possible or take swabs and count in scintillant using the appropriate channels on a scintillation counter. Always swab down the bench area with a suitable detergent, such as Decon 90, after each experiment.

3.3.2 Tissue handling and cell culture

- Handling fresh tissue or cells of human or animal origin is intrinsic to ISH. In the unfixed state, tissues should be considered as potentially hazardous.

- All procedures likely to generate aerosols from biological material must be performed with approved containment (i.e. in combined class I/III cabinets for tissue culture).

- All biological waste must be made safe by autoclaving before disposal. Bench surfaces should be wiped down with a suitable freshly prepared disinfectant.

- A record of cell lines, tissue samples, etc., should be kept, providing details of the source and storage conditions.

- Tissue and cells should be fixed immediately to achieve the best retention of nucleic acid and structural morphology for ISH. This also minimizes the handling time required for unfixed preparations substantially.

Clearly, these extra considerations for ISH must be carried out together with a general background of good laboratory discipline in handling all potentially hazardous material including solvents, oncogenic DNA, etc.

3.4 Preparation of target cells or tissues

Methods of collection and fixation of tissues or cells are an important and some-times overlooked preparatory step for ISH. The aim is to preserve the morphology of tissues and cells whilst ensuring maximal retention of target nucleic acid in a form which is still accessible to hybridization with a complementary probe.

When the target nucleic acid of interest is RNA, special care should be taken.

It is essential to fix tissue or cells with a minimum delay (30 min is a safe limit in most cases) after death or harvest of cells in order to minimize degradation of the RNA. The collection should be carried out using sterile instruments to minimize contact of the tissues with ribonucleases. In addition, the operator should wear disposable gloves.

A whole variety of fixatives has been used successfully for ISH by different researchers (28,37). It is generally believed that strongly cross-linking fixatives may prevent the penetration of probes whilst precipitating fixatives might be insufficient for RNA retention. Sometimes, 4% paraformaldehyde has been singled out as the best fixative for ISH (38). In our experience, each tissue type requires a different fixative for optimal hybridization efficiency (39). We recommend that the user optimizes the fixation procedure for each tissue type under investigation by screening with a panel of fixatives (see *Table 1*). Where retrospective studies are undertaken, it may be necessary to optimize permeabilization conditions. Both frozen and, more recently, paraffin-embedded tissue have been used successfully for ISH.

3.4.1 Frozen tissue blocks

Protocol 7 describes a method for preparing frozen tissue blocks.

Protocol 7

Preparation of frozen tissue blocks for *in situ* hybridization

Equipment and reagents

- 15% (w/v) sucrose in 1× PBS (see *Table 1*, footnote *c*)
- 15% (w/v) sucrose in 1× PBS plus 0.01% (w/v) sodium azide
- OCT compound (Merck)
- 10 μm cryostat
- Poly-L-lysine hydrobromide
- Silica gel
- Isoprene pre-cooled in liquid nitrogen

Method

1 Cut the tissue into small pieces (1 × 1 × 0.5 cm) and fix by immersion in the appropriate fixative.[a]

2 Cryoprotect the tissue blocks by rinsing in 15% sucrose in 1× PBS, for 4 × 1 h. Store excess tissue blocks at 4°C in 15% sucrose in PBS containing 0.01% sodium azide.

3 Orient the tissue block in a OCT compound then snap freeze it in isoprene pre-cooled in liquid nitrogen.

4 Cut 10 μm cryostat sections[b] and mount on to slides coated with poly-L-lysine hydrobromide. Dry the slides for at least 4 h in an oven set at 37°C prior to ISH.

[a] Animal brain tissue should be perfused with fixative.

[b] The sections may be used immediately or stored in the presence of silica gel at −70°C for at least 7 months without loss of target nucleic acid.

Table 1 A range of tissue fixation conditions

Fixative	Fixation time/temperature	
	Cultured cells	Tissue block
Ethanol–acetic acid (3:1 v/v) + ethanol	15 min at RT[a] + 5 min at RT	30 min at RT + 10 min at RT
Ethanol–acetic acid (95%:5% v/v)	15 min at RT	30 min at RT
Methanol–acetone (1:2 v/v)	4 min at $-20°C$	20 min at $-20°C$
Methanol–acetone (1:1 v/v)	4 min at $-20°C$	20 min at $-20°C$
Bouin's 74% picric acid + 25% formalin + 1% acetic acid	30 min at RT	1 h at RT
4% Paraformaldehyde[b] in PBS pH 7.2[c]	30 min at RT	1 h at RT
2% Glutaraldehyde in PBS pH 7.2	30 min at RT	1 h at RT
Paraformaldehyde–lysine–periodate (PLP)[d]	15 min at RT	1 h at RT
4% Paraformaldehyde–methanol (1:9 v/v)	30 min at RT	1 h at RT
4% Glutaraldehyde in 20% ethylene glycol	5 min at 4°C	30 min at 4°C

[a] Room temperature.

[b] Prepare paraformaldehyde solution in a fume-hood. Do not breath dust or fumes. Dissolve 12 g of electron microscopy grade paraformaldehyde in 300 ml PBS pH 7.2. Heat the solution to 50–60°C (do not boil) and stir rapidly until the paraformaldehyde is completely dissolved.

[c] Phosphate-buffered saline: 10 mM sodium phosphate pH 7.2, 130 mM sodium chloride.

[d] Paraformaldehyde–lysine–periodate: 4% (w/v) electron microscopy grade paraformaldehyde, 23.2 mM lysine, 2.58 mM sodium periodate in 0.1 M sodium phosphate pH 7.2.

3.4.2 Cultured cells

Both adherent and non-adherent cultured cells can be cytocentrifuged on to glass slides. These preparations contain rounded cells such that the cytoplasm forms a thin rim around the nucleus. *Protocol 8* provides a general method for preparating cultured cells for ISH.

Protocol 8

Preparation of cultured cells for *in situ* hybridization

Reagents

- 0.5% (w/v) trypsin, 0.02% (w/v) EDTA
- Slides coated with poly-L-lysine hydrobromide
- 10% (w/v) serum in medium
- 1× PBS (see *Table 1*, footnote *c*)

Method

1 Rinse adherent cells with PBS. Detach the cells from the flask using 0.5% trypsin, 0.02% EDTA (5 min at 37°C).

2 Wash the trypsinized cells in medium containing 10% serum (serum inhibits trypsin).

3 Resuspend cells at 1×10^5/ml.

4 Deposit 2000 cells on each slide (previously coated with poly-L-lysine hydrobromide) using a cytocentrifuge (Shandon).

5 Allow the cells to air dry for 5 min.

6 Fix the cells by immersion in the appropriate fixative.

7 Rinse the cytospin preparations in $1\times$ PBS (2×3 min).

8 Rinse the cytospin preparations in distilled water (2×5 min).

9 Dry the preparations at 37 °C for $\geq$4 h.

10 The cells may be used immediately for ISH or stored in the presence of silica gel at -70 °C for $\geq$7 months without loss of target nucleic acid.

Alternatively, adherent cells can be grown directly on slides or cover-slips and fixed *in situ*. This is useful when observing cytoplasmic distribution of RNA since the cells are spread out over the solid support. Seed sterile multi-well slides or glass cover-slips with 2000–3000 cells each and culture under appropriate conditions for $\geq$24 h. Rinse the preparations with three changes of isotonic PBS, then follow steps **7–10** of *Protocol 8*.

3.4.3 Preparation of glass slides and cover-slips

Loss of tissue sections or cells during the ISH procedure has presented itself as a problem. Several methods of coating or 'subbing' the slides have been developed (34,35). Poly-L-lysine hydrobromide (mol. wt 150 000–300 000) is efficient in retaining tissue sections and cytospin preparations of cultured cells, although more recently amino-modified silane compounds have become available commercially (e.g. from Vector Labs) and these appear to offer more reliable sample retention. Glass cover-slips can be siliconized, as described in *Protocol 9*, to prevent adherence of nucleic acid probe and cytological material (36).

Protocol 9

Procedure for siliconizing glass cover-slips

Equipment and reagents

- 5% (v/v) solution of dichlorodimethylsilane in chloroform
- Glass cover-slips
- Double-distilled de-ionized water
- 250 °C oven

Method

1 Immerse the glass cover-slips in a 5% solution of dichlorodimethylsilane in chloroform.

2 Rinse the cover-slips thoroughly in double-distilled de-ionized water.

3 Bake the cover-slips for $\geq$4 h at 250 °C to complete the process and render the cover-slips RNase-free prior to use.

3.4.4 Minimizing contamination with RNases

RNA molecules are rapidly degraded by RNases. Since both RNA probes and target RNA molecules are susceptible to degradation, it is important to take the following precautions to avoid contamination with RNases:

- Always wear disposable gloves.
- Use sterile disposable plasticware or baked glassware for solution preparation (bake glassware at 250 °C for ≥4 h to render it RNase free).
- For solutions that can be autoclaved, include 0.1% (w/v) diethylpyrocarbonate (DEPC), which destroys RNase activity. Excess DEPC must be converted to carbon dioxide and ethanol by autoclaving treated solutions, otherwise false-positive results can sometimes occur during subsequent ISH.
- Use exogenous inhibitors of RNases (for example, placental RNase inhibitor or vanadyl ribonucleoside complexes) in solutions that cannot be autoclaved.
- Use baked spatulas to avoid the introduction of RNases into stock solutions. Materials that cannot be baked should be treated with a commercially available RNase-inhibiting solution, e.g. RNase Away (Ambion Inc.).

3.5 Preparation of radiolabelled probe

3.5.1 Choice of radioisotope

^{32}P-, ^{33}P-, ^{35}S-, ^{3}H- and ^{125}I-labelled probes have been used for ISH. Unfortunately, there has to be a compromise between resolution and sensitivity, which is dependent on the energy of particles emitted from the radioisotope. For example, the low-energy beta particles emitted by tritium afford the best resolution in autoradiography but exposure times can be of the order of 4–6 weeks. At the other end of the scale, the beta particles from ^{32}P are high energy, affording low resolution but much shorter exposure times of the order of a few days. ^{33}P, ^{35}S and ^{125}I fall in the middle of this range.

The choice of radioisotope rather depends on the resolution required by the investigator; for example, ^{3}H-labelled probes generally offer intracellular localization whereas a ^{32}P-labelled probe could be used to indicate presence of a transcript over a cell or group of cells. Researchers often choose ^{33}P, ^{35}S or ^{125}I as a compromise, although autoradiographic backgrounds are sometimes a problem with the latter two.

3.5.2 Choice of probe

There is now a considerable body of evidence (40) to show that single-stranded RNA probes are more sensitive (10- to 15-fold) when used for ISH than double-stranded DNA probes. DNA probes may re-anneal before hybridization within a cell is accomplished. In addition, non-specifically bound RNA probes can be removed from the tissue preparation after hybridization by digesting with appropriate concentrations of RNase to improve the signal-to-noise ratio (new batches of RNase should be titrated on control samples prior to use to obtain an optimum signal-to-noise ratio).

Although many researchers continue to use double-stranded DNA probes for ISH, investigators requiring the highest possible sensitivity should consider inserting probe sequences into commercially available dual promoter vectors which will allow preparation of strand-specific probes.

Oligonucleotide probes also have the advantages of being single-stranded but are generally less sensitive because their small size means that a smaller target nucleic acid sequence is covered and hence less label is delivered to the target. There are, however, situations where short oligonucleotide probes are necessary to define sequence specificity, e.g. when probing for a family of related genes or precursors (41).

3.5.3 Alkaline hydrolysis of RNA probes

Synthesis of labelled RNA probes is detailed in Chapter 6 of this volume. For efficient hybridization in cytological preparations it is generally agreed that probes should be between 50 and 250 bases in length. This can be achieved by alkaline hydrolysis of the full-length transcripts as described in *Protocol 10*.

Protocol 10

Alkaline hydrolysis of RNA probes

Reagents

- 10 mM dithiothreiotol (DTT)
- 0.4 M $NaHCO_3$
- 0.6 M Na_2CO_3
- Probe dissolved in sterile water

- 3 M sodium acetate and 1.3 μl of glacial acetic acid
- Ethanol
- Hybridization buffer

Method

1 Calculate the hydrolysis time using the following formula (15):

$$t = \frac{L_o L_f}{K L_o L_f}$$

where L_o is the initial probe length (in kilobases), L_f is the final probe length (in kilobases) and K is the rate constant for hydrolysis (0.11 kb/min).

2 Prepare the hydrolysis mix:[a]

0.4 M $NaHCO_3$	20 μl
0.6 M Na_2CO_3	20 μl
H_2O	150 μl
Probe dissolved in sterile water	10 μl
Total volume	200 μl

3 Incubate the hydrolysis mix at 60°C for the appropriate time, calculated from step 1.

Protocol 10 continued

4 Stop the reaction by adding 6.6 μl of 3 M sodium acetate and 1.3 μl of glacial acetic acid.

5 Precipitate the hydrolysed probe with 2.5 volumes of absolute ethanol.

6 Pellet the probe in a microcentrifuge for 5 min.

7 Dry the pellet.[b]

8 Dissolve the probe in hybridization buffer at 0.5 ng/μl for use in the hybridization.

[a] The hydrolysis mix should contain 10 mM DTT if ³⁵S-labelled probes are used.

[b] At this stage, it is possible to check the original and final lengths of the transcript by electrophoresis in 2% agarose–formaldehyde gels followed by autoradiography. The above formula gives reproducible results provided all solutions are RNase free.

3.6 Hybridization

Before working through the procedures for tissue pre-treatment and the hybridization and subsequent stringency washes, it is important to ensure that the appropriate equipment has been prepared:

- Humid chambers are required for ISH because the hybridizations are carried out in small volumes, typically 10–20 μl. Airtight plastic sandwich boxes are ideal. The bottom of the sandwich box should be covered with a thin layer of incubation buffer. It is important that this solution has the same salt concentration as the medium currently on each slide to prevent distillation effects. Slides should be supported above the liquid, for example using pipettes taped to the bottom of the chamber.

- An orbital shaker should be coated with readily disposable absorbent material and stringency washes should be agitated at the lowest speed.

- A hot-air oven should be set to the correct hybridization temperature before starting the experiment. Alternatively, dedicated ISH workstations are now commercially available (e.g. OmniSlide system from Hybaid) which accurately control the temperature of hybridization and can handle ≤20 slides.

- Solutions should be pre-warmed as appropriate prior to use.

3.6.1 Tissue pre-treatment

Protocol 11 describes how to pre-treat slides before hybridization. This involves permeabilizing cells to allow access of probe to target and treating to reduce non-specific binding of probe.

Protocol 11

Pre-treatment of tissues before hybridization

Equipment and reagents

- Silica gel
- Permeabilization buffer: 0.3% (v/v) Triton X-100, 1× PBS, pH 7.2
- 1× PBS, pH 7.2 (see *Table 1* footnote *c*)
- Proteinase K
- 0.1 M glycine, 1× PBS, pH 7.2
- 0.25% acetic anhydride (w/v), 0.1 M triethanolamine, pH 8.0[c]

Method

1 Remove slides from storage at −70°C and allow them to come to room temperature in the presence of silica gel.[a]

2 Permeabilize the cells by immersing the slides in permeabilization buffer for 15 min.

3 Rinse the slides in 1× PBS, pH 7.2, for 2 × 3 min.

4 Improve access of the probe to cellular nucleic acid by limited proteinase K digestion.[b] Incubate all slides in the optimal proteinase K concentration at 37°C for 30 min.

5 Stop the action of proteinase K by immersing the slides in 0.1 M glycine, 1× PBS, pH 7.2 for 5 min.

6 Post-fix the tissues for 3 min to retain target nucleic acid.

7 Rinse the slides in 1× PBS, pH 7.2, for 2 × 3 min.

8 Reduce non-specific binding of the probe by acetylating the tissue sections. Immerse the slides in 0.25% acetic anhydride, 0.1 M triethanolamine pH 8.0 for 10 min.[c]

[a] Although the preparations are fixed at this stage it is important not to introduce RNases when using an RNA probe of defined length. In addition random fragmentation of target nucleic acids may result in altered stringency.

[b] Proteinase K differs from batch to batch. It is therefore necessary to titrate each batch for optimal results.

[c] Add the acetic anhydride to 0.1 M triethanolamine as it is stirring. Immediately place the slides in the staining jar and begin the incubation.

3.6.2 Pre-hybridization/hybridization procedure

Hybridizations are usually performed in formamide to prevent exposure of the tissues to high temperatures during the hybridization, which may affect the morphology. It is worth noting that the melting temperature of RNA–RNA duplexes is reduced by 0.35°C for each percentage point of formamide concentration (v/v). Hybridization is optimal at 25°C below the melting temperature, T_m (15). Ideally the T_m for each probe should be determined using northern or dot blots. Generally, when RNA probes are used to detect RNA, a hybridization temperature of 50–55 °C in 50% formamide is used. A pre-hybridization step is generally incorporated at the hybridization temperature. A widely applicable procedure is given in *Protocol 12*.

Protocol 12

Procedure for pre-hybridization and hybridization

Equipment and reagents

- 20× SSPE: 0.2 M sodium phosphate, pH 7.4, 3.0 M sodium chloride, 0.02 M EDTA
- Pre-hybridization buffer: 50% de-ionized formamide, 5× SSPE, freshly prepared and pre-warmed to appropriate temperature
- Probe
- Hybridization oven

- Hybridization buffer: 50% de-ionized formamide (v/v), 5× SSPE, 10% dextran sulfate (w/v), 0.25% (w/v) bovine serum albumin (BSA), 0.25% Ficoll 400 (w/v), 0.25% polyvinyl-pyrrolidone PVP 360 (w/v), 0.5% SDS (w/v), 100 μg/ml freshly denatured salmon sperm DNA (sonicated to an average size of 200 bp) (freshly prepared)

Method

1. Immerse the slides in pre-hybridization buffer (pre-warmed to the appropriate temperature) and incubate for ≥15 min.

2. Dissolve the hydrolysed probe in freshly prepared hybridization buffer at 0.5 ng/μl.

3. Drain the slides from pre-hybridization buffer. Apply 10–20 μl hybridization buffer containing probe to each slide.

5. For hybridization to double-stranded DNA in tissue/cell preparations, heat the slides in a mechanical convection oven at 100–150 °C for 10 min (42).

6. Cover the sections/cells with siliconized RNase-free cover-slips (see *Protocol 9*) where necessary. Cytospin preparations are usually well covered by 10 μl hybridization mix without the need for a cover-slip.

7. Incubate in the hybridization chamber at the appropriate temperature (typically 52 °C for RNA probes) for 2 h, or overnight.

3.6.3 Stringency washes

Stringency washes remove non-specifically bound probe from the cell/tissue preparations. Stringency is increased by raising the temperature and/or decreasing the salt concentration of particular washes. As with the hybridization temperature, stringency conditions should be evaluated for each probe using northern or dot blots. *Protocol 13* outlines the procedure. Note that it is important, when using [35]S- or ^{125}I-labelled probes, to include 10 mM DTT or 100 μM potassium iodide, respectively, in the stringency washes to reduce background signal.

Protocol 13

Procedure for stringency washes

Equipment and reagents

- 2× SSPE (see *Protocol 12*)
- 2× SSPE, 0.1% (w/v) SDS
- 0.1× SSPE, 0.1% (w/v) SDS
- 10 μg/ml RNase A, 2× SSPE
- 100%, 95% and 70% (v/v) ethanol containing 0.3 M ammonium acetate
- Staining jar

Method

1 After hybridization, remove the slides from the hybridization chamber and remove the cover-slip and hybridization buffer by dipping each slide into a beaker of 2× SSPE.

2 Place the slides in a staining jar containing 2× SSPE, 0.1% (w/v) SDS. Wash the slides for 4 × 5 min at room temperature with gentle shaking on an orbital mixer.

3 Wash the slides in 0.1× SSPE, 0.1% (w/v) SDS pre-warmed to the hybridization temperature for 2 × 10 min with gentle shaking.

4 RNA hybrids are relatively resistant to digestion by pancreatic RNase A, so low concentrations of RNase can be used to remove non-specifically bound RNA probes. Immerse the slides in a pre-warmed solution of RNase (10 μg/ml RNase A, 2× SSPE) at 37 °C for 15 min.

5 Dehydrate the tissues/cells in graded ethanols (70%, 95%, twice in 100%) containing 0.3 M ammonium acetate. The ammonium acetate is present to maintain the integrity of the hybrids (43).

3.7 Autoradiography (44)

Both X-ray film and cover slips coated with nuclear track emulsion have been used to detect ISH by opposing the tissues or cells against the film. X-ray film is useful for screening many samples or optimizing conditions for hybridization. However, these are not the most common methods of autoradiography. The method described here allows the worker to view the tissue or cells simultaneously with the exposed emulsion overlaid.

Autoradiography requires a completely light-tight dark room. It is an advantage to have a double-door system or a rotating door fitted so that the user can enter and leave the darkroom without exposing the emulsion during the drying period. If there is no double door, the slides can be dried in a light-tight well-ventilated box. The dark room should be fitted with a safe light containing a 25 W bulb and a Kodak Wratten series 11 filter when using Kodak NTB2 emulsion. Please note that different safe lighting conditions may be required with nuclear track emulsions from different manufacturers (e.g. Amersham Pharmacia Biotech or Ilford Ltd). The safe light should be positioned at least 1 m away from the working area. The dark room should also be equipped with a water bath set at 43–45 °C.

Nuclear track emulsions have a gelatin base and are therefore solid at room temperature. Since repeated melting of the emulsion increases the background level of grains, the emulsion should be initially aliquoted. Methods for aliquoting emulsion and dipping slides are provided in *Protocols 14* and *15*, respectively.

Protocol 14

Procedure for aliquoting new emulsion

Equipment and reagents

- Water bath set at 43–45 °C
- 0.6 M ammonium acetate
- humidity monitor
- Glass scintillation vials
- Nuclear track emulsion
- Light tight box

Method

A. Before adopting safe light conditions

1   Switch on the water bath set at 43–45 °C.
2   Ensure that the temperature in the room is 18–20 °C and that the humidity is ~60% (low humidity can cause stress grain formation during drying of emulsions (43).
3   Place an appropriate volume of distilled water containing 0.6 M ammonium acetate in a flask in the water bath. Leave the water to come to temperature.
4   Check the arrangement of apparatus. Extinguish all lights. Switch to the safe light when ready to proceed.

B. After adopting safe light conditions

5   Place the new container of emulsion in the water bath. Allow the emulsion to melt for 30 min.
6   Carefully pour the molten emulsion into the flask containing an equal volume of pre-warmed 0.6 M ammonium acetate. Gently mix the emulsion by turning the flask several times (too much agitation of the emulsion also increases the background grains).
7   Aliquot the emulsion (e.g. into 10 ml amounts in glass scintillation vials).
8   Store aliquots in a light-tight box at 4 °C in a refrigerator away from radioisotopes.

Protocol 15

Procedure for dipping slides in nuclear track emulsion

Equipment and reagents

- Glass dipping chamber
- Tray containing ice
- Metal or glass plate
- Pre-diluted nuclear track emulsion
- 43–45 °C water bath
- Absorbent tissue
- Light-tight boxes
- Silica gel

Protocol 15 continued

Method

1   Carry out steps **1** and **2** from *Protocol 14*. In addition, place a metal or glass plate on to a tray containing ice. This is used to facilitate gelling of the emulsion on the slide.

2   Place a suitable dipping chamber which will accommodate one slide at a time to warm in the water bath. We have used a glass dipping chamber constructed using blown glass (available commercially from Amersham Pharmacia Biotech).

3   Sort slides into sets for different exposure times. Under safe light conditions, melt an aliquot of pre-diluted emulsion at 43–45 °C for 10–15 min.

4   Pour the emulsion slowly down the side of the chamber ensuring that air bubbles do not form.

5   Check for the absence of air bubbles by repeated dipping of blank glass slides: hold the slide between thumb and forefinger, dip the slide slowly into the emulsion and withdraw. Drain the slide briefly on a pad of absorbent tissues and examine the slide for air bubbles under the safe light. Continue to dip slides until no air bubbles remain.

6   Dip the experimental slides carefully as described in step **5**. By withdrawing the different slides at as near a constant rate as possible from the emulsion, the uniformity of the emulsion coat between slides is more reproducible (41). After draining each slide briefly on a pad of absorbent tissue, wipe the back of the slide with a tissue and quickly place the slides on to the pre-cooled glass or metal plate.

7   Allow the emulsion to 'gel' for 10 min.

8   Remove the slides from the cooled plate and allow them to dry at room temperature for ≥2 h.

9   Pack the slides into light-tight boxes containing silica gel to maintain dryness. Seal the boxes with black tape and store at 4 °C for appropriate exposure times.

3.7.1 Developing autoradiographs

A procedure for developing autoradiographs is provided in *Protocol 16*. When nuclear track emulsion is 'wet', it is extremely fragile. Wrinkling and loss of emulsion from the slides will occur if the emulsion is subjected to different temperatures during the development or staining procedure. All solutions should therefore be equilibrated to room temperature (15–20 °C) in the dark room before development.

Protocol 16

Procedure for developing autoradiographs

Reagents

- Developer (e.g. Kodak D-19)
- Fixer (e.g. Kodak F24)
- Distilled water

Protocol 16 continued

Method

1   Remove the box containing slides from the refrigerator. Allow the slides to warm up to room temperature in the presence of desiccant.

2   Under safe lighting conditions, remove the slides from the box and immerse in developer (e.g. Kodak D-19 developer, 4 min). Do not agitate the slides.

3   Stop development by placing in distilled water for 5 min.

4   Fix the emulsion, e.g. using Kodak F24 fixer (5 min).

5   Rinse the slides in distilled water for 5 min.

6   At this stage the lights can be turned on.

3.7.2 Histological staining

The emulsion layer can result in variable and imprecise staining; however, counterstaining the tissue or cells through the photographic emulsion does produce acceptable results. In general, when carrying out histological staining, the staining solutions should be at the same temperature as the developing solutions. Silver grains can be lost through exposure to acidic solutions during staining procedures (e.g. for some haematoxylins and differentiation with acid alcohol). *Protocol 17* describes a procedure for counterstaining with Pyronin Y.

Protocol 17

A procedure for counterstaining with Pyronin Y

Equipment and reagents

- 2.5% (w/v) Pyronin Y (National Diagnostics) in distilled water, freshly prepared
- Acetone
- Xylene
- Synthetic resin, e.g. DPX (Merck)

Method

1   Immerse the slides in Pyronin Y for 30 sec.

2   Rinse in distilled water for a few seconds. This removes Pyronin stain. The optimal time should be determined by trial and error.

3   Dehydrate in acetone for 1 min.

4   Rinse in equal parts of acetone and xylene.

5   Clear in pure xylene.

6   Mount the slides in synthetic resin such as DPX.

3.8 Analysis of data

3.8.1 Microscopy

The tissue/cell autoradiographs are viewed using light microscopy. Bright-field illumination is adequate for viewing medium to strong hybridization signals. Weak hybridization signals may seem clearer if viewed using dark-field illumination. With dark-field illumination, the tissue or cells appear dark whilst the grains appear white or refractile *(Figure 2)*. For dark-field optics, the microscope should be fitted with a dark-field condenser. The best dark-field image is obtained by putting immersion oil between the condenser and the glass slide.

3.8.2 Quantification

One of the advantages of using radioisotopes in this application is the ability to quantify the signal by counting grains over particular cells. However, absolute quantification of signal from ISH is a contentious issue in several respects, since there are many variables both between different tissue sections and within the same section (e.g. uneven fixation and uneven probe penetration, autoradiographic efficiency, hybridization efficiency, etc.).

There is value in comparative grain counting (e.g. to establish signal-to-noise ratios). This is particularly important when assigning genes to chromosomes by ISH or when detecting rare transcripts in cells. In these cases, statistical significance must be established (43). Complex instrumentation is now available for image analysis so that comparative quantification by statistical analysis is less tedious than counting grains manually with the light microscope.

The task of quantification is eased by the availability of radioactive materials that are loaded with known amounts of activity (e.g. Microscales from Amersham Pharmacia Biotech). These can be applied to slides for exposure to emulsions or X-ray film. The grain density over the defined activity levels can then be used as a scale for quantification of the test samples, either by grain counting or by image analysis. Such scales are available treated with ^{3}H or ^{14}C; calculations have been made to allow ^{14}C scales to be used to assess ^{35}S- (45) and ^{33}P-labelled (46) probes.

3.9 Controls required for ISH

Controls for ISH experiments are designed to assess the reproducibility, specificity and sensitivity of the reagents and procedures used. For detection of mRNA, a comprehensive series of controls is detailed below:

- Northern blot analysis (see Section 3.9.1) should be carried out using *total* RNA extracted from the cells or tissues of interest. The blots are used to test the specificity of the particular antisense probe in question and any non-homologous or negative control probes used subsequently for ISH. It may be possible to eradicate spurious bands using altered stringency conditions or by using only part of the probe sequence. Northern blots or dot blots can also be used to establish the melting temperature (T_m) and therefore the optimal hybridization temperature for the probe of interest. Final stringency conditions are also determined in this way.

- ISH can be carried out using the probe of interest on a tissue or cell line known not to express the target transcript (checked by northern or dot blotting of total RNA).
- Many workers use sense transcripts of the sequence of interest as a negative control for ISH. This can be checked on blots.
- The specificity of the hybridization within the tissue section can be checked by competing out labelled antisense probe with unlabelled antisense probe.
- The overall hybridization procedure and solutions can be checked for each experiment by using a control probe (e.g. to α-actin) which is highly conserved and expressed in all cell types.
- Some workers use RNase pre-treatment of tissues to show that signal generation is due to the presence of target RNA. This may be viewed as contentious when RNA probes are used (E. J. Gowans, personal communication), as residual RNase may remain during the hybridization.
- The specific probe can be omitted to test for the presence of background grains following autoradiography.
- A blank slide should also be dipped to test the emulsion.

3.9.1 Northern blotting

A northern blot provides information about the size and concentration of specific RNA transcripts in a mixture. However, RNA molecules are extremely susceptible to degradation through cleavage with RNases at any stage during extraction, purification, separation and blotting (see Section 3.4 for appropriate procedures for minimizing RNase contamination).

An average mammalian cell contains ~10–20 pg of RNA including 80–85% ribosomal RNA (28S, 18S and 5S), 10–15% low molecular weight RNA species (small nuclear and transfer RNA) and 1.5% mRNA (34). mRNA comprises a range of molecular sizes. It is possible to isolate mRNA from total RNA prior to northern analysis by virtue of the 3' poly(A) tail on these molecules. The method of choice is chromatography on oligo(dT)–cellulose which can be obtained commercially. However, when northern analysis is used to assess probe specificity for ISH, it is important to use total RNA since this reflects the full range of target molecules *in situ*. There are several methods available to prepare total RNA (47,48).

In summary, northern analysis comprises the following basic steps:

(1) Separate transcripts according to size using denaturing gel electrophoresis (outlined in *Protocol 18*).
(2) Blot the transcripts from the gel on to a solid support (e.g. nylon membrane) (*Protocol 20*).
(3) Fix the molecules to the support (*Protocol 21*).
(4) Hybridize a specific radiolabelled probe to the blot (*Protocol 22*).

(5) Wash off unhybridized probe (*Protocol 22*).

(6) Visualize the hybrid using autoradiography.

(7) Interpret results.

(i) Denaturing gel electrophoresis

Denaturing conditions disrupt secondary structure within the RNA molecules during electrophoresis, enabling an accurate estimation of the length of the transcript. There are a number of denaturing gel systems currently in use, including glyoxal gels, formaldehyde gels and methyl mercuric hydroxide gels. The formaldehyde gel system is detailed here (*Protocol 18*) since, in our experience, these systems are effective in producing good quality northern blots which demonstrate high sensitivity.

Protocol 18

Denaturing gel electrophoresis of RNA

Equipment and reagents

- Total RNA (~20 μg)
- Formaldehyde
- Gel running buffer: 1× Mops (0.2 M 3-(*N*-morpholino)propane sulfonic acid), 50 mM sodium acetate pH 7.0, 5 mM EDTA
- De-ionized formamide[a]
- 5× Mops, pH 7, sodium acetate, EDTA buffer
- Agarose and electrophoresis apparatus
- Bromophenol Blue
- Sterile loading buffer: 50% (w/v) glycerol, 1 mM EDTA, 0.4% (w/v) Bromophenol Blue

Method

1   Incubate total RNA at 65 °C for 15 min in the following solution:

Total RNA (~20 μg)	4.5 μl
Formaldehyde	3.5 μl
De-ionized formamide[a]	10 μl
5× Mops buffer pH 7	2 μl

2   Chill on ice and add 2 μl of sterile loading buffer to each sample.

3   Prepare a 1.5% agarose gel by melting 1.5 g of agarose in 62.1 ml of water. Cool the gel mixture to 60 °C. Add 20 ml of 5× Mops buffer and 17.9 ml of formaldehyde. Mix and pour the gel immediately in a fume-hood.

4   Assemble the electrophoresis apparatus, cover the gel with Gel running buffer, pre-electrophorese the gel at 100V for 10 minutes.

5   Load the total RNA samples on to the gel alongside suitable DNA or RNA molecular weight markers (treat the markers in exactly the same way as the RNA samples prior to loading). Apply current to the gel until the Bromophenol Blue has migrated three-quarters of the way through the gel.

[a] See *Protocol 19*.

Protocol 19 details a procedure for de-ionizing formamide.

Protocol 19

Procedure for de-ionizing formamide

Equipment and reagents

- Formamide (AnalaR grade)
- Whatman no. 1 filter paper
- Mixed-bed ion-exchange resin (e.g. Bio-Rad AG 501-X8)

Method

1 Mix 50 ml AnalaR formamide with 5 g mixed bed ion-exchange resin until the pH is neutral.

2 Filter the de-ionized formamide twice through Whatman no. 1 filter paper.

3 Dispense the formamide into aliquots and store at $-20\,°C$ in tightly capped tubes ready for use.

(ii) Transfer of RNA to a membrane support by capillary blotting

A procedure for the capillary transfer of RNA to a membrane is provided in *Protocol 20*.

Protocol 20

Capillary transfer of RNA to a membrane support

Equipment and reagents

- Blotting buffer (20× SSPE, see *Protocol 12*)
- Whatman 3MM paper
- Clingfilm
- Absorbent paper towels
- Nylon membrane (e.g. Hybond (Amersham Pharmacia Biotech))
- Glass plate
- 0.75 kg weight

Method

1 Fill a tray or glass dish with blotting buffer.

2 Make a platform for the gel, e.g. using an inverted gel former, and cover it with a wick made from three sheets of Whatman 3MM paper saturated with blotting buffer.

3 Place the gel upside down on the wick; avoid trapping air bubbles beneath it. Surround the gel with clingfilm to prevent blotting buffer from being absorbed directly into the paper towels above the gel.

4 Cut a piece of nylon membrane to the exact size of the gel. Lower the membrane on to the gel, taking care not to introduce air bubbles between the membrane and the gel. If bubbles do appear, squeeze them out by rolling a sterile pipette over the membrane.

5   Place three sheets of Whatman 3MM filter paper cut to size and wetted with blotting buffer on top of the membrane.

6   Place a stack of absorbent paper towels (5 cm thick) on top of the 3MM paper.

7   Place a glass plate on top of the paper to spread the pressure from a 0.75kg weight.

8   Allow transfer to proceed for $\geq$12 h.

(iii) Fixing RNA molecules to the membrane

A procedure for fixing RNA to nylon is provided in *Protocol 21*. UV irradiation is the most effective means of fixing nucleic acids to nylon since a covalent linkage results. A wavelength of 312 nm is recommended. The power and wavelength of UV light from individual UV transilluminators may vary, so in order to achieve efficient UV cross-linking, it is essential to calibrate your transilluminator. Failure to do so can lead to very poor sensitivity of detection following hybridization. It should be noted that nitrocellulose membranes should never be irradiated due to the risk of fire.

Protocol 21

Fixing RNA to nylon membrane

Equipment and reagents
• UV transilluminator   • Saran Wrap[a]

Method

1   After blotting, carefully dismantle the apparatus. Before removing the membrane from the gel, mark the membrane with pencil to identify the tracks and loadings. Do not wash the filter.

2   Wrap the membrane in Saran Wrap and place, with RNA side down, on a standard UV transilluminator for 2–5 min.

3   Dry blots should be stored under vacuum in a desiccator. In this condition, they are stable for several months.

[a] Note that some brands of clingfilm are unsuitable for this application because they absorb UV light. We recommend the use of Saran Wrap.

(iv) Preparation of probe

RNA or DNA probes can be used. Their preparation is described in Chapter 6. It is recommended that RNA probes are used in this instance since they are subsequently used for ISH.

(v) Hybridization of the northern blot

Protocol 22 describes a procedure for hybridizing a northern blot.

Protocol 22

Hybridization of northern blot

Equipment and reagents

- Pre-hybridization/hybridization buffer: 50% de-ionized formamide (v/v), 5× SSPE (*Protocol 12*), 5× Denhardt's, 0.5% (w/v) sodium pyrophosphate, 100 µg/ml denatured, sheared herring sperm DNA, 10% (w/v) dextran sulfate, 2% (w/v) SDS[a]
- Plastic box or heat-sealable plastic bag
- RNA probe, radiolabeled
- 2× SSPE, 0.1% (w/v) SDS
- 0.1× SSPE, 0.1 % (w/v) SDS
- 10 µg/ml RNase A in 2× SSPE

Method

1 Immerse the northern blot in pre-hybridization buffer and incubate with shaking at 42–65 °C for 1 h. Pre-hybridization can be carried out either in a heat-sealed plastic bag or in a suitable plastic box using 5 ml of solution per 100 cm^2 of membrane.

2 Add the RNA probe to freshly prepared hybridization buffer to a final concentration of 5 ng/ml. Mix the buffer thoroughly to achieve an even distribution of probe before adding the membrane to the pre-warmed buffer.

3 Incubate the membrane with radiolabelled probe in hybridization buffer at 42–65 °C (depending on stringency required) for 12–16 h.

4 Wash the filter four times in 2× SSPE, 0.1% (w/v) SDS, for 5 min each at room temperature.

5 Wash the filter twice in 0.1× SSPE, 0.1% (w/v) SDS for 10 min each at 55–75 °C.

6 Treat the filter with 10 µg/ml RNase A in 2× SSPE at 37 °C for 15 min.

7 Wrap the damp filter in Clingfilm. (Do not allow the filter to dry completely if re-washing or re-probing is a possibility.)

[a] Alternatively, ready-to-use hybridization buffer for radioactive samples (Rapid-hyb) can be obtained commercially from Amersham Pharmacia Biotech.

(vii) Autoradiography

Expose the filter to X-ray film. The appropriate length of time for autoradiography will depend on the abundance of the target RNA molecule. For ^{32}P-labelled probes, autoradiograph at −70 °C using two intensifying screens and pre-flashed film for maximum sensitivity. For ^{35}S- and ^{33}P-labelled probes, autoradiograph dried filters at room temperature without Clingfilm (see Chapter 4).

(viii) Interpretation of results

In the simplest situation, a single band is observed when using a specific anti-sense RNA probe. However, this is not always the case when probing total RNA. Extra bands corresponding to spliced transcripts from different initiation sites or prematurely terminated transcripts are sometimes observed, depending on the probe of interest. It is important to have evidence that the extra bands are transcripts of the sequence of interest and not a closely related sequence from an entirely different gene. It may be possible to achieve this by varying the stringency of the hybridization and wash procedures, for example. It is useful to test a 'non-homologous' probe, e.g. a sense transcript of the sequence of interest for use, as a negative control in subsequent ISH experiments using a northern blot.

3.10 Simultaneous localization of proteins and nucleic acids

The development of the ISH technique has enabled researchers to question whether the localization of a protein at a specific site within a cell represents newly synthesized protein or stored sequestered protein. ICC has the power to detect subtle differences in molecules, for example, in different post-translational states, assuming that selective antibodies are available.

There are different procedures in the literature for simultaneous localization of protein and RNA transcripts (49,50). The major difference is whether the ICC is performed before or after the ISH. Problems arise from RNase contamination of antisera and also from non-specific binding of nucleic acid probe to diamino-benzidine (DAB) reaction products if ICC is performed before ISH. In this application, radioisotopes are extremely useful as an alternative detection system distinguishing transcript detection from antigen detection in the same cell or tissue section.

4 Conclusion

We have focused on the use of specific and sensitive antibody and nucleic acid probes for localization of biological molecules. For subcellular localization, the 'classical' techniques of organelle separation coupled with gel fractionation and blotting are particularly valuable since the size of molecules can be determined and subcellular preparations can be used for functional studies. Furthermore techniques are now being used to reconstitute *in vivo* systems based on subcellular fractions. Radioisotopes are widely used for high sensitivity (e.g. ^{125}I, ^{32}P) since resolution is not as critical on gels/blots as it is for microscopy. For the latter application ^3H-labelled probes are favoured for subcellular studies of mRNA since resolution is better, although development time is long.

It seems clear that the future development of techniques for precise location of molecules will lie with non-radioactive techniques that are sensitive, fast and have high resolution, and lessons may be learnt from developments that have occurred in non-radioactive ICC. However, conventional ISH without the need for high resolution may be quite adequate using other labels including ^{32}P, ^{33}P and ^{35}S. Although electron microscopy techniques give precise localization of molecules to organelles, they are time consuming and complex; in the future we may see the specificity of ICC/ISH coupled with more specific subcellular markers to identify the location of molecules using the light microscope. A parallel situation may occur in the field of cytogenetics where chromosome analysis may be speeded up using chromosome-specific probes.

Acknowledgements

The authors are grateful to their colleagues at Amersham Pharmacia Biotech, and in particular Martin Cunningham, for comments and data included in this chapter. Particular thanks are also extended to Chris Jones, Giorgio Terenghi, Judith Parke and Robert Burgoyne for experimental data and protocols.

References

1. Cumming, R. and Burgoyne, R.D. (1983). *Bio. Sci. Rep.* **3**, 997.
2. Rickwood, D. (ed.) (1989). *Centrifugation: A Practical Approach*. IRL Press, Oxford
3. Hames, B.D. (1990). In *Gel Electrophoresis of Proteins: A Practical Approach*. IRL Press, Oxford.
4. Rickwood, D., Chambers, J.A.A. and Spragg, S.P. (1990). In *Gel Electrophoresis of Proteins: A Practical Approach*. IRL Press, Oxford.
5. De Maio, A. (1994). In *Protein Blotting: A Practical Approach*. IRL Press, Oxford.
6. Mansfield, M.A. (1994). In *Protein Blotting: A Practical Approach*. IRL Press, Oxford.
7. Dunbar, B.S. (1987). *Two-dimensional Electrophoresis and Immunological Techniques*. Plenum Press, New York.
8. Durrant, I. and Fowler, S. (1994). In *Protein Blotting: A Practical Approach*. IRL Press, Oxford.
9. Latif, N. and Dunn, M.J. (1997). *Life Sci. News* **22**, 14.
10. Cumming, R. (1980). *J. Immunol. Methods* **37**, 301.
11. Hunt, S.P., Allanson, J. and Mantyh, P.W. (1986). In Polak J. M. and Van Noorden, S. (eds), *Immunocytochemistry*, p. 99. Wright, Bristol.
12. Cumming, R., Burgoyne, R.D., Lytton, N.A. and Gray, E.G. (1983). *Neurosci. Lett.* **37**, 215.
13. Drury, R.A.B. and Wallington, E.A. (1980). *Carlton's Histological Technique*. Oxford University Press, Oxford.
14. Ish-Horowicz, D. and Burke, J.F. (1979). *Nucleic Acids Res.* **7**, 1541.
15. Westermark, B. (1974). *Int. J. Cancer* **12**, 438.
16. Hames, B.D. and Higgins, S.J. (eds) (1985). *Nucleic Acid Hybridization: A Practical Approach*. IRL Press, Oxford.
17. Sommerville, J. and Scheer, O. (eds) (1987). *Electron Microscopy in Molecular Biology: A Practical Approach*. IRL Press, Oxford.
18. Gall, J.G. and Pardue, M.L. (1969). *Proc. Natl Acad. Sci. USA* **63**, 378.
19. Bentley Lawrence, J., Villnave, C.A. and Singer, R.H. (1988). *Cell* **52**, 51.
20. Rooney, D.E. and Czepulkowski, B.H. (eds) (1978). *Human Cytogenetics: A Practical Approach*. IRL Press, Oxford.
21. Davies, K.E. (ed.) (1986). *Human Genetic Diseases: A Practical Approach*. IRL Press, Oxford
22. Roberts, D.B. (ed.) (1986). *Drosophila: A Practical Approach*. IRL Press, Oxford.

23. Manuelides, L. (1985). *Ann. N.Y. Acad. Sci.* **450**, 250.

24. Bentley Lawrence, J. and Singer, R.H. (1986). *Cell* **45**, 407.

25. Hentzen, D., Renucci, A., Le Guellec, D., Benchaibi, M., Jurdic, P., Gandrillon, O. and Samaret, J. (1987). *Mol. Cell. Biol.* **7**, 2416.

26. Bloch, B., Popovice, T., Choucham, S. and Kowalski, C. (1986). *Neurosci. Lett.* **64**, 29.

27. Schuster, V., Mate, B., Wiegand, H., Traub, B., Kampa, D. and Neumann-Haefelin, D. (1986). *J. Infect. Dis.* **154**, 309.

28. Haase, A., Brahic, M., Stowning, L. and Blum, H. (1984). *Methods in Virology,* Vol. III, p. 189.

29. Maitland, N.J., Cox, M.F., Lynas, C., Prime, S., Crane, I. and Scully, C. (1987). *J. Oral Pathol.* **16**, 199.

30. Gowans, E.J., Burrell, C.J., Jilbert, A.R. and Marmion, B.P. (1983). *J. Gen. Virol.* **64**, 1229.

31. Gupta, J., Gendelman, H.E., Naghashfar, Z., Gupta, P., Rosenhein, N., Sawada, E., Woodtaff, J.D. and Shah, K. (1985). *Int. J. Gynaecol. Pathol.* **4**, 211.

32. National Radiological Protection Board. (1984). *Living With Radiation.* Her Majesty's Stationery Office, London.

33. Brady, M.A.W. and Finlan, M.F. (1990). In *In Situ Hybridization: Principles and Applications.* Oxford University Press, Oxford.

34. McAllister, H. and Rock, D. (1985). *J. Histochem. Cytochem.* **33**, 1026.

35. Bentley Lawrence, J. and Singer, R.H. (1985). *Nucleic Acids Res.* **13**, 5.

36. Terenghi, G., Cresswell, L. and Fallon, R. (1988). *Proc. R. Microscop. Soc.* **23**, 47.

37. Toutellotle, W.W., Verity, A.N., Schmid, P., Martinez, S. and Shapshak, P. (1987). *J. Virol. Methods* **15**, 87.

38. Maddox, P.H. and Jenkins, D. (1987). *J. Clin. Pathol.* **40**, 1256.

39. Sambrook, J., Fritsch, E.F. and Maniatis, T. (1989). *Molecular Cloning: A Laboratory Manual,* 2nd edn. Cold Spring Harbor Laboratory Press, Cold Spring Harbor, NY.

40. Cox, K., DeLeon, D.V., Angerer, L.M. and Angerer, R.C. (1984). *Dev. Biol.* **101**, 485.

41. Scott Young III, W., Mezey, E. and Siegel R.E. (1986). *Mol. Brain Res.* **1**, 231.

42. Unger, E.R., Budgeon, L.R., Myerson, D. and Brigatti, D.J. (1986). *Am. J. Surg. Pathol.* **10**, 1.

43. Brahic, M. and Haase, A.J. (1978). *Proc. Natl Acad. Sci. USA* **75**, 6125.

44. Miller, J.A. (1991). *Neurosci. Lett.* **121**, 211.

45. Eakin, T.J., Baskin, D.G., Breininger, J.F. and Stahl, W.L. (1994). *J. Histochem. Cytochem.* **42**, 1295.

46. Rogers, A.W. (1979). *Techniques for Autoradiography*, 3rd edn. Elsevier, Amsterdam.

47. Messe, E. and Blin, N. (1987). *Gene Anal. Tech.* **4**, 15.

48. Chan, V.T.-W., Fleming, K.A. and McGee, J. O'D. (1988). *Anal. Biochem.* **168**, 16.

49. Shivers, B.D., Harlan, R.E., Pfaff, D.W. and Schachter, B.S. (1986). *J. Histochem. Cytochem.* **54**, 39.

50. Hoeffler, H., Childers, H., Montminy, M.R., Gechan, R.M., Goodman, R.H. and Wolfe, H.J. (1986). *Histochem. J.* **18**, 597.

Chapter 8
Radioisotopes and immunoassay

ADRIAN F. BRISTOW and ROBIN THORPE
National Institute for Biological Standards and Control, Blanche Lane,
South Mimms, Potters Bar, Hertfordshire EN6 3QG, UK

1 Introduction

Immunoassay is the quantitative measurement of a substance of interest ('analyte') using antibodies which bind specifically to that analyte. For the rest of this chapter, these two main components of the assay will be referred to as ligand and antibody respectively. In order to perform such an assay, it is necessary to measure the binding of the ligand to the antibody at very low concentrations; this has most commonly been achieved with the use of radioisotopes. The first practical application of the principle was the techniques of radioimmunoassay (RIA), invented by Yalow and Berson in the late 1950s (1), for which the Nobel Prize was subsequently awarded. This technique, and subsequently developed related techniques such as immunoradiometric assays and enzyme-linked immunosorbent assays, have revolutionized the biological sciences. Using these methods the minute quantities found in biological materials of a whole range of analytes such as protein hormones, steroids, viral antigens, cytokines and clotting factors can be accurately measured without relying on difficult and sometimes imprecise measurements based on their biological activities. Indeed, it is hard to overestimate the impact that the development of immunoassays has had on the life sciences.

Although the basic principles of the radioimmunoassay and of the immunoradiometric assay have remained the same, both methods have been the subject of countless refinements and improvements, some of which have included the use of non-isotope labels (enzymes, luminescent labels), the development of novel methods of separating free from bound antigen such as magnetic particles, or of quantifying that ratio without separation (e.g. fluorescence depolarization) and the use of specific combinations of monoclonal antibodies. This chapter will attempt to cover the use of radioisotopes in immunoassays, with reference to a number of specific examples of assay methodology. This is a vast field, however, and the interested reader is referred to a number of excellent texts covering the various fields of immunoassay in the life sciences (2,3).

2 Immunoassays: theoretical considerations

The theoretical basis of immunoassays employing isotopes has been reviewed by Ekins (4). Two basic types of immunoassay can be recognized; 'limited reagent' assays, and 'reagent excess' assays, the two basic methods being typified by RIA and immunoradiometric assay (IRMA) respectively.

Limited reagent assays have been described by a number of terms such as 'saturation assay', 'competitive protein binding assay' and 'displacement assay'. Limited reagent assays employing isotopes and antibodies are almost invariably referred to as 'radioimmunoassays', a somewhat unfortunate term since it does not in any way describe the theoretical basis of the assay method. The underlying principle in all 'limited reagent' assays is that by limiting the concentration of one of the reagents, the system is saturable. In the technique of RIA, a radiolabelled ligand reacts with antibodies which bind specifically to that ligand. The amount of antibody available to react with the ligand is limited such that it is saturable, only a fraction of the total labelled ligand is bound to antibody. The reaction may be written as

$$Ab + {}^*L \rightarrow AB\text{--}{}^*L$$

where Ab is antibody and *: is radiolabelled ligand.

If the reaction is allowed to reach equilibrium in the presence of ligand that is not radiolabelled, two simultaneous reactions will take place:

$$Ab + {}^*L \rightarrow Ab\text{--}{}^*L$$

and

$$Ab + {}^*L \rightarrow Ab\text{--}L$$

where L is unlabelled ligand.

Under limited antibody conditions, therefore, unlabelled ligand will compete with radiolabelled ligand for available antibody, and the fraction of radiolabelled ligand that is antibody-bound will fall as the amount of unlabelled ligand present increases. In practice, the fraction of antibody-bound label is measured using a system that separates bound and free ligand. Increasing concentrations of unlabelled ligand generate a displacement curve of the type shown in *Figure 1*.

The parameters of the assay are the percentage bound in the absence of unlabelled ligand (B_0) and percentage bound in the absence of antibody (non-specific binding, NSB); the working range of the assay is the concentration range over which the displacement of radioactivity is linearly related to the log of ligand concentration.

In 'reagent excess' immunoassays, the antibody is present in excess, such that at equilibrium all available ligand is antibody-bound, and the majority of antibody remains unused. In practice, the amount of antibody that has ligand bound to it is determined. Measurement of antibody–ligand complex in this type of assay is most conveniently achieved by radiolabelling the antibody; the ligand-bound radiolabelled antibody is then separated from the free radiolabelled antibody by some convenient procedure. The most widely used immunoassay based

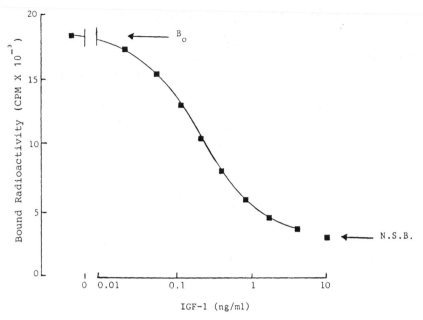

Figure 1. Radioimmunoassay of insulin-like growth factor-1 (IGF-1). Experimental procedure for this radioimmunoassay is given in *Protocol 4*. B_0 = per cent binding at zero IGF-1 concentrations. NSB (non-specific-binding) = per cent binding at zero antibody concentration.

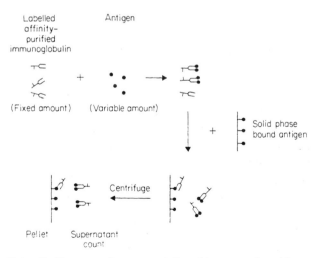

Figure 2. Diagrammatic representation of immunoradiometric assay (IRMA).

on the excess reagent principle is the immunoradiometric assay (IRMA). In the first IRMA procedures to be described, the free labelled antibody was removed from the reaction by using a large excess of the ligand coupled to a solid phase such as dextrose beads, which can easily be removed by centrifugation (this procedure is shown diagrammatically in *Figure 2*). This approach suffers from the serious drawback that it is only applicable when very large quantities of the

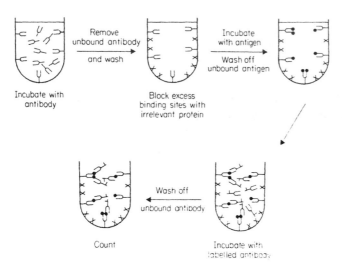

Figure 3. Diagrammatic representation of microtitre plate based two-site immunoradiometric assay (IRMA) for estimation of antigens.

ligand are available. A much more useful, and widely used approach, is the two-site IRMA, in which two separate antibodies are used, that react with two independent sites on the ligand. One of the antibodies is used to radiolabel the complex, and the other to remove the complex from the radiolabelled pool. The latter is achieved most commonly by coupling the second antibody to a solid phase. The procedure is schematically illustrated in *Figure 3*.

In the two-site IRMA a linear response is obtained, in which bound radio-activity is directly proportional to the concentration of ligand in the reaction mixture. The two-site IRMA offers a number of advantages over the 'limited reagent' RIA:

- The IRMA is generally more sensitive than corresponding RIAs for the same ligand;

- In the RIA the ligand is radiolabelled by a process of chemical modification; the antibody may then not react with the labelled and unlabelled ligand to the same extent;

- The two-site IRMA requires a specific interaction between two different anti-bodies and their binding sites; the specificity of such methods is therefore greater than in the RIA, in which immunologically similar but distinct molecules may interfere.

Despite these advantages, however, the usefulness of the two-site IRMA is often restricted by the requirement for two antibody-binding sites. Small molecules, or molecules that are poorly antigenic, may simply not produce the right combination of antibodies, and the RIA remains a powerful and widely used analytical tool.

3 Radioimmunoassays

3.1 Reagents

3.1.1 Antibodies

Currently, two basic types of antibody are available to the biologist, monoclonal antibodies and polyclonal antibodies. Monoclonal antibodies, although widely used in IRMAs, are very often of little use in RIA. Instead, the analyst has to obtain polyclonal antisera, raised in such animals as the rabbit, guinea pig, sheep or goat. A complete summary of methods of producing antisera for use in RIA is beyond the scope of this book, and is available elsewhere (5).

(i) Immunogens

The immunogen is the preparation to be injected into the animal to elicit antibody formation. Generally speaking, proteins with molecular weight >5000 will readily elicit antibody formation in experimental animals. Smaller molecules such as steroids or thyroid hormones are non-immunogenic without chemical modification and conjugation to carrier proteins, such as albumin or thyroglobulin, enabling the animal to see them as 'foreign', and so produce antibodies. It should also be noted that even for larger proteins, immunogenicity varies in different species of animal. Generally speaking, the more foreign a protein is the better will be the immune response. For instance, human insulin is a poor immunogen in the rabbit, whose own insulin is very similar in structure, but is a much better immunogen in the guinea pig, which has insulin of rather different structure.

(ii) Evaluation of antisera

Antisera for use in RIA are evaluated in terms of titre, avidity and specificity. Determination of the titre of an antibody is the most useful primary screening technique. The antibody titre, usually expressed as a dilution (e.g. 1/200 000) is the concentration of antibody that binds a certain percentage (usually 40–60%) of radiolabelled tracer under a given set of experimental conditions. Although such measurements are not absolute in that they vary with the experimental conditions used, they do represent the easiest way of monitoring the production of antibody during the immunization schedule.

The avidity, or association constant, is an absolute measurement of the extent to which an antibody binds to its ligand at equilibrium. In an RIA, equilibrium is reached between labelled and unlabelled ligand binding to a limited concentration of antibody (Ab). The association constant (K_{ass}) for the reactions is given by

$$K_{ass} \frac{[Ab-L]}{[Ab] \times [L]}$$

It can be readily seen that the higher the association constant, or avidity, the lower the concentration of ligand required to form the antibody–ligand complex. Hence there is a direct relationship between the sensitivity of an RIA (the concentration of ligand that can be measured) and the avidity of the antibody. In practice, the most common method for determining the avidity of an antibody is Scatchard analysis, in which the percentage of radiolabelled ligand bound is measured as a function of antibody concentration. Data are then plotted on a graph where the y-axis (ordinate) is the ratio of bound to free ligand and the x-axis (abscissa) is the concentration of ligand in mol/l. The slope of the line gives K_{ass} (1/mol) (5).

Antisera for use in RIA should have appropriate specificity. This means that while the substance to be measured should bind to the antibody, there should be no binding to any other substances present in the test sample. Many biochemical analytes such as steroids share common structural features, and even more complex structures such as proteins may exhibit immunological similarities. In practice few antibodies exhibit absolute specificity for one ligand, and most will show a decreased level of binding to structurally related ligands. Under these circumstances it needs to be shown that any other substances that bind to the antibody are not present in the test samples in sufficient concentrations to affect the assay significantly.

3.1.2 Tracers

Radiolabelled ligands used in RIA are generally referred to as tracers. Preparation of the tracer is one of the most important aspects of an RIA and indeed the use of poor tracers is the most common cause of assay variability, or lack of sensitivity and/or precision. The most common isotope used in RIA is ^{125}I, although other isotopes such as ^3H have been used, particularly for smaller molecules such as steroids. The major point of concern in considering tracers for use in RIA is that it should be remembered that the mechanism of RIA assumes that the antibody does not discriminate between the unlabelled ligand and the tracer (radio-labelled ligand). Although incorporation of tritium into steroids, or ^{125}I into thyroxine, may be done synthetically without appreciably altering the chemical structure of the ligand, for the most part incorporation of radioiodine into bio-logical macromolecules involves a process of chemical modification, and it should be recognized that the possibility always exists that this chemical modification alters the immunological properties of the ligand such that the antibody will discriminate between antibody and tracer.

Detailed procedures for radioiodination have been covered in Chapter 6. For the purposes of radioiodination, two basic methods have been used: those employing ^{125}I in the presence of oxidizing agents, and those employing radio-iodinated alkylating agents that do not require oxidizing conditions. Generally speaking, for robust ligands, oxidizing conditions such as the chloramine-T method are suitable. The side reactions that accompany such procedures, how-ever, often modify amino acid side-chains other than those being iodinated, and this can occasionally modify or destroy antibody-binding. The alternative, non-oxidizing procedures, such as the Bolton–Hunter method, are often suitable.

Such methods, however, introduce quite bulky iodine-containing chemical groups, which may themselves modify the immunological properties of the tracer.

Although the radioiodination procedures described in Chapter 6 serve as a useful guide, detailed methods need to be developed for each ligand studied. Since the underlying principle of an RIA requires displacement of tracer with equal concentrations of unlabelled ligand, it follows that the molar concentration of tracer should be as low as possible, and, therefore, that the specific activity of the tracer should be as high as possible. In practice, therefore, tracers for use in RIA should be radiolabelled to as high a specific activity as possible without affecting the antibody binding. Evaluation of the suitability of a tracer for use in an RIA is achieved using the parameters set out in Section 3.5.1.

3.1.3 Choice of working concentrations

In order to establish an RIA, the working concentrations of antibody and of tracer must be optimized. RIA as a 'limiting reagent' technique, in which the sensitivity of the method increases as the concentration of reagents approaches zero. The precision of radioactivity measurements, however, decreases as the amount of radioactivity to be measured approaches zero. Optimization of working concentrations of antibody and tracer is therefore a balance between increasing the sensitivity and decreasing the precision by reducing the concentration of reagents.

(i) Tracer

For an RIA in which the tracer is labelled with ^{125}I, typical working concentrations would be 10 000–20 000 d.p.m./tube. Higher levels of radioactivity do not appreciably improve the precision, but reduce the sensitivity. At lower levels of radioactivity experimental error becomes unacceptably high.

(ii) Antibody

The antibody concentration should be chosen such that some fraction of the radioactivity is bound to antibody at equilibrium. Using 20 000 d.p.m./tube of tracer, 20% binding would correspond to 4000 d.p.m./tube. This addition of increasing concentrations of unlabelled ligand would displace this bound radioactivity. In practice, therefore, the assay would be measuring d.p.m./tube between 0 and 4000. Such a working range represents a reasonable compromise between sensitivity and precision, although with well-optimized conditions and experienced operators working ranges of 0–1000 d.p.m./tube or even lower can sometimes be used.

The concentration of antibody giving 20% binding needs to be determined under the conditions of the assay from an antibody dilution curve. A typical procedure is illustrated in *Protocol 1*.

215

Protocol 1

Determination of an antibody dilution curve for anti-interleukin-1α

Reagents

- Human interleukin-1α, radiolabelled to a specific activity of 3.07×10^6 Bq/µg (83 µCi/µg)
- Sheep anti-interleukin-1α antiserum
- Bovine γ-globulins (Sigma)
- Polyethylene glycol 6000 (BDH)
- Diluent: 0.05 M sodium phosphate buffered isotonic saline (0.15 M NaCl) pH 7.4, containing 0.2% bovine serum albumin

Method

1 Prepare twofold dilutions of antibody in diluent to cover the range 1/1000 to 1/1 024 000.

2 Prepare LP-3 (Luckam's) plastic tubes in triplicate containing:

   100 µl diluent

   100 µl tracer (10 000 c.p.m./100µl)

   100 µl antibody dilution

   Set up set of triplicates containing zero antibody.

3 Incubate at 4°C for 16 h.

4 Add to each tube, 300µl of PEG/γ-globulin reagent:

   5 g PEG 6000/12.5 ml H_2O

   30 bovine γ-globulin/2.5 ml H_2O

   5.0 ml 0.2 M Tris–HCl, pH 8.5

5 Vortex and incubate for 1 h at 4°C.

6 Centrifuge (2000 g, 30 min), carefully aspirate off the supernatants and determine the radioactivity in the pellets by gamma counting.

7 Plot counts precipitated against log_{10} antibody concentration.

3.1.4 Standards

RIA is not an absolute technique. Concentrations of ligand in test samples are determined by reading the values from displacement curves obtained with standard preparations. Although for well-optimized RIAs the dose–response range will remain reasonably constant, there will inevitably be day-to-day drift, and a separate standard curve needs to be included in each assay. The assay standard needs to meet certain requirements. Most important, it must be immunologically identical with the ligand present in the test sample. Although this may seem a relatively easy criterion to meet, it should be remembered that many molecules of interest to the biologist are complex structures that are only isolated from biological tissues after long and painstaking purification procedures. They may not even be present in homogeneous form in the body or cell to start with. As a result, few preparations of proteins or other biological macromolecules are actually homogeneous in biochemical terms. The different isoforms present may react differently with the antibody, especially when one considers that the antibody may have been raised to a different preparation of the ligand in the first place. The standard must also be stable in its stored form. An unstable standard will result in a gradual erroneous increase in RIA values as the standard degrades.

3.1.5 Samples

The component of an RIA over which the operator often has the least control is the sample for analysis. The binding of antibodies to their specific ligands may be perturbed by a number of factors present in samples. These factors are generally known as matrix effects and are described below.

(i) Solvent effects

RIA is frequently used to monitor fractionation procedures such as chromatography, or other biochemical procedures. Commonly, such procedures utilize conditions of high or low pH, chaotropic or organic solvents. Any of these may disrupt antibody–ligand binding, and samples would have to be appropriately treated, by neutralization, desalting, lyophilization or simple dilutions, before they could be examined by RIA.

(ii) Proteases

Biological samples, including blood and tissue extracts, frequently contain proteases that under the conditions of the RIA, may degrade the tracer, resulting in artefactual results. The presence of such activity can be revealed by gel filtration of the tracer following incubation with the sample. Proteases, if present, can be blocked with protease inhibitors, or the ligand of interest may be extracted from the sample before assay (see *iv* below).

(iii) Binding proteins

Many ligands exist in biological samples coupled to specific binding proteins. Occasionally these present a problem in that they will mask the antibody-binding site and prevent the assay from working. The example given in *Protocol 2* utilizes low pH to separate the ligand (insulin-like growth factor-1) from its binding protein.

(iv) Sample extraction

The most common cause of sample unsuitability is simply that the ligand of interest is not present in high enough concentration in the sample. Many procedures have been developed for extracting and concentrating the ligand from large volumes of sample. The method given in *Protocol 3* describes extraction and concentration of growth-hormone releasing factor from plasma.

Protocol 2

Sample pre-treatment for the removal of insulin-like growth factor-1 (IGF-1) binding proteins

Reagents

- Clinical plasma samples
- Formic acid, 2.4 M
- Ethanol

- RIA buffer: 0.1M EDTA, 0.145M NaCl, 0.1% Tween 20, 0.1% sodium azide, 0.2M sodium phosphate, pH 7.4

Method

1 To 100 μl plasma, add 25 μl of 2.4 M formic acid and 500 μl ethanol.
2 Mix and incubate for 30 min at room temperature.
3 Centrifuge (2000 g, 30 min) and collect the supernatants.
4 Neutralize 100 μl of supernatant with 600 μl RIA buffer.

Protocol 3

Extraction of growth-hormone releasing factor from serum

Equipment and reagents

- Serum or plasma samples
- Vycor glass
- Methanol, 80%
- Hydrochloric acid, 1.0 M

- Assay diluent: 0.1 M phosphate, 0.025 M EDTA, 0.15 M NaCl, 0.01% thiomersal, 0.5% human serum albumin pH 7.4

Method

1 Dilute 200 μl plasma samples to 1 ml with assay diluent.
2 Add 1 ml of a 10 mg/ml suspension of activated Vycor glass in distilled water, and rotate tubes for 30 min at 4°C.
3 Wash glass successively with 2 ml H_2O and 2 ml 1.0 M HCl.
4 Elute growth hormone from the glass with 1.0 ml 80% methanol, evaporate to dryness and reconstitute in assay diluent.

3.1.6 Assay matrix

The solvent, or matrix, in which the assay is performed, contains two main components: buffer salts and an agent to prevent assay reagents from binding non-specifically to other proteins or the walls of the tube. The most common buffer used in immunoassay reagents is sodium phosphate-buffered isotonic saline (pH 7.4), although many other neutral buffers can be used. Two types of agent may be used to prevent non-specific binding: proteins, such as serum albumin, at 0.1% to 1% (w/v), and detergents such as Tween. Occasionally, immunoassay matrices may contain additional reagents such as protease inhibitors (e.g. benzamidine, trasylol), chelating agents (e.g. EDTA) or inhibitors of bacterial growth (e.g. thiomersal). The detailed immunoassay procedures given in *Protocols 4* and *5* use typical immunoassay matrices exhibiting many of these features.

Clinical immunoassays, in which the ligand of interest is present in serum, may suffer quite markedly from matrix effects due to other substances present in the sample. A common way round this problem is to use as an assay diluent 'stripped' serum, i.e. serum from which all the ligand of interest has been removed by some suitable procedure, such as charcoal or immunoaffinity absorption.

3.2 Assay conditions

An RIA consists of three stages:

(1) pre-incubation (sample/standard + antibody);

(2) incubation (sample/standard + antibody + tracer); and

(3) separation (separation of bound from free tracer).

Increased sensitivity is usually achieved by increasing the pre-incubation and incubation times, bringing the reactants closer to equilibrium. Periods of 3 days each for the pre-incubation and incubation stages have been used. Where ultimate sensitivity is not a requirement, the pre-incubation may be reduced to a few hours, or even incorporated into the main incubation, which can itself be reduced to 16 h or shorter. Incubations are usually carried out at 4°C, but higher temperatures may be used to reach equilibrium more quickly.

Most RIA methods uses plastic or glass test tubes, and incubation volumes of around 0.5 ml. There are, however, many alternative procedures. The assay may be performed in microtitre-well plates or using antibody-coated tubes or on filter-paper spots.

3.3 Separation techniques

The final stage of an RIA is to separate the bound antibody from the free radio-labelled tracer. Many different procedures have been developed for different ligands. Some of the more commonly used methods are described below:

- *Adsorption*: low molecular weight ligands may be adsorbed on to charcoal, leaving antibody-bound ligand in solution.

- *Solvent precipitation*: where the ligand is soluble in organic solvent, the antibody-bound ligand may be removed by ethanol precipitation.

- *PEG*: PEG forms specific precipitating complexes with immunoglobulins (antibodies). It may be used, therefore, to separate free from bound tracer either alone, in combination with non-specific immunoglobulin to form a better precipitate (see *Protocol 1*) or in combination with secondary antibody (see below) to accelerate the separation.

- *Secondary antibodies*: immunoglobulins possess a constant region that, within a given species, contains several common epitopes. Anti-antibodies (usually referred to as secondary antibodies) can therefore be prepared in a different animal which will precipitate the primary antibody. Secondary antibodies are widely used in RIA, either alone or with PEG (see *Protocol 5*) or coupled to a solid phase (see below).

- *Solid-phase secondary antibody*: a particularly useful approach is to couple the secondary antibody to a solid phase such as microcrystalline cellulose, allowing easy and rapid precipitation of the antibody-bound tracer fraction (*Protocol 6*).

Protocol 4

Radioimmunoassay for IGF-1

Reagents

- Serum or plasma samples, pre-treated to remove IGF-1-binding proteins (*Protocol 2*)
- ^{125}I-labelled IGF-1, radiolabelled using chloramine-T to a specific activity of 7.4×10^6 Bq/μg (200 μCi/μg)
- Rabbit anti-IGF-1 antibody
- Cellulose-linked anti-rabbit antibody (see *Protocol 6*)
- Purified IGF-1 standard
- Assay diluent: 0.1 M EDTA, 0.145 M NaCl, 0.1% Tween 20, 0.1% sodium azide, 0.2 M sodium phosphate, pH 7.4

Method

1 Carry out the assay in triplicate in LP3 (Luckham's) plastic tubes.

2 Prepare a standard curve containing IGF-1 standard at the following concentrations (ng/ml): 40, 20, 10, 5, 2.5, 1.25, 0.625, 0.313, 0.156, 0.075, 0.039, 0.02 and 0

3 Prepare tubes containing 100 μl sample or standard, plus 100 μl anti-IGF-1 antibody at a dilution of 1/4000. An additional set in triplicate should be prepared containing 200 μl assay diluent only. These are used for measurement of non-specific binding.

4 Incubate at 4 °C for 6 h.

5 Add to each tube 100 μl [^{125}I]IGH-1 (10 000 c.p.m./100 μl). Include one set of triplicates which contains tracer only for determination of total counts.

6 Incubate for a further 16 h at 4 °C.

7 To each tube except those for the total counts add 100 μl stirred cellulose-linked anti-rabbit antibody. Vortex and incubate at room temperature for 30 min. Add 1 ml water to each tube, centrifuge (1500g, 10 min), aspirate off the supernatant and determine the radioactivity in the pellets by gamma counting.

8 Typical results are given in *Figure 1*.

- *Solid-phase primary antibody*: The need for a separation stage is overcome if the primary antibody is coupled to a solid support. A particularly attractive method is to couple the antibody to the walls of the test tube, separation being achieved simply by removing the incubation medium. Alternatively, cellulose-linked primary antibody may be used and removed by centrifugation.

3.4 Specific protocols

Detailed protocols for specific RIAs are given in *Protocols 4, 5* and *7*. Although identical in fundamental principles, each of the assays uses different sample treatment, incubation conditions and separation procedures, illustrating many of the principles and procedures described earlier in this chapter.

Protocol 5

Radioimmunoassay for growth-hormone releasing factor (GRF)

Reagents

- Serum samples, extracted as described in *Protocol 3*
- [^{125}I]-GRF radiolabelled using chloramine-T to a specific activity of 1.295×10^7 Bq/μg (350 μCi/μg)
- Rabbit anti-GRF antibody
- Purified GRF standard
- Assay diluent (see *Protocol 3*)
- Donkey anti-rabbit antiserum
- Normal rabbit serum
- PEG 6000

Method

1 Prepare standard concentrations covering the range 0.01–10ng/ml, as described in *Protocol 4*.

2 Pre-incubate standards or samples with antibody (1/100 000) for 24 h at 4 °C. Include tubes to measure non-specific binding (see *Protocol 4*)

3 Add [^{125}I]-GRF (10 000 c.p.m. in 100 μl) and incubate for a further 48 h at 4 °C.

4 Add 100 μl of secondary antibody reagent:
   - donkey anti-rabbit anti-serum 1/200
   - normal rabbit serum 1/2400
   - PEG 4%

5 Incubate for a further 1 h, centrifuge (3,300 g, 30 min) and determine the radio-activity in the pellets by gamma counting.

Protocol 6

Preparation of cellulose-linked secondary antibody

Equipment and reagents

- Thin-layer chromatography grade microcrystalline cellulose (Whatman)
- Donkey anti-rabbit antiserum
- 0.1 M trisodium citrate pH 6.5
- Cyanogen bromide: **Caution:** cyanogen bromide is extremely toxic by skin contact and inhalation. All procedures should be done in a fume-hood, and all equipment soaked in 4 M NaOH before washing.

- Resuspension buffer: 0.2 M sodium phosphate, 0.1 mM EDTA, 0.145 M NaCl, 0.1% sodium azide, 0.5% BSA, 0.5% Tween 20, pH 7.4
- 0.5 M sodium carbonate pH 10.5

- 2 M ethanolamine
- 4 M sodium hydroxide
- 0.05 M sodium phosphate, 0.15 M NaCl, pH 7.4

Method

1 Dialyse 15 ml donkey anti-rabbit antiserum exhaustively against 0.1 M trisodium citrate, pH 6.5.

2 Suspend 250 g cellulose in 1 litre of 0.5 M sodium carbonate, pH 10.5.

3 Dissolve 60 g cyanogen bromide in 250 ml distilled water (this may take 2–3 h).

4 Add the cyanogen bromide in 20 ml aliquots to the cellulose suspension. The mixture should be continuously stirred and the pH maintained at 10.5 with 4 M NaOH.

5 When the pH has ceased to fall, filter the cellulose through a Buchner funnel and wash the cake with 2 litres of ice-cold citrate buffer.

6 Transfer the washed cake to a beaker, add the dialysed antiserum and 45 ml 0.1 M trisodium citrate, pH 6.5 and incubate overnight at 4°C.

7 Next morning, add 8 ml ethanolamine solution and continue mixing at room temperature for 1 h.

8 Wash the cellulose in 0.05 M sodium phosphate buffered isotonic saline (0.15 M NaCl) pH 7.4 and resuspend in 600 ml of resuspension buffer.

Protocol 7

Microtitre plate IRMA

Equipment and reagents

- Phosphate-buffered saline (PBS): 10 mM sodium phosphate, 130 mM sodium chloride, pH 7.2
- PBS containing 3% bovine haemoglobin (Hb-PBS)
- Purified antibody 1 (for capture)
- Purified standard antigen

- ^{125}I-labelled purified antibody 2 (for detection)
- Samples containing antigen
- 96-well microtitre plate
- hot-wire plate cutter
- Gamma counter

Method

1 Prepare antibody 1 in PBS at 1–20 μg/ml (try 4 μg/ml initially). Add 50–200 μl to each well of a 96-well flexible microtitre plate and incubate overnight at 4°C.

2 Discard antibody solution by holding plate over the sink and tapping briskly. Wash the plate three times with Hb-PBS by filling wells and discarding contents as above.

3 Block wells with haemoglobin by filling wells with Hb-PBS and incubating for 30 min to 1 h at room temperature. Wash once with Hb-PBS.

4 Prepare dilutions of standard antigen in Hb-PBS (see *Figures 4-7* for typical ranges). Add 45–195 µl of dilutions to each well and include a blank (Hb-PBS); carry out triplicate assays for each point. Include unknown solutions in triplicate. Incubate overnight at 4°C or for 2–4 h at room temperature or 37°C.

5 Discard well contents and wash three times with Hb-PBS. Add 45–195 µl ^{125}I-labelled antibody 2 (3×10^5 c.p.m./well) diluted in Hb-PBS and incubate for 1–2 h at room temperature.

6 Discard well contents and wash three times with Hb-PBS. Cut out wells using a hot-wire plate cutter and determine the radioactivity bound to the wells by gamma counting.

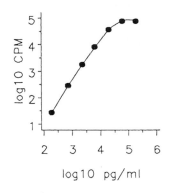

Figure 4. Dose–response curve for a two-site IRMA or rDNA-derived human IL-1α using monoclonal antibodies which recognize different epitopes.

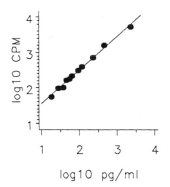

Figure 5. Dose–response curve for IRMA of rDNA-derived human IL-1α using a monoclonal antibody to capture antigen and ^{125}I-labelled sheep antibody to develop.

3.5 Interpretation of results

3.5.1 Assay validation

An RIA displacement curve, as illustrated in *Figure 1*, lies between two values, the B_0 value, or the fraction of tracer bound in the absence of unlabelled ligand, and the NSB, or fraction of the tracer which is apparently bound in the absence of antibody. The log dose-displacement curve is sigmoidal, and the slope may be measured over the linear part of the curve. In practice, these three experiment-

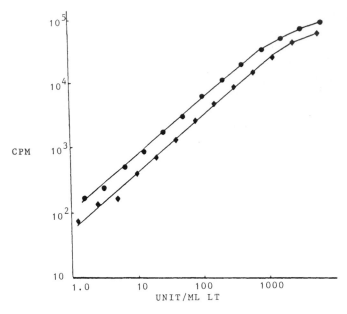

Figure 6. Bead-based IRMA for human lymphotoxin using two different mAbs. ◆, rDNA-derived lymphotoxin (LT). ●, Natural LT. Courtesy of Dr Tony Meager.

ally determined variables form useful diagnostic criteria for the performance of the RIA in measuring analyte in biological samples.

(i) Assay slope

Samples should always be measured at more than one dilution, in order that the slope of the displacement curve may be compared with that of the standard. Differences in slope indicate immunological differences between the assay standard and the ligand being measured in the samples. They also effectively invalidate the assay since different answers will be obtained depending on which parts of the standard curve are used.

(ii) Non-specific binding

The NSB value is the amount of tracer non-specifically precipitated or removed by the separation technique used. In practice, the displacement curve should, at high ligand concentration, reach the NSB value. If maximum displacement in the standard curve is higher than the NSB, it usually indicates that the tracer and the unlabelled ligand are not immunologically identical; this may arise from modification to the tracer during the iodination process, or, more commonly, may reflect ageing of the tracer, either by radioactive decay or radiolysis. If the maximum displacement obtained with the assay samples is lower than the NSB, it indicates that the tracer is being degraded under the assay conditions and that appropriate sample pre-treatment is necessary.

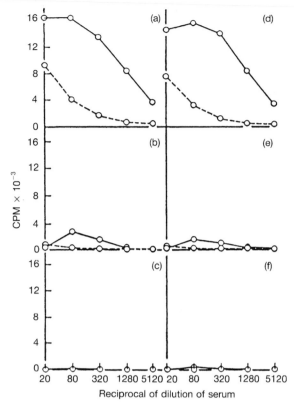

Figure 7. Two-site immunoradiometric assays for human IgE using mAbs and a polyclonal antiserum. (a), (b) and (c) using mAb 102 as capture antibody and (a) polyclonal anti-IgE serum, (b) mAb 102 and (c) mAb 117 as developing antibodies. (d), (e) and (f) Using mAb 117 as capture antibody and (d) polyclonal anti-IgE serum, (e) mAb 102 and (f) mAb 117 as developing antibodies (○—○) using the international reference preparation for human serum IgE (85/502) (100 000 IU/ml): (○- - - -○) using pooled normal human serum.

(iii) Zero binding, B_0

If the bound fraction in the assay samples is higher than B_0, it indicates the presence in the samples of components that are affecting the antibody–ligand reaction. If the sample displacement curve is lower than B_0 at high dilution, it may also indicate tracer degradation.

3.5.2 Analysis of data

The simplest method of data analysis is to plot the standard curve as percentage binding against log ligand concentration and to read the sample values off the curve. For some applications a more rigorous statistical analysis of the data is required. A number of approaches have been developed further, most of which use some form of data transformation to generate linear displacement curves, which are then analysed as parallel line assays. A complete description of the statistical analysis and validation of RIAs is beyond the scope of this book. The interested reader is referred to a number of excellent texts on this subject (for example, ref. 6).

4 Immunoradiometric assays

Originally, IRMAs were carried out using antibodies (or ligands) covalently coupled to a solid support in bead form, for example Sepharose or cellulose. Currently, this approach is rarely used as it has been shown that antibodies and the majority of protein antigens stick particularly well by physical processes to most plastic and some types of glass. Not only is this easier and quicker than covalent coupling, it also does away with chemical procedures which may inactivate some antibodies, particularly the less stable murine immunoglobulins. The two-site IRMA technique is particularly well-suited for use with monoclonal antibodies (mAbs) or mAb–polyclonal antibody combinations. This overcomes the limitations of most mAbs when used in conventional RIAs and by careful selection of appropriate antibody combinations both extreme specificity and high sensitivity can often be achieved.

4.1 Choice of solid support

The most common solid supports currently used to immobilize the first ('capturing') antibody are the walls of plastic tubes and plastic microtitre plates or the surfaces of spherical plastic beads. Some manufacturers make plates specially for immunoassays; these plates are supposed to be particularly good for binding proteins to the well surfaces. The authors have found that the flexible types of microtitre plates are perfectly adequate for IRMAs, and wells can be readily cut out from these using a hot wire, which is much less tedious than using the shears or band-saw necessary for rigid plates.

The use of tubes for IRMAs is obviously more time consuming than using microtitre plates (particularly in the washing steps) but a larger volume can be accommodated, and this, in some cases, can increase sensitivity. However, the use of large volumes obviously consumed larger volumes of reagents and requires a greater amount of sample, so this is often not a real advantage. Plastic tubes are available with moulded internal projections that increase the surface area available for antibody binding (for example, 'star' tubes) and these can increase sensitivity in some cases without excess consumption of reagents and requirement for a large sample volume.

Use of plastic beads for immobilizing antibody can also increase sensitivity as a fairly large volume of sample can be used. Beads are available with finely etched surfaces (for example, from Northumbria Biologicals) that increase the surface area available for protein binding.

4.2 Antibodies for IRMAs

Antibodies for IRMAs should be purified to as near homogeneity as possible. This is necessary as the sensitivity of the assay is directly related to the number of suitable binding sites on the solid support occupied by primary (capturing) antibody. If the coating antibody is impure, contaminants may compete for the binding sites and hence reduce sensitivity. The secondary or 'developing' antibody must also be highly purified as it must be radiolabelled; the use of impure preparations will lead to these also becoming radioactive, resulting in poor labelling efficiency of the antibody, decreased sensitivity and high assay backgrounds.

For IgG monoclonal antibodies it is usually sufficient to isolate the IgG fraction of ascitic fluid or culture fluid (if fetal calf serum or serum-free medium is used during production). For this combination of ammonium sulfate precipitation (40–45% saturation) and some form of ion-exchange chromatography is usually best and the authors have found the high performance liquid chromatography (HPLC) system described by Clezardin *et al.* (1985) most satisfactory (7). IgM mAbs can be problematical for use in IRMAs, but if used they can be purified by a variety of techniques including gel filtration on Sepharose 4B or equivalent.

Polyclonal antisera should also be highly purified, and in some cases affinity purification may be necessary. (See refs 8 and 9 for details of antibody purification.)

4.3 Radiolabelling antibodies

Radiolabelling antibodies with ^{125}I is almost always used for IRMAs. This is usually quite easily achieved using chloramine-T as the catalyst (see Chapter 6), but some antibodies (particularly murine mAbs) radioiodinate poorly or are impaired in ability to bind antigen by the radioiodination procedure. If this is the case, less harsh catalysts such as Iodo Gen or lactoperoxidase may overcome the problem.

4.4 Assay design

4.4.1 Choice of conditions and non-antibody reagents

A general protocol for a two-site IRMA using microtitre plates is given in *Protocol* 7 and a 'bead' version in *Protocol 8*. Tube-based versions are merely adaptations of plate assays. In general two-site IRMAs consist of five operations:

(1) coating solid phase with primary (capturing) antibody;

(2) blocking unoccupied protein binding sites on a solid phase;

(3) incubation with sample;

(4) incubation with radiolabelled secondary (developing) antibody;

(5) counting bound radioactivity.

Washing procedures are inserted between each operation to remove excess reagents, etc. In general it is best to coat plates/beads/tubes with antibody overnight at 4°C. A coating concentration of 1–20 μg/ml IgG is optimal in most cases; higher concentrations often produce less sensitive assays. Blocking of unoccupied binding sites is best achieved by incubation for ~30 min with an irrelevant protein; the very cheap and effective bovine haemoglobin works well for this. Ovalbumin, bovine serum albumin, human serum albumin and many other proteins can be substituted, but the authors caution against the use of non-protein blocking agents such as Tween, which may block inefficiently in some cases.

Protocol 8

Bead IRMA

Equipment and reagents

- PBS (see *Protocol 7*)
- PBS containing 2% BSA (BSA-PBS)
- PBS containing 0.1% BSA (washing solution)
- Purified antibody 1 (for capture)
- ^{125}I-Labelled purified antibody 2 (for detection)

- Purified standard antigen
- Samples containing antigen
- Plastic tubes (e.g. Luckham P4)
- Plastic beads (e.g. from Northumbria Biologicals)
- Filter paper

Method

1 Prepare antibody 1 in PBS in 100–200 μg/ml. Coat beads in antibody by incubating in this solution overnight at 4°C; 20 ml of the solution is enough for ~100 beads of 6.5 mm in diameter.

2 Block plastic tubes (e.g. Luckham LP4) with albumin by completely filling with BSA-PBS and incubating overnight at 4°C.

3 Aspirate and discard the BSA-PBS. Prepare dilutions of the standard antigen in BSA-PBS (see *Figures 4–7* for typical ranges). Add 200 μl of the dilutions to the tubes and include triplicate assays for each point. Include a blank (BSA-PBS) and unknown solutions also in triplicate.

4 Remove beads from antibody solution and wash five times with washing solution. Blot to dryness on filter paper and add one bead to each tube. Incubate overnight at 4°C or for 4 h at 4°C or room temperature.

5 Remove sample by aspiration and wash five times with washing solution. Add 200 μl of ^{125}I-labelled antibody 2 in BSA-PBS containing 1–3 × 10^5 c.p.m. to each tube and incubate overnight at 4°C or for 3–4 h at room temperature.

6 Remove labelled antibody and wash five times with washing solution. Determine the radioactivity bound to the beads by gamma counting.

Incubation with antigen-containing sample is often most conveniently carried out overnight at 4°C, but a shorter period, e.g. 3 h at room temperature or at 37°C, usually produces no loss of sensitivity. This approach may be valuable if a single-day assay is required.

Incubation time and amount of developing antibody required to produce the best IRMA must be determined empirically, but 1–2 h with ~300 000 c.p.m./well is optimal for most microtitre plate IRMAs.

4.4.2 Choice of antibodies

Careful evaluation of mAb or mAb–polyclonal antibody combinations for optimal performance in IRMAs is probably the most important factor in assay design (see *Figures 4–7*). In most cases it is necessary to select combinations that bind to different epitopes on the antigen; this is often difficult if polyclonal antisera are used. Immunochemical procedures can be used to establish that mAbs recognize different antigenic determinants and the reader is referred to a practical immunochemical text, for example (8), for such methods. If mAb–polyclonal combinations are used it is usually preferable to use the mAb to 'capture' antigen. If the antigen exists predominantly as a multimer, for example immunoglobulins, then it may be possible to use the same antibody to capture antigen and develop the assay (see *Figure 7*).

5 Non-radioisotopic immunoassays

As outlined in earlier sections of this chapter, the role of radioisotopes in quantitative immunoassays is that of a 'tag' or 'label', which allows small quantities of the labelled analyte or antibody (depending on assay type) to be accurately measured and quantified. Although a detailed account of developments is outside the scope of this article, it should be noted that in the three decades since the widespread introduction of RIAs, there has been a more or less constant development of non-isotopic tags as alternatives. Thus, immunoassays based on enzyme labels, chemiluminescent, fluorescent and electron spin labels have all been described. Such procedures offer a number of advantages, including:

- safety—the use of ionizing radiation is avoided;
- practical advantages—the life of non-isotopic labels is not restricted by radioactive decay or radiolysis;
- sensitivity—many non-isotopic methods are either intrinsically more sensitive (e.g. those using luminescent labels) or potentially amplifiable using secondary or indirect detection end-points (e.g. those using enzymes);
- ease of automation.

The most common non-radioisotopic RIAs remain enzyme-based systems usually called enzyme-linked immunosorbent assays (ELISAs). A detailed discussion of ELISA and other non-radioactive immunoassays is beyond the scope of this book (for a general discussion of ELISA and protocols, see ref. 10), some features are provided in *Table 1* and a brief description of the commonest type of assay is given below.

5.1 Enzyme immunoassays

Any assay configuration, direct or competitive binding or two-site immunoradiometric, can be modified to use enzymes. Enzymes are normally chosen which catalyse with simple kinetics the formation of stable, coloured or fluorescent products from colourless or non-fluorescent substrates. In many applications a positive/negative distinction can be read by eye, for example in primary hybridoma screening or in commercial pregnancy tests. For quantitative work,

Table 1. Advantages and disadvantages of ELISA

Advantages

- No radiation precautions needed
- Can be read by eye or simple plate readers
- Many samples can be processed in automated systems in a short time
- Reagents are cheap and readily available
- Enzyme conjugates are stable

Disadvantages

- Some reagents are toxic
- Enzyme reactions are prone to interference by biological samples
- Conjugation of enzymes to proteins is not as easy or efficient as for radioisotopes
- Enzyme groups are large and can interfere with antigen–antibody reactions
- Requires determination of optimal conditions for enzyme reaction

purpose-built spectrophotometers (ELISA readers) can read a 96-well microtitre plate in ~1 min.

A wide range of commercial anti-species specific immunoglobulins (i.e. IgGs against species-specific antibodies) or protein A products is available conjugated to the most commonly used enzymes, particularly horseradish peroxidase (HRPO) or alkaline phosphatase (AP). Conjugation of enzymes to antibodies can satisfactorily be performed in the laboratory using periodate- or glutaraldehyde-based methods (10, 7). Once formed, these conjugates are very stable if kept at high concentration. For HRPO the most common substrate is orthophenylene diamine, and that for AP is P-nitrophenyl phosphate. Both systems use cheap reagents that generate a stable product. The choice of enzyme–substrate combination depends on the type of sample to be analysed: some samples (e.g. BSA) contain endogenous enzyme activity. Milk, or some cells such as macrophages, contain peroxidase, while B lymphocytes contain AP. Other samples can contain inhibitor of enzyme activity for example azide. Alternative enzymes, such as β-galactosidase, urease or glucose oxidase, can be used if HRPO or AP is unsuitable.

For all enzyme-based immunoassays it is essential to establish conditions for the enzyme-catalysed reactions that give the optimal dynamic range for measurement of the analyte. The time and temperature of incubation for a given amount of antibody–enzyme conjugate are usually the most significant parameters. In a multi-plate assay it is also important to keep the temperature constant for every plate, and either to read the plates within a short time of each other, or to inactivate the enzyme with low (HRPO) or high pH (AP). The inclusion of negative controls and a standard titration of analyte on each plate should reduce problems of plate variation.

5.2 Radioisotopes versus non-isotopic methods in immunoassay: future trends

Non-isotopic labels usually introduce a large chemical group into either the analyte or the antibody being labelled, such as complex fluorophores, enzymes or enzyme-binding co-factors. In practice, whilst both immunoglobulins and high molecular weight analytes may be successfully tagged with a variety of non-isotope labels and still retain their antigen/antibody properties, small molecules cannot usually be labelled without destroying their antibody-binding properties. The extent of the trend away from the use of isotopes in immunoassays may be understood by considering this aspect of non-isotopic labelling in parallel with the comparison of the features of limited reagent versus reagent excess assays outlined in Section 2. Two main types of immunoassay application may be envisaged: those for high molecular weight analytes and those for low molecular weight ones.

5.2.1 High molecular weight analytes

For large molecules it is almost always possible to find a pair of antibodies that bind simultaneously to the molecule. Since there is no limitation to the development of a two-site, immunometric (reagent excess) assay, the advantages of sensitivity, speed and specificity outlined in Section 2 have led to the almost complete replacement of the older (limited reagent) RIAs for these molecules. For labelling antibodies the advantages of non-isotopic labelling are overwhelming, and the use of both competitive immunoassays and isotopes in assay of large molecules is declining. For example, in 1997 some 95 clinical laboratories were performing immunoassays for thyroid-stimulating hormone (TSH) (a 30 kDa glycoprotein) in the UK, of which 78 were carrying out non-isotopic two-site assays. This use of non-isotopic immunoassays for TSH has risen from 0% in 1980, to 60% in 1993 and 82% in 1997. Figures for many other high molecular weight clinical analytes are comparable.

5.2.2 Low molecular weight analytes

For low molecular weight analytes it is usually not possible to find two non-exclusive antibody-binding sites, either as a result of limited antigenicity or simply due to steric hindrance preventing two large antibody molecules binding to a single small molecule. The development of two-site methods for such analytes has, therefore, been restricted. As a result, the use of competitive (limited reagent) assays is still widespread. Chemical modification of the ligand by non-isotopic labelling limits the usefulness of these methods for small molecules; indeed in many cases they cannot even be radiolabelled by standard methods, and radio-labels have to be introduced during synthesis. For these reasons the use of RIAs often remains predominant. Thus only 4% of current clinical immunoassays in the UK for the steroid androstenedione are non-isotopic. Even in the case of small molecules, however, the introduction of non-separation immunoassay methods (outside the scope of this review) has for many analytes allowed the use of non-isotope immunoassays. Currently, for example, 100% of cortisol immunoassays in UK clinics are non-isotopic.

As a summary of these trends, it can be seen that whilst the development of immunoassay methods was critically dependent on the use of radioisotopes, the combined advantages of reagent excess assay methods and non-isotopic labels are continuing to reduce the need to use isotopes, a trend which is likely to continue to the point at which RIAs are of little more than historical significance.

References

1. Yalow, R. S. and Berson, S. A. (1959). *Nature (Lond.)* **184**, 1648.
2. Hunter, W. M. and Corrie, J. E. T. (eds) (1983). *Immunoassays for Clinical Chemistry*, 2nd edn. Churchill Livingstone, Edinburgh.
3. Thorell, J. I. and Larson. S. M. (1978). *Radioimmunoassay and Related Techniques*. C. V. Mosby, St Louis, MO.
4. Ekins, R. P. (1976). General principles of hormone assay. In *Hormone Assays and their Clinical Application* (eds J. A. Loraine and E. T. Bell), pp. 1–72. Churchill Livingstone, Edinburgh.
5. Munro, A. C., Chapman, R. S., Templeton, J. G. and Fatori, D. (1983). Production of primary sera for radioimmunoassay. In *Immunoassays for Clinical Chemistry* (eds W. M. Hunter and J. E. T. Corrie), pp. 447–456. Churchill Livingstone, Edinburgh.
6. Finney, D. J. (1978). *Statistical Methods in Biological Assay*, 3rd edn. Griffin, London.
7. Clezardin, P., McGregor, J. L., Manach, M., Boukerche, H. and Dechavanne. M. (1985). One-step procedure for the rapid isolation of mouse monoclonal antibodies and their antigen binding fragments by fast protein liquid chromatography on a mono Q anion-exchange column. *J. Chromatog.* **319**, 67–77
8. Johnstone, A. and Thorpe, R. (1996). *Immunochemistry in Practice*, 3rd edn. Blackwell Scientific, Oxford.
9. Baines, M. G., Gearing, A. J. H. and Thorpe, R. Purification of murine monoclonal antibodies. In *Methods in Molecular Biology*, Vol. 5, pp. 647–668. Humana Press, Totowa, NJ, USA.
10. Nakamura, R. M., Voller, A. and Bidwell, D. E. (1986). In *Handbook of Experimental Immunology*, Vol. 1, *Immunochemistry* (ed. D. M. Weir) pp. 271–2720. Blackwell Scientific Publications, Oxford.
11. Tijsenn, P. and Kurstak, E. (1984). *Ann. Biochem*, **136**, 451.

Chapter 9
Pharmacological techniques

ISABEL J. M. BERESFORD and MICHAEL J. ALLEN

Stroke and Migraine, Neurology, GlaxoSmithKline, New Frontiers Science Park, Third Avenue, Harlow, Essex CM19 5AW, UK

Systems Research, GlaxoSmithKline, Gunnels Wood Road, Stevenage, Herts SG1 2NY, UK

1 Introduction

Radioligands are essential for many types of pharmacological assay. A vast array of radiolabelled agents is now commercially available, including agonists and antagonists for a huge number of receptors, markers of G protein activation, signal transduction molecules, enzyme substrates and ions. Suppliers of these agents include Amersham Pharmacia Biotech, American Radiolabeled Chemicals, Cambridge Research Biochemicals, ICN Pharmaceuticals Ltd, NEN Life Science Products and Sigma–Aldrich Co. Ltd (for addresses see Appendix). In this chapter, we provide protocols for receptor binding and functional assays for enzymes, ion channels and G-protein-coupled receptors.

2 Radioligand binding assays

2.1 Introduction

Radioligand binding assays are a valuable means of studying many aspects of receptor behaviour, such as receptor distribution, characterization, classification and mechanisms of ligand–receptor interaction. In addition, radioligand binding assays are widely used to identify novel compounds which bind to specific receptors. Using recently available robotics, it is now possible to screen many thousands of compounds in such assays (1).

In brief, a radioligand binding assay involves:

(1) incubating a tissue or cell preparation containing the receptor of interest with a radiolabelled ligand;

(2) separating the bound from the free ligand using an appropriate separation technique (or using new homogeneous techniques to measure only bound radioligand); and

(3) measuring the concentration of bound radioligand.

An unlabelled compound that is able to bind to the receptor of interest will compete with the radiolabelled ligand. Thus, competition curves using increasing concentrations of the unlabelled ligand can be constructed and the affinity of the unlabelled ligand for the receptor determined. A comprehensive account of radioligand binding is given in a companion book in this series, *Receptor–Ligand Interactions: A Practical Approach* (2).

2.2 Choice of reagents

2.2.1 Receptor preparation

A number of different receptor preparations can be employed, such as whole animals, slices of tissue, whole cells, membrane fractions, cytosolic extracts, solubilized preparations and purified proteins. The majority of receptor binding assays for membrane-bound receptors employ membrane preparations. These provide controlled conditions in which drug–receptor interactions can be studied in the absence of intracellular substances such as guanine nucleotides. Protocols for the preparation of crude membrane fractions are given below. Further details about alternative preparations are given in refs 3 and 4.

Protocol 1 describes the preparation of crude membranes from brain tissue. A similar method can be employed to prepare membranes from other tissues. Tougher tissues such as heart, salivary glands and smooth muscle should be filtered before use by passing through a nylon filter (300 μm) after homogenization. *Protocol 2* describes how to prepare crude membranes from cell lines. Such crude preparations are sufficient for most radioligand binding assays. If a purer preparation is required, a protocol for the preparation of synaptosomes is given in ref. 3.

Protocol 1

Preparation of crude brain membranes

Equipment and reagents

- 50 mM Hepes (Sigma), adjusted to pH 7.4 using potassium hydroxide
- Ultra-Turrax T25 homogenizer (Jencons)

Method

1 Carefully remove the brain, dissect if required and place in ice-cold Hepes buffer.

2 Weigh the tissue and homogenize in 15 vols buffer (w/v) using an Ultra-Turrax homogenizer (three bursts of 5 sec at 24 000 r.p.m.), e.g. homogenize 1 g tissue in 15 ml buffer.

3 Centrifuge the homogenate at 48 000g for 20 min at 4 °C. Discard the supernatant and resuspend the pellet in 15 vols buffer (w/v, as in step **2**).

4 Homogenize the pellet as above, centrifuge again at 48 000g for 20 min at 4 °C and resuspend the resultant pellet in assay buffer (e.g. the type used in *Protocol 3*). Resuspend at a suitable concentration to be used directly in the binding assay or, if the receptor is stable,[a] to be stored at −80 °C until use.

[a] Most membrane preparations can be stored for ≥6 months. Protease inhibitors can be included to improve receptor stability. Tablets ('Complete') containing a cocktail of protease inhibitors can be obtained from Boehringer Mannheim. However, note that protease inhibitors can inhibit some receptor binding.

Protocol 2

Preparation of membranes from Chinese hamster ovary (CHO) cells

Equipment and reagents

- Access to cell culture laboratory
- CHO cells stably expressing the receptor of interest
- Phosphate-buffered saline (PBS)
- 50 mM Hepes, adjusted to pH 7.4 using potassium hydroxide
- PBS containing 5 mM EDTA
- Waring blender (Christison Scientific Equipment)
- Ultra-Turrex homogenizer (Jencons)
- 35 and 50 ml plastic centrifuge tubes

Method

1 Grow CHO cells in 175 cm² flasks in a suitable growth medium.

2 Once cells are confluent, remove medium from the flasks and wash with 5 ml PBS per flask.

3 Remove PBS and add 5 ml PBS containing 5 mM EDTA per flask. Leave the cells to detach for ~10 min at 37 °C. Tap the flasks sharply two or three times to aid cell detachment.

4 Transfer the cells to 50 ml plastic centrifuge tubes and centrifuge at 500g for 5 min at room temperature.

5 Remove the supernatant and resuspend the pellet in 10–20 vols (v/v) ice-cold Hepes (i.e. ~20 ml for 0.5 ml cell pellet) and leave on ice for 20 min to allow cells to swell.

6 Homogenize the pellet using a Waring blender (three 15 sec bursts, placing tube on ice for 2 min between bursts) and re-centrifuge at 500g for 20 min at 4 °C.

7 Transfer the supernatant[a] to 35 ml plastic centrifuge tubes and re-centrifuge at 48 000 g for 30 min at 4 °C.

8 Resuspend the pellet in Hepes (1 ml/flask) and homogenize using an Ultra-Turrax homogenizer (10 sec at 13 500 r.p.m.).

9 Aliquot (1 ml) into microcentrifuge tubes and freeze (–80 °C).[b]

[a] It is essential to discard the pellet in order to eliminate nuclei and unbroken cells.

[b] Most membrane preparations can be stored for ≥6 months. Protease inhibitors can be included to improve receptor stability. Tablets ('Complete') containing a cocktail of protease inhibitors can be obtained from Boehringer Mannheim. However, note that protease inhibitors can inhibit some receptor binding.

2.2.2 Choice of radioligand

Tritium (^{3}H) and iodine-125 (^{125}I) are the most frequently used isotopes for radioligand binding assays. Factors to consider in the choice of radioligand are the specific activity of the radioligand (unit radioactivity per mole), affinity and specificity of the radioligand for the receptor of interest and the level of receptor expression. Tritium is the isotope of choice as its half-life (12.43 years) is much longer than that of iodine-125 (60 days). Furthermore, tritium-labelled molecules are usually chemically identical to their unlabelled counterpart and are safer to handle. Drawbacks are reduced sensitivity compared with iodine-125 and the increased possibility of self-radiolysis. Sensitivity is less important if the tissue preparation contains a high level of receptors; this is frequently the case for recombinant cell lines. Sensitivity is maximized by using a ligand with high specific activity. Self-radiolysis during storage is minimized by dissolving the ligand in a suitable solvent (e.g. ethanol) that is unable to react with secondary species (e.g. free radicals), freezing rapidly and storing diluted aliquots at low temperature. Iodine-125 is the ligand of choice where high sensitivity is needed, for example when receptor expression is low. However, introducing iodine into a molecule may alter the binding properties of the ligand. The smaller the molecular size of the ligand, the more likely it is that the incorporation of iodine-125 will disturb the interaction of the ligand with the receptor. Thus, whilst large proteins and peptides are frequently labelled with iodine-125, low molecular weight ligands are better labelled with tritium. When a ^{125}I-labelled ligand is used, experiments should be performed to show that the incorporation of the iodine group has not altered the nature of the ligand–receptor interaction. In particular, the affinity (K_D; equilibrium dissociation constant; determined by saturation analysis; see Section 2.5.2) of the radiolabelled ligand should be similar to the affinity (K_i; determined by competition analysis; see Section 2.5.1) of the unlabelled molecule.

For some G-protein-coupled receptors, both agonist and antagonist radiolabelled ligands are available. The choice of ligand is dependent on the purpose of the studies. For receptor distribution studies, a radiolabelled version of the endogenous ligand is preferable. For screening to identify novel ligands, a radioligand which is highly selective for the receptor of interest is required. However, selectivity is less important when using recombinant cell lines that do not express other receptors which may bind the same ligand. G-protein-coupled receptors exist in two affinity states. Antagonist ligands bind to both states of the receptor with equal affinities, whilst agonists preferentially bind one state, which has high affinity for agonists. Therefore, if the radioligand binding assay is to be used to screen for agonist molecules, an agonist radioligand is preferred.

2.3 Separation techniques

2.3.1 Introduction

Radioligand binding assays usually necessitate separating radioligand which is bound to the receptor from free radioligand. Gel filtration, precipitation or charcoal adsorption assays can be used for solubilized preparations. In addition, charcoal adsorption assays are frequently used for naturally soluble cytosolic receptors, such as steroid hormone receptors. Details of these techniques can be obtained from ref. 2. For particulate preparations such as membranes, separation is usually achieved by centrifugation or filtration (see Section 2.3.2). Centrifugation assays involve sedimenting the receptor-bound radioligand, removing the unbound ligand in the supernatant and counting the radioactivity in the pellet. Centrifugation assays are useful for low affinity ligands (which may dissociate too rapidly from the receptor for a filtration assay to be used) and for ligands where high levels of radioligand bind non-specifically to filters. The disadvantages are high non-specific binding, low throughput and little opportunity for automation. Further details of centrifugation assays are given in ref. 5. For high affinity ligands ($K_D < 20$ nM), filtration is the method of choice (see Section 2.3.2).

Homogeneous radioligand binding assays do not require separation of bound and free ligand. Such techniques are amenable to high-throughput screening (see Section 1) and are discussed in more detail in Section 2.3.3.

2.3.2 Filtration assay

Rapid filtration is the principal method used to separate bound from free radioligand. A filter of suitable pore size to trap the receptor-bound ligand is required. The filters are washed to reduce the free ligand trapped by the membrane particles (i.e. the non-specific binding). Filtration is rapid, easy, efficient and reproducible. Drawbacks are that it is not suitable for low affinity ligands ($K_D > 20$ nM), and some radioligands bind to the filters. There are a number of ways in which filters can be treated to reduce filter binding (see Section 2.4). If filter binding cannot be reduced to a workable level, centrifugation should be used instead (see Section 2.3.1). A comprehensive account of the filtration technique is given in ref. 6. *Protocol 3* describes a filtration assay to measure the binding of ^{3}H-labelled substance P to rat cerebral cortex.

Protocol 3

Measurement of the binding of ^{3}H-labelled substance P in membranes prepared from rat cerebral cortex

Equipment and reagents

- 96-well plates
- Assay buffer: 50 mM Tris pH 7.4,[a] containing 3 mM MnCl$_2$, 40 μg/ml bacitracin (Sigma), 4 μg/ml leupeptin (Sigma), 2 μg/ml chymostatin (Sigma) and 0.02% (w/v) bovine serum albumin (BSA; Sigma)[b]
- Wash buffer: 50 mM Tris pH 7.4, containing 3 mM MnCl$_2$[a]
- 1.5 nM ^{3}H-labelled substance P (Cambridge Research Biochemicals; specific activity 5.11 TBq/mmol) in assay buffer (2.5× final concentration)
- Test compounds (serial dilutions prepared at 5× final concentrations in assay buffer)

- 5 μM substance P (can be obtained from several suppliers, e.g. Neosystem Laboratoire) in assay buffer (5× final concentration; to determine non-specific binding (NSB))[c]
- 80 mg/ml freshly prepared rat cortex membranes in assay buffer (8 mg/well), prepared as described in *Protocol 1*
- Cell harvester (Brandel from Semat) and vacuum pump
- GF/B grade glass-fibre filters (Semat) soaked in 0.3% (w/v) polyethyleneimine and 0.5% (v/v) Triton X-100 for ≥1 h
- Scintillation vials, scintillation fluid and liquid scintillation counter

Protocol 3 continued

Method

1 Prepare serial dilutions of test compounds in assay buffer at 5× final concentrations.

2 Add test compounds (50 μl) and ^{3}H-labelled substance P (100 μl) to 96-well plates. Total and non-specific binding are determined by replacing test compound with buffer (50 μl) or Substance P (50 μl; final concentration 1 μM) respectively.

3 Initiate the reaction by addition of membranes (100 μl) and incubate for 40 min at room temperature.

4 Separate bound radioligand by harvesting through GF/B grade glass-fibre filters using a Brandel cell harvester (or similar).

5 Wash filters with 3 × 1 ml wash buffer.

6 Place filters in scintillation vials, add 4 ml scintillation fluid, leave for ≥4 h and count in a liquid scintillation counter.[d]

[a] Adjust pH to 7.4 before adding $MnCl_2$.

[b] Bacitracin, leupeptin and chymostatin are peptidase inhibitors, to prevent degradation of the radioligand. Aliquots of these can be prepared in advance and stored at −20 °C. Chymostatin should be dissolved in dimethylsulfoxide. Tablets ('Complete') containing a cocktail of protease inhibitors can be obtained from Boehringer Mannheim.

[c] It is preferable to use a structurally unrelated ligand to define non-specific binding, such as a non-peptide antagonist, e.g. Tocris catalogue no. L-732138.

[d] The method for using a Brandel cell harvester is given. Other filtration and radioligand counting methods are also available (see Table 1).

Protocol 3 describes trapping bound radioactivity using a Brandel cell harvester and counting filters by liquid scintillation spectrometry. While such harvesters can harvest between 24 and 96 samples simultaneously, filter mats must then be punched and individual filters transferred to scintillation vials to which scintillation fluid is added. While equipment to punch filters into scintillation vials and add scintillation fluid is available, the procedure is still time-consuming and generates substantial amounts of radioactive, organic waste. Developments in scintillation technology have led to techniques in which the filter is sealed into a bag with liquid or solid scintillant. The filter is then counted in a β-plate counter, which usually counts simultaneously from a number of filter positions. This saves considerable time but suffers from lower efficiency of detection. In addition, harvesters that are compatible with this technology typically use smaller filters that have lower protein loading capabilities than the Brandel harvester (see below). For these reasons, the technique is better suited to membrane binding assays that employ high-expressing recombinant cell lines.

While the amount of protein which can be successfully harvested depends on the type of membrane preparation, filter type, wash protocol and vacuum pressure, Brandel harvesters should successfully harvest ≥2 mg protein/well. Ninety-six-well microplate harvesters are available from Tomtec and Packard, which are suitable for harvesting ≥100 μg protein/well for GF/B grade filters and ≥500 μg

Table 1 Equipment to harvest and count filters

Harvester	Filters	Scintillant	Scintillation counter
Brandel (24, 48 or 96 samples)	Whatman filters	liquid scintillation fluid	liquid scintillation counter
Tomtek (48 or 96 samples; different filter sizes for different protein requirements)	Wallac Filtermats[a]	Wallac Meltilex solid scintillant or Betaplate Scintillation fluid	Wallac 1204 BigSpot Betaplate (for larger filters); Wallac 1450 Microbeta TriLux
Packard FilterMate (24 or 96 samples)	Packard UniFilter or FlexiFilter plates[a]	Packard solid FlexiScint or MicroScint cocktail	Packard TopCount

[a] Wallac and Packard filters can be counted in either Wallac or Packard betaplate counters.

protein/well for GF/C grade filters. If larger amounts of protein are being used, the Tomtec Mac IV harvester (which harvests 12×4 samples) in conjunction with the Wallac 1204 BigSpot Betaplate counter combines the ability to harvest similar quantities of protein to the Brandel cell harvester with the advantages of plate counting technology. *Table 1* details some of the harvesting and counting equipment that is currently available.

2.3.3 Homogeneous radioligand binding assays

The techniques described above all require physical separation of bound from free radioligand. In order to increase ease of automation and, therefore, to increase throughput of binding assays, there is a need to remove this separation step. Two proprietary technologies, Scintillation Proximity Assay (SPA) and Flash-plate, allow the quantity of bound ligand to be measured without the need for physical separation. To date, there are few publications that detail their uses for radioligand binding assays. Examples of their use in assays for G-protein-coupled second messengers and ion channel function are given in *Protocols 6 and 10*.

The SPA (Nycomed Amersham) relies on the interaction between a radiolabel that emits weak beta particles (e.g. ^{3}H or ^{125}I) and receptor (in membrane fragment) attached to a scintillant (SPA) bead (by wheatgerm agglutinin) (*Figure 1*). When the radiolabelled ligand binds to the receptor, the radioactive energy it emits is absorbed by the SPA bead and light is generated. In contrast, radiolabel that is not bound to the receptor does not generate a signal as it is not close enough to the bead for the energy to be absorbed. Hence, a specific signal resulting from the interaction between the membrane and the ligand is generated (7).

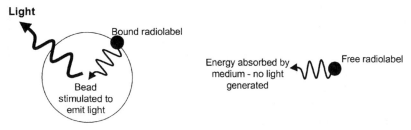

Figure 1 Principle of Scintillation Proximity Assay.

Flashplate (NEN) uses the same principle as SPA, except that instead of a scintillant bead, scintillant is incorporated into the bottom of a 96- or 384-well plate. Membrane is attached to the bottom of the well by wheatgerm agglutinin and, again, only radioactivity bound to the membrane (or receptors in the membrane) is close enough to the scintillant to stimulate light generation.

2.4 Validation and optimization of assay

Whilst radioligand binding assays are usually quick and simple to perform, their optimization, characterization and validation can be very complex. Key parameters to be optimized are described below. Further information can be obtained from ref. 8.

For a valid binding assay, the following criteria must be met:

- specific binding must be saturable and reversible;
- the assay must be performed at equilibrium;
- agonist and antagonist ligands should compete with the radioligand with the expected pharmacology; and
- free radioligand concentration should not be signficantly depleted by binding to cells or membranes.

In order to meet these criteria, a number of assay parameters must initially be optimized:

(1) *Choice of ligand to define non-specific binding.* Non-specific binding of the radioligand (for example to the receptor preparation and filters) is determined by measurement of radioligand binding in the presence of an excess of an unlabelled ligand which binds to the receptor of interest. Typically, a concentration of the unlabelled ligand that is ~100 times higher than its IC_{50} (concentration of the drug which inhibits specific binding by 50%) for the receptor is used. Ideally, the unlabelled ligand should be structurally dissimilar from the radioligand to avoid displacement from non-specific sites.

(2) *Incubation time.* The time taken for total binding to reach equilibrium should be determined. Note that this is dependent on the concentration of radioligand that is used. Since some competing agents may have slower kinetics, a routine incubation time should be chosen which is greater than the time taken for the radiolabelled ligand to reach equilibrium.

(3) *The concentration of membrane protein.* Total and non-specific binding should be determined over a range of membrane protein concentrations. The binding should increase in proportion to the increase in receptor (protein) concentration. If binding decreases with increasing protein concentration, this may suggest that the ligand is being degraded during the incubation or that there is an endogenous ligand present which interferes with the radioligand binding. A protein concentration should be chosen which gives a good signal:noise ratio, is on the linear part of the protein–response curve and is such that <10% of the total added radioligand is bound.

(4) *Ligand concentration (saturation analysis)*. Total and non-specific binding should be determined over a range of ligand concentrations. Specific radioligand binding should be saturable, whilst non-specific binding should increase in a linear manner. From these data, K_D (equilibrium dissociation constant) and B_{max} (number of receptors per unit protein) can be determined (see Section 2.5.2). For competition curves (incubation of a single concentration of radioligand with a range of concentrations of test compound), a radioligand concentration should be chosen which does not saturate receptors, gives a good signal:noise ratio and is such that <10% radioligand is bound (binding of >10% significantly reduces free radioligand concentration, an affect known as ligand depletion). Ideally, the ligand concentration should be as low as possible, both to minimize radioligand usage and so that the IC_{50} of the competing ligand is a reasonable approximation of its affinity (K_i) (see Section 2.5.1).

(5) *Temperature*. For practical reasons, it is preferable to perform assays at room temperature. However, if one is not working in an air-conditioned laboratory, it is advisable to control the temperature. If the radioligand or receptor is particularly unstable, it may be necessary to work at 4°C. Alternatively, if the radioligand is slow to reach equilibrium, a temperature of 37°C may be necessary.

(6) *Filtration parameters (if appropriate)*. Dissociation of the bound ligand from the receptor must be minimized by using ice-cold buffer and a high vacuum pressure, to achieve rapid separation. Different wash protocols should be assessed to maximize the window between total and non-specific binding. Typically, three or four washes of 1 ml buffer are used with a Brandel cell harvester. Different filters and filter treatments should also be examined to minimize non-specific binding. Commonly used filter treatments are polyethyleneimine (~0.1% (w/v)) or BSA (~0.5% (w/v)).

(7) *Buffer composition*. Tris- or Hepes-based buffers at pH 7.4 are usually appropriate. Hepes is preferable, as its pH is independent of temperature. Binding to G-protein-linked receptors is modified by the presence or absence of ions within the buffer. For example, divalent cations such as magnesium (typically 10 mM) are often included in buffers for agonist radioligands as they stabilize the receptor–G-protein complex. This increases the proportion of receptors in the high affinity state and hence the amount of agonist binding. Under certain circumstances, guanine nucleotides such as guanosine 5′-triphosphate (GTP) or its non-hydrolysable analogues guanosine-5′-O-(3-thiotriphosphate (GTPγS) or 5′-guanylimidodiphosphate (GppNHp) may be included (typically at 0.1–100 μM). These act to dissociate the G protein from the receptor and therefore convert receptors to the low affinity state for agonists.

2.5 Data analysis

2.5.1 Competition analysis

Competition (also known as displacement or inhibition) studies enable the determination of the affinity (K_i; units moles/litre) of an unlabelled ligand for the receptor. Such studies are carried out by incubating a range of concentrations of the competitor ligand (typically eight to 12 concentrations using three- to four-fold serial dilutions) with a fixed concentration of the radioligand, ideally in duplicate. Competition curves are plotted as bound ligand (y-axis) versus log drug concentration (x-axis). They are frequently analysed by fitting a four-parameter logistic equation (*Equation 1*) to the data:

$$B = ((max(1 - [x^n]))/(IC_{50n} + [x^n])) + NSB \qquad (1)$$

where B is the level of specific binding, max is the maximum specific binding (total − non-specific binding), $[x]$ is the ligand concentration, IC_{50} is the concentration of drug which inhibits specific binding by 50%, n is the Hill slope coefficient and NSB is the minimum, i.e. non-specific binding. This analysis can be performed by the curve-fitting packages listed in Section 6.

The affinity of a drug, K_i, is calculated from the IC_{50} using the Cheng–Prusoff equation (see *Equation 2*):

$$K_i = IC_{50}/(1 + ([L]/K_D)) \qquad (2)$$

where $[L]$ is the concentration of radioligand used in the assay and K_D is the equilibrium dissociation constant of the radioligand.

2.5.2 Saturation analysis

Saturation analysis is an essential part of the characterization and validation of assays. It involves incubating a fixed amount of receptor protein with increasing concentrations of radioligand in the absence (total binding) and presence (non-specific binding) of the unlabelled ligand that is being used to define non-specific binding. Saturation analysis enables measurement of the K_D (the concentration of radioligand which occupies 50% of the receptors; units moles/litre) and B_{max} (maximum amount of radioligand bound by the receptors; units moles/mg protein). If one molecule of radiolabelled ligand binds to one receptor, B_{max} will also represent the amount of receptor expressed by the protein. Saturation analysis is also used to determine the optimum concentration of the radioligand to use in competition analysis, to demonstrate that radioligand binding is saturable and that non-specific binding increases in a linear manner with radioligand concentration.

Saturation analysis involves measurement of total and non-specific binding over a wide range of ligand concentrations (at least from one tenth to ten times the estimated K_D), usually in duplicate or triplicate. Specific binding is calculated by subtracting non-specific binding from total binding. In addition, each ligand concentration must be accurately determined by scintillation counting and a sample of the receptor preparation retained for protein determination. The data are graphically displayed by plotting data as bound ligand (y-axis) versus free ligand concentration (x-axis). Historically, data were analysed using Scatchard analysis, which involves plotting the ratio of specifically bound/free radioligand) (y-axis) versus specifically bound radioligand (x-axis). A straight line indicates that the radioligand is binding to a single receptor population. If so, the intercept on the x-axis gives the B_{max} and the slope of the line is equal to $(-1/K_D)$. A curvilinear plot indicates that the radioligand is binding to more than one receptor. Whilst Scatchard analysis remains a popular way to assess graphically whether the radioligand is binding to single or multiple sites, this technique has now been superseded by non-linear regression analysis (*Figure 2*), which can be performed using the curve-fitting packages listed in Section 6. Raw total binding data are fitted to *Equation 3*:

$$Bound = B_{max}[L]^n/(K_D{}^n + [L]^n) + NSB[L] \qquad (3)$$

where *Bound* is the total binding, B_{max} is the maximum amount of radioligand bound by the receptors, $[L]$ is the free radioligand concentration, n is the Hill slope coefficient, K_D is the equilibrium dissociation constant and *NSB* is the non-specific binding.

The above equation simultaneously fits specific binding as a hyperbolic function $(B_{max}[L]^n/(K_D{}^n + [L]^n))$ and non-specific binding as a linear function $(NSB[L])$ to the raw total counts. B_{max}, K_D and *NSB* are estimated by non-linear regression.

An alternative method is to subtract measured non-specific binding (counts in the presence of a saturating concentration of unlabelled ligand) from total counts and simply fit the resultant specific binding only to the hyperbolic function. If n

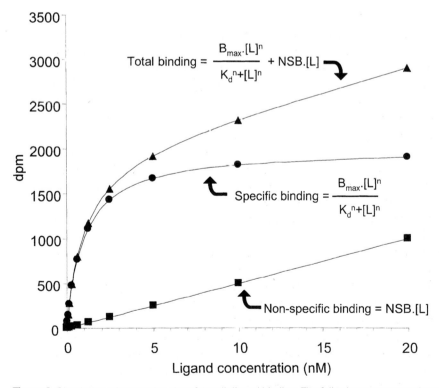

Figure 2 Simulation of saturation data for radioligand binding. The following parameters were fixed: ligand $K_d = 1$ nM; Hill slope (n) = 1.0; $B_{max} = 2000$; *NSB* = 50. Graph shows total binding (triangles), specific binding (circles) and non-specific binding (squares).

is not significantly different from 1, the mid-point of the hyperbola represents the K_D of the radioligand. *NSB* is the coefficient for linear, non-specific binding. The validity of using a non-specific binding parameter should be verified by checking that the non-specific binding data fit to a straight line (i.e. by fitting $y =$ *NSB*[L]). If specific binding data are used, *NSB* is equal to zero.

There are a number of potential explanations for fits that result in the Hill slope (*n*) being significantly different from unity. A *steep* curve ($n > 1$) may be explained by positive co-operativity (binding of one molecule facilitating the binding of further molecules), ligand depletion or non-equilibrium of the system. A *shallow* curve ($n < 1$) may be explained by negative co-operativity (binding of one molecule hindering the binding of further molecules) or binding to more than one receptor type.

2.5.3 Kinetic analysis

Association experiments are essential to establish that equilibrium has been reached. Total and non-specific binding are determined at different times following the addition of protein. In practice, this is most easily accomplished by adding protein to assay wells at different time points and then simultaneously separating bound from free. The observed association rate constant (k_{obs}) is derived from fitting an exponential curve to a plot of bound counts versus time, or can be given from the slope of the line of $\ln(B_o/(B_o - B_t))$ versus time, where B_o is counts at equilibrium and B_t is counts at time t. The association rate is dependent on ligand concentration. Thus, to determine the time to reach equilibrium, experiments should be performed at the ligand concentration that will be used in future experiments. The association rate constant (k_{+1}) can be calculated from k_{obs} using *Equation 4,* where k_{-1} is the dissociation rate constant (see below).

$$k_{+1} = (k_{obs} - k_{-1})/[L] \qquad (4)$$

The association rate constant is usually expressed as $M^{-1}\ min^{-1}$. Ideally the observed rate constant should be determined at a range of ligand concentrations. A plot of observed rate constant versus ligand concentration has a slope of k_{+1} and an intercept of k_{-1}.

Dissociation experiments are used to assess whether radioligand binding is reversible. In addition, they can be used to investigate the nature of interaction between the radioligand and an unlabelled ligand, since ligands which bind competitively will not affect the dissociation rate of the radioligand, whilst an allosteric modulator will alter the dissociation rate. To perform the assay, binding is allowed to reach equilibrium and then dissociation is initiated by the addition of an excess of the unlabelled ligand (>100 times IC_{50}) or by a ≥ 50-fold dilution of the reaction mixture. In practice, the former is easier. As for association experiments, the assay should be designed such that total and non-specific binding tubes at all time-points can be harvested simultaneously. In addition, some tubes should be included which do not receive non-radioactive displacer to demonstrate that steady-state binding is maintained throughout the dissociation timecourse. The dissociation rate constant (k_{-1}) is determined by fitting an exponential decay curve to the plot of bound counts versus time, or can be given from the slope of the line of $\ln(B_o/B_t)$ versus time. Convenient units of k_{-1} are min^{-1}.

K_D can be determined from both saturation analysis (see Section 2.5.2) and kinetic analysis, using *Equation 5*.

$$K_D = k_{+1}/k_{-1} \qquad (5)$$

Half-lives for association and dissociation are equal to $(\ln 2)/k$. Incubation times of five and seven times the association half-time will achieve 97% and 99% receptor occupancy, respectively.

2.6 Non-radioactive assays

There are currently few methods for examining ligand–receptor interactions in non-radioactive formats. Below are brief descriptions of currently emerging technologies:

2.6.1 Homogeneous time-resolved fluorescence (HTRF; Packard)

Ligand and receptor are both labelled with fluorescent molecules. The absorption and emission spectra are chosen such that if the two labels are in close proximity, the emitted light from one labelled molecule (europium cryptate) is able to excite the second labelled molecule (XL665) (*Figure 3*). This principle is generally known as fluorescence resonance energy transfer (FRET). Since the XL665 labelled molecule is not excited by the wavelength used to excite the europium cryptate, it will only emit light when bound close to the europium label. Additionally, europium cryptate fluoresces over a much longer timescale (microseconds) than background fluorescence and so the FRET-based fluorescence can be distinguished from background by reading at a time when the background fluorescence has decayed. In this way a non-radioactive measurement of binding may be obtained with no need for a separation step.

2.6.2 Fluorescence polarization

A ligand is labelled with a fluorophore (e.g. fluorescein) and then excited with polarized light. When the ligand is bound to receptors, the rotational velocity of

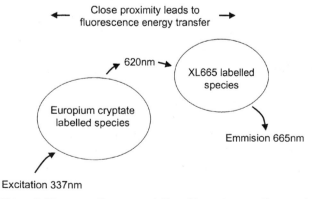

Figure 3 Diagrammatic representation of homogeneous time resolved fluorescence

the fluorophore is reduced and the emitted light is maintained in the same plane as the exciting light. If the fluorophore is unbound then it has a high rotational velocity and the emitted light will vary in its plane of polarization. Hence, by measuring the amount of light emitted in the two planes, the quantity of ligand bound to its receptor may be determined with no need for a separation step. This method has been used to examine the binding of substance P to its receptor (9). A plate reader which will count both 96- and 384-well plates is now available from LJL Biosystems.

2.6.3 Fluorometric Microvolume Assay Technology (Biometric Imaging/Perkin Elmer)

This technology again utilizes a fluorescently labelled ligand. Labelled ligand binds to cells or membrane-coated beads (10). The beads or cells are allowed to settle in each well of a 96- or 384-well plate. Confocal microscopy is used to image the bottom 50 μm of the well (hence excluding background fluorescence in the solution of the well). Computer software analyses the image of the well and is able to quantify the fluorescence associated with each bead or cell. The amount of ligand bound may be quantified with no need for a separation step.

3 G-protein-coupled receptor second messenger assays

3.1 Introduction

The availability of cultured cells and recombinant receptors expressed in immortalized cell lines has challenged the pharmacologist to develop robust and rapid assays to measure receptor activation. In the absence of a functional response (e.g. contraction or secretion), a common approach is to measure second messengers within the cells. This section describes methods commonly used to measure second messengers associated with activation of G-protein-coupled receptors. These receptors commonly couple to one of three major classes of G protein. The second messengers stimulated by each class and some of the radioactive assays associated with these are summarized in *Table 2*.

Table 2 Radioactive assays suitable for different classes of G-protein-coupled receptor

G protein family	Second messengers	Suitable radioactive assay
G_s	increased cAMP	cAMP radioimmunoassay
G_i	reduced cAMP	cAMP radioimmunoassay
G_q	increased inositol phosphate production; increased intracellular calcium	[^{3}H]inositol phosphate turnover; inositol phosphate radio-receptor assay; intracellular [^{45}Ca^{2+}] release
All	guanine nucleotide exchange	[^{35}S]GTPγS binding (best results obtained with G_i-coupled receptors)

3.2 Measurement of [^{35}S]GTPγS binding

3.2.1 Introduction

While most assays to measure G protein activation measure the downstream production of second messenger molecules (see Sections 3.3 and 3.4), it is also possible to measure this event directly by determination of agonist-induced guanine nucleotide exchange using [^{35}S]GTPγS. Agonist stimulation of a G-protein-coupled receptor causes guanosine 5'-diphosphate (GDP), bound to the α-subunit of the G protein, to be exchanged for GTP, allowing α-GTP to dissociate from the $\beta\gamma$ subunit. Subsequent downstream events are then stimulated by both α-GTP and $\beta\gamma$. Incubation of membrane preparations containing the receptor of interest with the non-hydrolysable GTP analogue, [^{35}S]GTPγS, effectively causes irreversible activation of the G protein since GTPγS is not readily susceptible to the intrinsic GTPase activity of G_α. This activation is observed as an increase in binding of [^{35}S]GTPγS to the membrane preparation. This quantitative technique measures the primary response in the signalling pathway following receptor activation, which is the only step in the pathway that is directly regulated by ligands. Stimulation of [^{35}S]GTPγS binding has been demonstrated following activation of a number of G-protein-coupled receptors, including adenosine A_1, muscarinic, dopamine D_2, somatostatin sst_5, metabotropic glutamate, melatonin mt_1, mu (μ) opioid and cannabinoid receptors (e.g. 11). Best results are generally obtained for G_i-coupled receptors. Assays can use membranes prepared from recombinant cell lines, as described below, or native tissues, or can employ autoradiographical techniques in tissue slices.

Protocol 4

Measurement of [^{35}S]GTPγS binding in membranes prepared from Chinese hamster ovary (CHO) cells stably expressing human melatonin mt_1 receptors[a]

Equipment and reagents

- Assay buffer: 20 mM Hepes pH 7.4, containing 10 mM $MgCl_2$, 100 mM NaCl[b]
- 1 nM [^{35}S]GTPγS (Nycomed Amersham; specific activity >37 TBq/mmol) in assay buffer (10 $\times$ final concentration)
- 50 μM GDP (sodium salt, Sigma, catalogue no. G-7127) in assay buffer[c] (5 $\times$ final concentration)
- 500 μM GTP (sodium salt, Sigma, catalogue no. G-8752) in assay buffer[c] (5 $\times$ final concentration)
- Cell harvester (Brandel or similar)

- 100 μg/ml CHO-mt_1 membranes in assay buffer (2 $\times$ final concentration), prepared as described in *Protocol 2*
- Test compounds (serial dilutions prepared at 5 $\times$ final concentrations in assay buffer, or 10 $\times$ final concentrations if testing an antagonist in the presence of an agonist concentration–response curve)
- GF/B grade glass-fibre filters (Semat)
- Scintillation vials (4 ml)
- Scintillation fluid
- Liquid scintillation counter

Protocol 4 continued

Method

1   Incubate test compounds (50 μl)[d] with GDP[c] (50 μl; final concentration 10 μm) cell membranes (125 μl) for 120 min at 30°C in a shaking water-bath. Non specific binding is defined by placing GDP with GTP[c] (50 μl; final concentration 100 μm).

2   Add [^{35}S]GTPγS (25 μl; final concentration 100 μm) to the wells and incubate in the shaking water-bath for a further 30 min at 30°C.

3   Separate bound radioligand by harvesting through GF/B grade glass-fibre filters using a Brandel cell harvester (or similar).

4   Wash filters with 4 × 1 ml distilled water.

5   Place filters in 4 ml scintillation vials, add 4 ml scintillation fluid and count the following day in a liquid scintillation counter.[e]

[a] The assay is performed in duplicate in 96-well plates in a final assay volume of 250 μl.

[b] Can be stored at 4°C for ≤1 month. Saponin (≤1 mg/ml) can be included. This permeabilizes vesicles in the membrane preparation and may improve access of [^{35}S]GTPγS.

[c] Can be stored as frozen aliquots.

[d] Or 25 μl agonist (10 × final concentration) and 25 μl antagonist (10 × final concentration).

[e] Method for using a Brandel cell harvester are given. Other filtration and radioligand counting methods are also available (see *Table 1*).

Protocol 4 describes the measurement of [^{35}S]GTPγS binding to membranes containing human melatonin mt$_1$ receptors. This protocol involves separating bound and free [^{35}S]GTPγS by filtration. Recently, a homogeneous method using Flashplate technology has been described (11), which is amenable to automation and hence increased throughput. Flashplates (NEN) are 96-well polystyrene plates which are coated with scintillant (see Section 2.3.3). The Flashplate is coated with membranes containing the receptor of interest and incubated with [^{35}S]GTPγS, GDP and test compound as described in *Protocol 4*. In the method described (11), the reaction is stopped by aspirating the contents from each well. Alternatively, the Flashplate could be centrifuged. The plate is counted in a β-plate scintillation counter (e.g. Wallac Microbeta or Packard Topcount counter). A kit for measuring [^{35}S]GTPγS binding using SPA technology (see Section 2.3.3) is also available from Nycomed Amersham (catalogue no. RPNQ0210).

3.2.2 Data analysis

Agonist-stimulated [^{35}S]GTPγS binding can be expressed in three ways:

- as pmole/mg protein;
- as a percentage of basal binding;

- basal binding can be subtracted from all data points and data expressed as a percentage of the response generated by a fixed or a maximum concentration of a standard agonist.

The potency of an agonist to evoke [^{35}S]GTPγS binding is measured by determining the concentration of the agonist which evokes half-maximal increase in [^{35}S]GTPγS binding (EC_{50}). EC_{50} values are best determined by constructing concentration–response curves and fitting a four-parameter logistic equation (see *Equation 6*) to the data:

$$y = ((max[x^n])/(EC_{50}{}^n + [x^n])) + min \qquad (6)$$

where y is the response, *max* is the maximum response, $[x]$ is the drug concentration, n is the Hill slope coefficient and *min* is the minimum (basal) response. This analysis can be performed by the curve-fitting packages listed in Section 6.

The ability of an antagonist to inhibit agonist-stimulated [^{35}S]GTPγS binding can be determined in a number of ways. One of the best methods is to calculate agonist EC_{50} (concentration of drug which evoked 50% of the maximum response) values in the presence of increasing concentrations of the putative antagonist and to fit data to *Equation 7* (12), which is a modified form of the Schild equation, to determine the K_B (dissociation equilibrium constant) of the antagonist.

$$-pEC_{50} = \log_{10}(EC_{50}{}^c) + \log_{10}(1 + [B]^n/K_B) \qquad (7)$$

where pEC_{50} is logio of the EC_{50} value, $EC_{50}{}^c$ is the control EC_{50} value, $[B]$ is the concentration of antagonist and n is the slope parameter (equivalent to the Schild slope). If n is not significantly different from unity, consistent with simple competition, it can be constrained to unity to obtain an estimate of K_B.

3.2.3 Assay optimization

The following assay parameters should be optimized:

- *The concentration of GDP.* Detection of agonist-stimulated [^{35}S]GTPγS binding requires the presence of GDP, which reduces basal levels of [^{35}S]GTPγS binding to a greater extent than it lowers binding in the presence of agonist, enabling agonist-specific effects to be observed. Therefore, the ability of a maximally effective concentration of a known agonist to stimulate [^{35}S]GTPγS binding in the presence of a range of concentrations of GDP (e.g. 0.1–100 μM) should be examined and the concentration of GDP which results in maximum agonist-evoked increase in binding over basal determined.

- *The concentration of membrane protein.* The ability of a maximally effective concentration of a known agonist to stimulate [^{35}S]GTPγS binding in the presence of a range of membrane protein concentrations (e.g. 1–100 μg/ml) should be determined. A membrane protein concentration should be chosen which evokes good signal:noise ratio, is on the linear part of the membrane protein concentration–response curve and is such that <10% of total added radioligand is bound in the presence of a maximally effective concentration of the agonist (to prevent artefacts caused by ligand depletion).

- *Time to equilibrium.* The length of time to incubate drugs with membranes prior to addition of $[^{35}S]GTP\gamma S$ should be determined in order that equilibration of drugs is achieved (typically 30–120 min).
- *The incubation time with $[^{35}S]GTP\gamma S$.* The signal:noise ratio should be optimized whilst remaining on the linear portion of the stimulated $[^{35}S]GTP\gamma S$ association time-course. Typically an incubation period of 30 min is used.

3.3 Measurement of 3',5'-cyclic adenosine monophosphate (cAMP)

3.3.1 Introduction

For G_i- and G_s-coupled receptors, the measurement of intracellular cAMP provides a convenient measure of receptor activation. G_s-coupled receptors couple to increase the activity of adenylate cyclase which acts to convert 5'-AMP to cAMP. G_i-coupled receptors couple to reduce the activity of adenylate cyclase. As the rate of breakdown of cAMP (by phosphodiesterase) is not acutely regulated by the cell, the levels of cAMP reflect activation of G_s and G_i (cellular phosphodiesterases in the cell are usually inhibited in cAMP assays to increase the sensitivity of the system). *Protocols 5* and *6* describe a method for measuring cAMP in a typical cell line (CHO cells expressing recombinant β_3-adrenoceptors, which couple through G_s). The method described utilizes 96-well plates and measures intracellular cAMP using a commercially available kit. The kit described (SPA; available from Nycomed Amersham) provides a homogenous method for measurement of cAMP (i.e. no separation step is required), and allows for rapid reading in a 96-well scintillation counter (see Section 2.3.3 for further information). An alternative method to the one described is the use of a Flashplate available from NEN (see Section 2.3.3 for further information). Other second messengers (e.g. guanosine 3',5'-cyclic monophosphate (cGMP) or inositol triphosphate) may be measured using similar methods.

Protocol 5

Preparation of CHO cells stably transfected with human β_3-adrenoceptor and extraction of cAMP for measurement of cAMP by ^{125}I Scintillation Proximity Assay

Equipment and reagents

- Access to cell culture laboratory with CO_2 incubator
- CHO cells stably expressing human β_3-adrenoceptor receptors
- DMEM/F12 cell culture medium, containing 10% (v/v) fetal calf serum (FCS) and selection agents (if required)
- DMEM/F12 medium, containing 300 mM isobutylmethylxanthine (IBMX; Sigma)
- Sterile PBS
- Sterile PBS containing 5 mM EDTA
- Test compounds (serial dilutions prepared at $10 \times$ final concentrations in DMEM/F12)
- cAMP SPA assay buffer (Nycomed Amersham; provided in the cAMP SPA kit)
- 50 ml plastic centrifuge tubes
- 100% Ethanol ($4°C$)

Protocol 5 continued

Method

1   Grow cells in DMEM/F12 medium containing 10% FCS in 175 cm^2 flask until approaching confluency.

2   Once cells are confluent, remove media from the flasks and wash with 5 ml PBS per flask.

3   Remove PBS and add 5 ml PBS containing 5 mM EDTA per flask. Leave the cells to detach for ~10 min at 37°C. Tap the flasks sharply two or three times to aid cell detachment.

4   Transfer the cells to 50 ml plastic centrifuge tubes and centrifuge at 100 g for 5 min. Resuspend the pellet in 70 ml DMEM/F12 containing 10% FCS.

5   Pipette 100 μl of cell suspension into each well of a 96-well plate (sufficient for seven plates) and incubate overnight (37°C, 5% CO_2).[a]

6   Aspirate medium from all wells and replace with 90 μl DMEM/F12 (37°C) containing 300 mM IBMX and no fetal calf serum. (If antagonists are required, add 80 ml of above medium and 10 ml medium containing antagonist). Incubate for 30 min at 37°C.

7   Add agonist (in 10 ml of medium). Leave for 15 min at 37°C. Note, agonist and antagonist contact times may need to be optimized (typically 15–30 min).

8   Aspirate medium from all wells. Add 100 ml of 100% ethanol (cold, from refrigerator) and leave for 30 min at 4°C.

9   Transfer 20 ml ethanol from each well to a fresh plate and add 80 ml of SPA assay buffer or water. The cAMP content of an aliquot (in 20% ethanol) can then be determined using the SPA *assay* (see *Protocol 5*).[b]

[a] Shake suspension periodically during addition to plates, to ensure that cells do not settle out.

[b] If, using the above method, the cAMP content of samples is insufficient for assay, cAMP can be concentrated by transferring the entire sample into a fresh plate and drying down the ethanol. The cAMP remaining can then be dissolved in 25 or 50 ml SPA assay buffer for assay.

Protocol 6

[^{125}I]cAMP SPA of cAMP from samples generated in *Protocol 5*

Equipment and reagents

- SPA plate (Wallac rigid sample plate, 1450-512).
- 96-well plate scintillation counter (e.g. Packard Topcount or Wallac Microbeta).
- cAMP SPA kit (Nycomed Amersham RPA 556). The kit is sufficient for ten 96-well plates[a]
- Plate sealer (e.g. Wallac sealing tape)

Protocol 6 continued

Method

1 Dilute assay buffer to 1250 ml with purified (e.g. distilled) water.

2 Dilute cAMP standard in 2 ml assay buffer (12.8 pmol/well standard). Serially dilute standard (0.5 ml standard + 0.5 ml assay buffer) to give 6.4, 3.2, 1.6, 0.8, 0.4 and 0.2 pmol/well.

3 Dilute ^{125}I tracer in 14 ml assay buffer per vial (one vial is sufficient for five plates). Gently mix by hand.

4 Dilute antiserum in 15 ml assay buffer per vial (one vial is sufficient for five plates). Gently mix by hand.

5 Dilute SPA anti-rabbit reagent in 30 ml assay buffer.

6 Dilute samples in assay buffer if required.

7 For five plates, mix together one vial of tracer (14 ml), one vial of antiserum (15 ml) and 15 ml of SPA reagent. Stir with a magnetic stirrer (adjust volumes accordingly for different numbers of plates).

8 Pipette 25 μl of standard (duplicates) or unknown into each well of an SPA plate (Wallac).[a]

9 Add 75 μl of tracer–antiserum–SPA mix (continue stirring bottle throughout).

10 Seal plate with plate sealer. Incubate overnight at room temperature (in lead-impregnated Perspex box) and count in β-plate scintillation counter (1 min).

[a] This method uses lower quantities of each reagent than the method described in the kit (25 μl instead of 50 μl). Remember that the standard curve will therefore contain 50% of the quantity per well than is given in the kit.

3.3.2 Data analysis

cAMP content of samples is assayed against a standard control curve. Computerized interpolation of the standard curve allows for rapid conversion of measured counts per minute to cAMP quantity. Responses to agonists are typically plotted as concentration–response curves and fitted to a four-parameter logistic equation (see Section 3.2.3). The ability of an antagonist to inhibit agonist-evoked cAMP production can be determined as described in Section 3.2.3.

3.3.3 Assay optimization

The following parameters should be optimized:

- *Incubation time in the presence of agonist.* The time taken to reach maximal cellular cAMP varies for different cells and agonists. The time course should be investigated to ensure that the response has reached equilibrium. This should also be performed when antagonists are being investigated, as the presence of an antagonist may slow down the time to equilibrium response.

- *cAMP levels.* For G_i-coupled receptors, cAMP levels must be elevated above basal levels. The agonist should then induce a reduction of cAMP levels. The most commonly used method for elevating cAMP is to incubate with the adenylate cyclase activator, forskolin (Sigma). A concentration–response curve to forskolin (1–100 μM) should be performed in the cell type in question, and a sub-maximal concentration selected (a concentration giving 50–75% of maximum would normally give a large enough signal for accurate measurement without reducing the sensitivity to G_i-coupled receptors that could occur with supra-maximal stimulation). The incubation time in the presence of agonist and forskolin should then be investigated. Commonly, both agents may be added simultaneously, but for agonists with a slow onset of action it may be necessary to pre-incubate in the presence of agonist before addition of forskolin.

- *Acetylation of samples.* For cells that produce very low levels of cAMP, it is possible to enhance the sensitivity of the assay by acetylation of the samples. The method and reagents required are typically included in commercially available cAMP assay kits.

3.4 Measurement of inositol triphosphate

3.4.1 Introduction

Section 3.3 described the measurement of intracellular cAMP, which may be used to investigate the actions of G_s- and G_i-coupled receptors. The third major group of G-protein-coupled receptors, those that couple through G_q, do not directly modulate intracellular cAMP. G_q-coupled receptors stimulate phospholipase C, which leads to the breakdown of membrane phospholipids (particularly phosphatidylinositol), to generate two major second messengers, inositol 1,4,5-trisphosphate and diacylglycerol. The activity of phospholipase C may be measured indirectly by measuring the effect of one of the second messengers (e.g. intracellular calcium release stimulated by inositol 1,4,5-trisphosphate) or by direct measurement of inositol phosphate generation. *Protocol 7* describes a method for measurement of total inositol phosphate accumulation in CHO cells stably transfected with a human oxytocin receptor. Alternative radioactive assays for G_q-coupled receptors include direct measurement of inositol 1,4,5-trisphosphate by commercially available radioimmunoassay kits. Intracellular calcium release may also be measured directly (and distinguished from entry of extracellular calcium entry) by pre-loading cells with [^{45}Ca^{2+}] and measuring release from intracellular stores on activation by an agonist.

Protocol 7

Measurement of total inositol phosphate accumulation in CHO cells transfected with human oxytocin receptor

Equipment and reagents
- CHO cells stably transfected with human oxytocin receptor (CHO-OT cells)
- Sterile PBS
- Access to cell culture laboratory with CO_2 incubator

Protocol 7 continued

- DMEM/F12 cell culture medium, containing 10% FCS and selection agents (if required)
- Modified Krebs buffer: 10 mM Hepes, pH 7.4, containing 118 mM NaCl, 4.7 mM KCl, 4.2 mM $NaHCO_3$, 1.2 mM $MgSO_4$, 11 mM glucose, 1.2 mM KH_2PO_4, 2 mM $CaCl_2$, 10 mM LiCl
- Test compounds (serial dilutions prepared at twice the final concentrations in modified Krebs buffer)
- Sterile PBS, containing 5 mM EDTA
- Glass chromatography columns (Fisher)
- 50 ml plastic centrifuge tubes
- 24-well plates
- Dowex-1 chloride resin (8% cross-linked, 200–400 mesh, Sigma)
- HCl (1 mM)
- Myo[2-^{3}H]inositol (Nycomed Amersham)
- Trichloroacetic acid (TCA, 1 mM, Aldrich)
- 1,1,2-Trichlorotrifluorethane (Aldrich)
- Tri-n-octylamine (Aldrich)
- $NaHCO_3$ (60 mM)
- Ammonium formate (25 mM)

A. Preparation of columns[a]

1 Stir 60 g Dowex into 50 ml distilled water overnight (room temperature).

2 Make up to 200 ml with distilled water.

3 Add 0.8 ml Dowex per column.

4 Wash column with 10 ml HCl (1 mM) followed by two washes with distilled water.

B. Method

1 Grow cells in 175 cm^2 flasks in DMEM/F12 medium containing 10% FCS, until approaching confluency.

2 Once cells are confluent, remove medium from the flasks and wash with 5 ml PBS per flask.

3 Remove PBS and add 5 ml PBS containing 5 mM EDTA per flask. Leave the cells to detach for ~10 min at 37°C. Tap the flasks sharply two or three times to aid cell detachment.

4 Transfer the cells to 50 ml plastic centrifuge tubes and centrifuge at 100 g for 5 min. Resuspend the pellet in DMEM/F12 (10% FCS) containing [^{3}H]myo-inositol (1.5 μCi/ml), using ~100 ml for a confluent 175 cm^2 flask.[b] Plate cells out in 24-well plates (0.5 ml per well). Incubate cells for 48 h (37°C, 5% CO_2).

5 Remove medium from plate and wash cells twice with 0.3 ml modified Krebs buffer. Incubate at 37°C for 15 min.

6 Add agonist (300 μl; e.g. oxytocin) and incubate for 15 min. Add 300 μl of ice-cold TCA (1 mM) and incubate on ice for 15 min.

7 Remove 460 μl from each well and place in microcentrifuge tubes (1.5 ml) containing 40 μl EDTA (10 mM). Add 500 μl 1,1,2-trichlorotrifluorethane/tri-n-octylamine (1:1 mix, freshly prepared in fume cabinet prior to experiment). Cap microcentrifuge tube and mix by vortexing. Centrifuge tubes at 5000g for 2 min.

Protocol 7 continued

8 Remove 250 μl of the upper (aqueous) phase and transfer to microcentrifuge tubes (1.5 ml) containing 50 μl NaHCO$_3$ (60 mM).

9 Apply 300 μl extract obtained from step **8** to prepared Dowex columns and allow to run under gravity.

10. Wash each column with 2 ml distilled water,[c] followed by 10 ml ammonium formate[d] (25 mM).

11. Elute inositol phosphates by adding 10 ml HCl (1 mM) to column. Mix 5 ml of eluent with 15 ml scintillation cocktail and count with a liquid scintillation counter.

[a] Columns may be regenerated by washing with 10 ml HCl (2 M) and 10 ml distilled water. Resin should be replaced after 15 uses.

[b] Volume should be such that cells are plated out at ~50% confluency when plated at 0.5 ml/well in 24-well plates.

[c] To remove free [^{3}H]inositol.

[d] To remove glycerol phosphoinositides.

3.4.2 Data analysis

To account for differences in labelling between experiments, total [^{3}H]inositol phosphate accumulation should be expressed as per cent above basal accumulation. Agonist and antagonist data can then be analysed as described in Section 3.2.3.

3.4.3 Assay optimization

The time course of the response to agonist, either alone or in the presence of antagonist, should be investigated and responses measured once equilibrium has been reached.

3.5 Non-radioactive assays

There are non-radioactive alternatives for measurement of second messengers generated, as summarized in *Table 3*. cAMP may be measured by enzyme immunoassay. However, this has a lower throughput than SPA or Flashplate assays

Table 3 Non-radioactive assays suitable for different classes of G-protein-coupled receptor

G protein family	Second messengers activated	Suitable non-radioactive assay
G$_s$	increased cAMP	cAMP enzyme immunoassay; cAMP reporter cell line
G$_i$	reduced cAMP	cAMP enzyme immunoassay; cAMP reporter cell line
G$_q$	increased inositol phosphate production; increased intracellular calcium	Fura-2 fluorescent measurement of intracellular calcium
All	guanine nucleotide exchange	None currently available

and is more difficult to automate. cAMP reporter cell lines (e.g. cAMP response element promoter coupled to luciferase production) may also be generated, though this requires specialized techniques and considerable work-up time. Intracellular calcium levels may be conveniently measured using fluorescence of a dye such as Fura-2; this is the method of choice unless measurement of inositol phosphate production or intracellular calcium release is specifically required. At the time of writing, there is no alternative to [^{35}S]GTPγS assays for measurement of guanine nucleotide exchange.

4 Enzyme assays

4.1 Introduction

While spectrophotometric and fluorimetric methods are probably the most widely used procedures for assaying enzymes (13), radiometric enzyme assays continue to be used due to their high sensitivity, specificity and freedom from interference. A comprehensive account of radiometric enzyme assays is given in a companion book in this series, *Enzyme Assays: A Practical Approach* (13,14). Radiometric enzyme assays generally involve determination of the rate of conversion of radiolabelled substrate to labelled product. Requirements are the availability of a suitable labelled substrate of known specific activity, and of a simple, rapid and accurate method to separate substrate and product. Commonly used separation methods include:

- ion-exchange techniques, using either resin (e.g. *Protocol 9*) or phosphocellulose filter paper (e.g. measurement of p34^{cdc2} kinase (15) by transfer of radioactive phosphate from [^{32}P]adenosine triphosphate to a peptide substrate, which is captured by adsorption on to phosphocellulose paper);

- solvent extraction methods (e.g. measurement of acyl coenzyme A:cholesterol transferase (ACAT) by enzyme-catalysed formation of [^{14}C]cholesterol oleate from [^{14}C]oleoyl-coenzyme A and subsequent extraction using chloroform/methanol (16));

- release of radioactivity (e.g. measurement of tryptophan hydroxylase by enzyme-catalysed formation of 5-hydroxy-[4-^{3}H]tryptophan from [5-^{3}H]tryptophan and the subsequent acid-dependent release of ^{3}H$_2$O and removal of un-reacted substrate with charcoal (17)).

In addition, transport systems are usually measured using radiolabelled substrates, for example measurement of transport of radiolabelled aromatic amino acids into rat liver cells (18).

4.2 Measurement of nitric oxide synthase activity using [^{14}C]arginine

4.2.1 Introduction

Protocol 8 describes how to extract enzymes from tissues or cells. Further methods, including extraction from subcellular fractions, are given by Price (19).

If the purpose of the assay is to quantify the ability of a substance to inhibit enzyme activity, rat brain cysotol can be used as a crude source of neuronal nitric oxide synthase (NOS) (see *Protocol 8* for further details). Inducible NOS can be obtained by stimulating rat or human hepatocytes (19) or a suitable cell line (e.g. human colorectal adenocarcinoma cell line DLD-1 or murine macrophage cell lines) with a mixture of cytokines such as interferon γ, interleukins-6 and 1β, and tumour necrosis factor α (20). Alternatively, it may be more convenient to obtain NOS by expression using the baculovirus–insect cell system (21)

Protocol 9 describes a method to measure NOS activity by conversion of radio-labelled arginine to citrulline using ion-exchange separation. Earlier forms of this assay used rather slow and labour-intensive separation methods. This protocol employs a simple method of separating arginine from citrulline using the Na^+ form of the strongly acidic cation exchanger, Dowex 50.

Protocol 8

Extraction of nitric oxide synthase from tissue or cell extracts

Equipment and reagents

- Extraction buffer: 20 mM Hepes, pH 7.2, containing 320 mM sucrose, 1 mM EDTA, 1 mM dithiothreitol (Sigma), 10 μg/ml leupeptin (Sigma), 10 μg/ml soya bean trypsin inhibitor (Sigma) and 2 μg/ml aprotinin (Sigma)[a]
- For cell extracts, PBS containing 5 mM EDTA
- PBS

- 10 mg/ml phenylmethylsulfonyl fluoride (PMSF, Sigma), prepared in absolute ethanol and stored at −20 °C[a]
- Ultra-Turrax T25 homogenizer (Jencons) or Vibra-cell VC100 ultrasonic processor (Jencons)
- Access to a cold room is desirable

A. Preparation of tissue extracts[b]

1 Place tissue in 10 ml pre-cooled plastic tube. Add ice-cold extraction buffer (5 ml/g). Homogenize for 10 sec using an Ultra-Turrax homogenizer (pre-cool probe in ice/water). Add PMSF (10 μl per ml buffer) and homogenize for a further 10 sec.[a]

2 To extract NOS from the post-mitochondrial supernatant (cytosol and microsomes, containing most of the NOS in cells), centrifuge the homogenate at 10 000g for 30 min and retain the supernatant for assay.

3 To extract NOS from the cytosol (rat brain cytosol is a convenient source of NOS for inhibitor experiments), centrifuge the homogenate at 100 000 *g* for 60 min and retain the supernatant for assay.

B. Preparation of cell extracts[b]

1 For adherent cells in a 175 cm^2 flask, wash cells with 5 ml PBS,[c] remove PBS and add 5 ml PBS containing 5 mM EDTA per flask. Leave the cells to detach for ~10 min at 37 °C. Tap the flasks sharply two or three times to aid cell detachment. Transfer the

cells to a 50 ml plastic centrifuge tube and centrifuge at 100g for 5 min. Resuspend the pellet in 2 ml ice-cold extraction buffer.

2   For cells in suspension, centrifuge (400g, 5 min), resuspend in PBS (2 ml per ml pellet), re-centrifuge (400g, 5 min) and resuspend the cell pellet in extraction buffer (2 ml per ml pellet).

3   Disrupt the cell extract by sonication. For example, sonicate using a Vibra-cell VC100 ultrasonic processor with a pre-cooled probe with 6 mm tip. Sonicate twice for 10 sec at 10 μm amplitude with 30 sec cooling in ice/water in between. Add PMSF (10 μl per ml buffer) between sonication steps.[a]

4   It may be necessary to remove endogenous arginine[c] from tissue or cell extracts by treating the extract with ion-exchange resin, prepared as described in *Protocol 9* and washed in extraction buffer. Mix 2 vols enzyme extract with 1 vol ice-cold resin in microfuge tubes. Centrifuge the mixture at 10 000 g for 1 min at 4°C and collect the supernatant for assay. Ensure that no resin is carried forward into the assay as it will adsorb the labelled substrate.

[a] Extraction buffer can be stored in aliquots (typically 50 ml) at −20°C. The composition of the buffer is designed to permit extraction of NOS without breaking intracellular organelles and minimizing proteolysis by chelating divalent cations with EDTA and inclusion of protease inhibitors. Since it is unstable in aqueous solution, 10 mg/ml PMSF is added to the buffer during the extraction procedure (step **2**).

[b] Extraction of tissue must be carried out at 0–4°C, preferably in a cold room, to avoid loss of enzyme activity. Tissue or cells must be assayed immediately or rapidly frozen and stored at −70°C until use.

[c] The concentration of arginine in the assay (20 μM) has been chosen as a compromise between having a high concentration (to minimize the effect of endogenous arginine) and having a low concentration (to maximize assay sensitivity). Since cell culture media contain high concentrations of L-arginine (typically >0.5 mM), cells must first be washed with an arginine-free medium (e.g. PBS).

Protocol 9

Measurement of nitric oxide synthase activity using [^{14}C]arginine

Equipment and reagents

- [^{14}C]arginine assay buffer: 50 mM potassium phosphate, pH 7.2, including 1.67 mM L-citrulline (Sigma), 1.67 mM $MgCl_2$, 0.33 mM $CaCl_2$, 167 μM NADPH and L-[U-^{14}C]arginine (Nycomed Amersham; specific activity >11 GBq/mmol) plus L-arginine (to give a total concentration of labelled and unlabelled arginine in the buffer of 33 μM)[a]

- NOS inhibitors: S-ethylisothiourea (SEITU; Aldrich; also known as 2-ethyl-2-thiopseudourea) is recommended to determine basal level of radioactivity.[b] 0.5 mM SEITU in 50 mM potassium phosphate (pH 7.2), to give a final concentration of 100 μM

Protocol 9 continued

- 5 mM [ethylene-bis(oxyethylenenitrilo)] tetracetic acid (EGTA) (to give a final concentration of 1 mM); set pH to 7.2 using potassium hydroxide and store at 4 °C

- Ion-exchange resin: Dowex 50W, 200–400 mesh, 8% cross-linked (Sigma), converted to the Na$^+$ form^c

- Test compounds (if applicable); serial dilutions prepared at 5 × final concentrations in 50 mM potassium phosphate (pH 7.2)

Method

1 To 10 ml clear plastic tubes, add 20 μl of one of the following: water (control), test compound (if applicable), 5 mM EGTA (calcium-free), or 0.5 mM SEITU (basal).

2 Add 60 μl assay buffer and pre-warm tubes to 37 °C in a water-bath.

3 Initiate the assay by adding 20 μl enzyme extract and incubate for 10 min at 37 °C.^{d}

4 Terminate the reaction by adding 1.5 ml ion-exchange resin at room temperature.

5 Add 5 ml water and mix. Allow the resin to settle for ~10 min. Remove 4 ml supernatant, mix with 10 ml scintillation fluid and count in a liquid scintillation counter.

6 To determine total [^{14}C]arginine added to wells, count the radioactivity in 60 μl assay buffer.

^{a} Assay buffer can be stored at −20 °C. Final concentrations in assay will be 60% of the concentrations in this buffer (60 μl assay buffer in 100 μl final assay volume), e.g. 20 μM L-arginine. [^{3}H]Arginine can be used instead of [^{14}C]arginine. However, [^{14}C]arginine gives lower blanks and hence a larger signal:noise ratio. If blank (no enzyme) values start to increase, [^{14}C]arginine can be purified by adding the radiolabel in 1 ml of 20 mM Hepes buffer (pH 5.5) to a 1 ml column of Dowex resin, washing with 8 ml water and eluting with 4 ml of 0.5 M ammonia. The eluent is freeze-dried, re-dissolved in 2% ethanol and stored at 4 °C.

^{b} Basal ^{14}C not bound to resin independently of NOS activity is determined in the presence of an excess of a NOS inhibitor. SEITU is recommended. Alternative NOS inhibitors are N^G-monomethyl-L-arginine acetate (L-NMMA; Tocris) and N(G)-nitro-L-arginine (L-NNA; Tocris). To assess the calcium-dependence or -independence of the measured NOS, 1 mM EGTA is added to some incubations to chelate free-calcium ions.

^{c} This must be converted from the H$^+$ form to the Na$^+$ form by washing the resin four times with at least 10 vols 1 M sodium hydroxide. Add sodium hydroxide to the resin, mix, allow the resin to settle under gravity (~5 min) and pour off supernatant. Then wash the Na$^+$ form of the resin is with water until it reaches a pH of <8 and bring it to 50% resin in water (estimate using the height of the settled resin). Can be prepared in bulk and stored at 4 °C.

^{d} It is crucial to ensure that the assay is linear with both time and concentration of enzyme extract added (see Section 4.2.4).

4.2.2 Data analysis

The amount of citrulline generated is expressed in pmol/min/mg protein or pmol/min/g tissue weight. To determine NOS-dependent citrulline formation, basal radioactivity in the presence of an excess of the NOS inhibitor (e.g. 100 μM

SEITU) is subtracted from that present in the other incubations. Citrulline content is determined using *Equation 8*.

$$\text{Citrulline (pmol/min per mg protein)} = \frac{((\text{test or control dpm-based dpm})/\text{Specific activity}^{b}) \times (6.6\text{ml}/4.0\text{ ml}^{a})}{\text{Incubation time} \times \text{tissue or protein weight in 20 } \mu\text{l}^{c}}$$

Total and calcium-independent NOS activities can be determined from incubations in the absence and presence of 1 mM EGTA to chelate free calcium ions, respectively.

The potency of a compound to inhibit NOS activity is measured by determining the concentration of the compound that inhibits 50% of citrulline generated in control incubations (IC_{50}). IC_{50} values are best determined using non-linear curve-fitting analysis, such as a four-parameter logistic equation (see Section 2.5.1).

4.2.3 Assay optimization

With some sources of enzyme, especially purified NOS, other additions are required for maximal activity. These are (6R)-5,6,7,8-tetrahydro-L-biopterin (BH$_4$; Shirks Laboratories; 10 μM, prepared fresh in dilute HCl), flavin adenine dinucleotide (Sigma) and flavin mononucleotide (Sigma) (both 1 μM, 50–100× stocks can be stored frozen), and calmodulin (Sigma; 100 nM, 100× stock can be stored frozen). Experiments should be performed to determine if the presence of these agents increases enzyme activity.

If basal levels of NOS activity in the presence of an excess of a NOS inhibitor are substantially higher than blanks at time zero, it suggests that endogenous arginase is present, which is competing for the arginine substrate. This occurs particularly in liver tissue or cells. If this is the case, the arginase inhibitor L-valine (Sigma; 60 mM, dissolved directly in the assay buffer on the day of assay) should be added. To determine if endogenous arginase is present, initial experiments should be performed in the absence and presence of L-valine.

It is crucial to ensure that the assay is linear with both time and concentration of enzyme extract added. These can vary widely. For example, at 37°C, under the conditions described in *Protocol 9*, the neuronal NOS assay is linear for ~10 min, whereas the inducible NOS assay is linear for ~60 min. At 25°C, the neuronal NOS assay is linear for ~60 min.

When initially establishing the assay, a number of controls should be determined, including:

* blank in the absence of enzyme;
* activity at time zero;

---

[a] Adjustment to take into account that 4 ml sample is assayed from a total assay volume of 6.6 ml.

[b] Specific activity expressed as d.p.m./pmol is calculated by dividing total d.p.m. added to assay tube (d.p.m. in 60 μl assay buffer) by the total amount of arginine (100 μl of 20 μM).

[c] Tissue or protein present in the 20 μl volume of enzyme extract.

- basal activity in the presence of an excess of an NOS inhibitor;
- assessment of calcium-independent NOS activity by determining activity which remains in the presence of 100 mM EGTA.

4.3 Homogeneous assays

Recently, homogeneous methods utilizing either SPA or Flashplate technologies (see Section 2.3.3 for further information) have been developed in which no separation of substrate and product is required. This obviously has tremendous potential for radiometric enzyme assays, offering a simple assay which is amenable to automation and high-throughput screening. While kits for such enzyme assays are currently quite limited and fairly expensive, their ease of use makes it likely that these assay technologies will become more widespread. *Protocols* 6 and 7 described extraction and assay of cAMP using a ^{125}I SPA. Flashplates can also be used in a similar manner for these assays. SPA kits to measure cAMP- (catalogue no. TRKQ7090) and cGMP-dependent (catalogue no. TRKQ7100) phosphodiesterase activity and phospholipase A$_2$ activity (catalogue no. TRK7040) are available from Nycomed Amersham. Spencerfry and colleagues (15) compared tyrosine kinase assays using phosphocellulose filter binding and SPAs. NEN have described the use of Flashplates to measure protein kinases using [γ-^{33}P]ATP (see NEN website for further details: http://www.nenlifesci.com/). In addition, basic Flashplates (NEN catalogue no. SMP200) can be used to develop your own enzyme assay.

5 Ion channel assays

5.1 Introduction

The above sections have dealt with some of the common pharmacological targets, namely receptors, second messengers and enzymes. This section deals with the last common pharmacological target, ion channels. Modulation of ion channels has proved beneficial in a wide range of clinical settings such as anaesthesia (Na$^+$ channels), angina (Ca^{2+} channels) and diabetes (K$^+$ channels). Here we describe a simple method for measuring influx of a radioactive ion into cells.

5.2 Measurement of Na$^+$ channel function by [^{14}C]guanidine flux

5.2.1 Introduction

The assay described here uses sterile Flashplate plates (NEN). These plates, which are suitable for cell culture, have scintillant incorporated into the base and utilize the 'proximity effect' principle (see Section 2.3.3). Thus, radioactivity that enters the cell accumulates and is held close enough to the scintillant base to stimulate light production. Radioactivity in the medium above the cells, however, is not in close enough proximity to stimulate scintillation. Without any separation of intracellular and extracellular label, the quantity of a label entering a cell may be measured. Additionally, the time-course of this entry may be measured by simply repeating the measurements at intervals. In the assay we describe here, the function of the Na^+ channel is measured using [^{14}C]guanidine as a marker for sodium. Guanidine has the advantage over using sodium itself in that it is not subject to the normal cellular mechanisms of sodium transport, so changes in signal reflect entry through the sodium channels. Additionally, since guanidine is not subject to transport out of the cell, it will accumulate and give greater sensitivity than using sodium itself. The method described measures the activity of compounds that open the channel. The activity of compounds that inhibit channel opening can be investigated by opening channels with toxins (e.g. scorpion venom and veratrine for type II sodium channels) and then examining inhibition of [^{14}C]guanidine influx. This protocol may also be used to measure influx of ^{45}Ca into cells.

Protocol 10

Measurement of [^{14}C]guanidine flux into CHO cells stably expressing recombinant Na^+ channels

Equipment and reagents

- Access to cell culture laboratory and CO_2 incubator
- CHO cells stably expressing recombinant Na^+ channels
- DMEM/F12 cell culture medium, containing 10% FCS and selection agents (if required)
- Sterile PBS
- Sterile PBS containing 5 mM EDTA
- Test compounds (serial dilutions prepared at $20 \times$ final concentrations in assay buffer)

- Assay buffer: 50 mM Hepes containing 125 mM choline chloride, 5.5 mM glucose, 0.8 mM $MgSO_4$, 5 mM KCl, adjusted to pH 7.4 with Tris base
- Sterile Flashplate (NEN)
- [^{14}C]Guanidine (Nycomed Amersham) in assay buffer (307 kBq/ml)
- Plate sealers (Titertek)
- 96-well plate scintillation counter (e.g. Packard Topcount or Wallac Microbeta)

Method

1   Culture cells transfected with recombinant sodium channels (or a cell line containing channel of interest) in 175 cm^2 flasks.

2   Once cells are confluent, remove medium from the flasks and wash with 5 ml PBS per flask.

3   Remove PBS and add 10 ml PBS containing 5 mM EDTA per flask. Leave the cells to detach for ~10 min at 37°C. Tap the flasks sharply two or three times to aid cell detachment.

4   Transfer the cells to 50 ml plastic centrifuge tubes and centrifuge at 100g for 5 min. Resuspend pellet in 70 ml DMEM/F12 containing 10% FCS.

Protocol 10 continued

5   Pipette 100 μl of cell suspension into each well of a 96-well sterile Flashplate (sufficient for seven plates).

6   Remove medium and add 130 μl assay buffer to each well.

7   Add 10 μl of test compound to wells. Incubate for 10 min (37 °C, 5% CO_2).

8   Add 60 μl (18.5 kBq) of [^{14}C]guanidine solution to each well. Seal plate with plate sealer.

9   Read in β-plate scintillation counter (e.g. Wallac Microbeta or Packard Topcount). The time-course can be followed, or influx of [^{14}C]guanidine can be halted by addition of 10 μl of tetrototoxin (1 μM, **caution**: toxic) to each well after a pre-set time.

5.2.2 Data analysis

Data are expressed as c.p.m./well (proportional to quantity of [^{14}C]guanidine in cells. Agonist and antagonist data can be analysed as described in Section 3.2.3.

6 Some commercially available curve-fitting programs

Data Analysis Toolbox, MDL Information Systems (UK) Ltd, Ground Floor, Building 4, Archipelago, Lyon Way, Camberley, Surrey GU16 5ER, UK (http://www.mdli.com/tech/lsw.html).

Prism, GraphPad Software, Inc., 5755 Oberlin Drive, #110, San Diego, CA 92121, USA (http://www.graphpad.com/www/welcome.html).

XlFit, IDBS, 5 Huxley Road, The Surrey Research Park, Surrey, GU2 5RE, UK (http://www.idbs.co.uk/).

Acknowledgements

We would like to thank Drs Amanda Martin and Richard Knowles for assistance with preparation of protocols and Drs Michael Sheehan, David Hall, Ann Mills-Duggan and Heather Giles for critical review of the manuscript.

References

1. Mellor, G.W., Fogarty, S.J., O'Brien, M.S., Congreve, M., Banks, M.N., Mills, K.M., Jefferies, B. and Houston, J.G. (1997). *J. Biomol. Screening*, **2**, 153.
2. Hulme, E.C. (ed.) (1992). *Receptor–Ligand Interactions: A Practical Approach*. IRL Press, Oxford.
3. Hulme, E.C. and Buckley, N.J. (1992). In *Receptor–Ligand Interactions: A Practical Approach* (ed. E.C. Hulme), p. 177. IRL Press, Oxford.

4. Haga, T., Haga, K. and Hulme, E.C. (1990). In *Receptor Biochemistry: A Practical Approach* (ed. E.C. Hulme), p. 1. IRL Press, Oxford.

5. Hulme, E.C. (1992). In *Receptor–Ligand Interactions: A Practical Approach* (ed. E.C. Hulme), p. 235. IRL Press, Oxford.

6. Wang, J-X., Yamamura, H.I., Wang, W. and Roeske, W.R. (1992). In *Receptor–Ligand Interactions: A Practical Approach* (ed. E.C. Hulme), p. 177. IRL Press, Oxford.

7. Bosworth, N. and Towers, P. (1989). *Nature*, **341**, 167.

8. Hulme, E.C. and Birdsall, N.J.M. (1992). In *Receptor–Ligand Interactions: A Practical Approach* (ed. E.C. Hulme), p. 63. IRL Press, Oxford.

9. Tota, M.R., Daniel, S., Sirotina, A., Mazina, K.E., Fong, T.M., Longmore, J. and Strader, C.D. (1994). *Biochemistry*, 33, 13079–13086.

10. Shulakoff, E. (1998). *Genet. Eng. News*, **18**, 18.

11. Watson, J., Selkirk, J.V. and Brown, A.M. (1998). *J. Biomol. Screening*, **3**, 101.

12. Lew, M.J. and Angus, J.A. (1995). *Trends Pharmacol. Sci.*, **16**, 328.

13. John, R.A. (1992). In *Enzyme Assays: A Practical Approach* (ed. R. Eisenthal and M.J. Danson), p. 59. IRL Press, Oxford.

14. Oldham, K.G. (1992). In *Enzyme Assays: A Practical Approach* (ed. R. Eisenthal and M.J. Danson), p. 93. IRL Press, Oxford.

15. Spencerfry, J.E., Brophy, G., Obeirne, G. and Cook, N.D. (1997). *J. Biomol. Screening*, **2**, 25.

16. Nagata, Y., Yonemoto, M., Iwasawa, Y., Shimizu-Nagumo, A., Hattori, H., Sawazaki, Y. and Kamei, T. (1995). *Biochem. Pharmacol.*, **49**, 643.

17. Beevers, S.J., Knowles, R.G. and Brown, C.I. (1983). *J. Neurochem.*, **40**, 894.

18. Salter, M.S., Knowles, R.G. and Pogson, C.I. (1986). *Biochem. J.*, **233**, 499.

19. Price, N.C. (1992). In *Enzyme Assays: A Practical Approach* (ed. R. Eisenthal and M.J. Danson), p. 255. IRL Press, Oxford.

20. Johnson, M.L., Shapiro, R.A. and Billiar, T.R. (1998). In *Nitric Oxide Protocols* (ed. M. Titheradge), p. 43. Humana Press, Totowa, NJ.

21. Garvey, E.P. (1998). In *Nitric Oxide Protocols* (ed. M. Titheradge), p. 33. Humana Press, Totowa, NJ.

22. Charles, I.G., Foxwell, N. and Chubb, A. (1998). In *Nitric Oxide Protocols* (ed. M. Titheradge), p. 51. Humana Press, Totowa, NJ.

Chapter 10

Summary of legislation in radiological protection in the UK

DAVID PRIME and BARRY FRITH

Radiological Protection Service, University of Manchester, Oxford Rd., Manchester M13 9PL, UK

1 Introduction

The use of equipment or materials that emit radiations is governed by two pieces of legislation—the Radioactive Substances Act (1993) and the Ionising Radiations Regulations (1999). This legislation imposes a system of radiological protection that is almost the same in all premises using radioactive materials or radiation-emitting equipment. This system will be outlined in two sections dealing with the respective items of legislation.

2 The Radioactive Substances Act (1993)

This act controls the acquisition, storage and disposal of radioactive materials.

2.1 Acquisition

Before obtaining radioactive materials it is necessary to obtain a registration under this Act. This registration will specify both the radionuclides or class of radionuclides and the total activities allowed.

2.2 Storage

Radioactive materials must be stored under certain conditions as outlined in the Act. This will always mean a lockable store sufficiently shielded, normally to reduce the radiation dose to <7.5 μSv/h outside the store. Storage involves labelling and security.

2.3 Radioactive waste

The disposal of waste is often particularly difficult. There are usually several different categories of waste of all phases generated. Every one of these disposal pathways is subject to an authorization under the Act.

Aqueous waste can often be disposed of down special sinks directly to the sewers. Using this method of disposal requires knowledge of the total flow rate in the sewer to confirm that dilution is adequate. It is normal to incinerate organic solvents and other organic wastes, although animal carcasses can sometimes be macerated and then disposed of as aqueous waste.

Since premises are often in centres of population, the disposal of gaseous radioactive waste is rarely permitted directly from the site.

Solid waste disposal routes are a function of activity and nuclide. Low levels can usually go, under special conditions, to local waste dumps where the radioactive material is immediately buried under normal refuse. Higher levels must be taken to an approved contractor, where special arrangements can be made for temporary or permanent storage.

3 The Ionising Radiations Regulations (1999)

This set of regulations controls work with materials or equipment emitting ionizing radiations. It imposes working practices and a management structure. The regulations consist of eight separate parts, each of which will be briefly discussed below.

3.1 Interpretation and general

The first four regulations are grouped together in this section. The main features of these particular regulations are that all work involving radiation must be notified to the Health and Safety Executive (HSE) so that the premises can be inspected appropriately and that there should be close co-operation between different employers. This last point is of particular importance in premises where a worker may spend time in more than one establishment and it is important to record all radiation doses.

3.2 General principles and procedures

3.2.1 Authorization of specified practices

This particular regulation (5) was enacted later than the rest of the legislation. It requires certain practices to be authorised by the HSE. These are mainly of concern to industrial users of radiation-emitting isotopes and equipment and there is general approval for most practices. It is unlikely that any user of radioactive isotopes in research will be required to obtain prior authorization.

3.2.2 Prior risk assessment etc.

There is a much greater emphasis on risk assessment in these regulations. Before any new activity involving work with ionizing radiation starts, the employer will have to make a risk assessment. Such risk assessments should include:

- the nature of the sources;
- estimated radiation dose rates to persons;
- the likelihood of contamination arising and being spread;
- the results of previous monitoring or dosimetry;

- advice from equipment manufacturers or suppliers about safe use or maintenance;
- engineering controls and design features;
- any planned systems of work;
- estimated levels of surface and airborne contamination;
- effectiveness and suitability of protective clothing;
- the extent of unrestricted access;
- the likelihood and potential severity of accidents;
- the consequence of the failure of control measures;
- accident prevention measures.

3.2.3 Restriction of exposure

Many design features must be incorporated to reduce radiation exposure. Shielding and containment are of special importance; the general philosophy is that all radiation doses should be kept as low as is reasonably practicable. This means that it is not merely necessary to keep within the dose limits, but to keep as far below these limits as is practicable. Also, priority must be given to keeping individual doses to as low as is reasonably practicable. This means that sometimes a sharing out of radiation exposure must be carried out by increasing the numbers of persons carrying out a particular practice. Engineering controls and design features have to be augmented by protective clothing and written systems of work in order to restrict exposure. Strangely, the dose limit to women declared pregnant is also in this section. In the eyes of the law, a pregnant woman is only considered pregnant when she has informed her employer in writing; after this time the fetus is restricted to a dose of 1 mSv.

3.2.4 Personal protective equipment and the maintenance and examination of engineering controls etc.

The employer must supply these types of protection where it is necessary and ensure that there is a suitable maintenance routine. Facilities must be provided for the storage of protective clothing.

3.2.5 Dose limitation

The new dose limits are detailed in schedule 4 of the regulations. It is unlikely that research workers will get greater than one-tenth of any limit under normal working conditions.

3.2.6 Contingency plans

The risk assessment identifies potential hazards, and a contingency plan (or plans) must be made to deal with the hazard if it occurs. Details of these plans have to be included in the departmental Local Rules. Where 'appropriate', the plans have to be rehearsed at suitable intervals.

3.3 Arrangements for the management of radiation protection

3.3.1 Radiation Protection Adviser

The employer of radiation workers should normally provide a Radiation Protection Adviser (RPA), an expert in the theory and practice of radiation protection.

3.3.2 Information, instruction and training

Appropriate courses in radiological protection for workers and training in handling techniques must be provided by the employer.

3.3.3 Co-operation between employers

The employer must come to satisfactory arrangements with other employers whose workers are on his or her sites and vice versa.

3.4 Designated areas

3.4.1 Designation of controlled and supervised areas

It is likely that fewer controlled areas will be needed than were required under the Ionising Radiations Regulations 1985. This is because the regulations are less prescriptive on the dose rates or contamination levels necessary for the designation of controlled areas.

3.4.2 Local Rules and Radiation Protection Supervisors

The Radiation Protection Supervisor (RPS) must draw up Local Rules in consultation with the RPA, and should have a copy available for inspection by the HSE. Any modifications should be discussed with the RPA before implementation. The RPS must ensure that workers in his or her department have read and understood these rules and that they follow them.

3.4.3 Monitoring of designated areas

All supervised and controlled areas must be regularly monitored at an interval specified in the Local Rules. Records must be kept for inspection by the HSE. The RPA should provide advice on all aspects of monitoring.

3.5 Classification and monitoring of persons

There is not likely to be any reason to classify radioisotope workers. If however, this is thought necessary after consultation with the RPA, an Approved Dosimetry Service must carry out appropriate personal dosimetry of each worker and full records must be kept of such measurements. Medical examinations of such classified radiation workers should be carried out annually. Such examinations are used not to confirm dosimetric results, since medical effects would be undetectable at normal dose levels, but to check that the health of the worker is compatible with the working conditions.

Monitoring of non-classified workers can be carried out although it should be emphasized that there is no point in providing film badges or other dosimeters for workers using low-energy beta emitters and usually, hand monitoring with thermoluminescent dosimeters is more appropriate for high-energy beta use.

3.6 Arrangements for the control of radioactive substances, articles and equipment

This section deals first with the keeping and moving of sources. The whereabouts of all sources should be known at all times and accounting methods should be employed to keep careful records of receipts and disposals. Special equipment and methods may be required for the movement of sources and it may even be necessary to declare special radiation areas during transport operations. It is necessary for the RPS to know on a daily basis the whereabouts of every source of radiation in his or her department.

3.7 Duties of employees and miscellaneous

This final part deals mainly with a timetable of implementation of the regulations and information about general duties of workers with regard to their fellows.

Chapter 11

Federal regulations on use of radionuclides in the USA

EILEEN D. HOTTE and NEAL NELSON
GlaxoSmithKline 1250 S. Collegeville Road, Collegeville PA 19426, USA

1 Introduction

The use of radioactive materials and radiation-emitting devices in the USA is regulated by a number of agencies. In 1967, The World Health Organization reported that regulation of devices and materials emitting ionizing radiation was divided among 50 state governments, several city governments, six federal agencies, a Federal Council and a Federal Commission (1). The situation in the early 2000s is not much different. In the USA, regulation is by laws, statutes, standards, regulations and ordinances. Laws are the rules of conduct applied and enforced under the authority of state or federal government. They establish proper, permitted, denied and penalized behaviour and statements of goals or intent (2). Statutes are acts of legislature, administrative regulations or any enactment given the force of law by the state (2).

While laws and statutes give legislative requirements, interpretation is often necessary. Regulations are issued by the government agencies administering the laws. Such regulations define what is required by the law and how the legal requirements of the law may be met (2). Standards, like regulations, explain compliance with the law and provide a means of comparison or evaluation to show compliance with the law (2). Ordinances are legal requirements issued by a city or municipality.

It would be impossible to itemize all state and local regulations on control of ionizing radiation, so only federal regulations will be addressed here. It should be noted, however, that state and local regulations must be at least as strict as federal and may be more strict. Individuals wishing to work with radionuclides or other sources of ionizing radiation must comply with all regulations, federal, state and local.

When a federal agency wishes to propose or change a regulation or standard, there is an extended public process. This may include an advanced Notice of Proposed Regulations and a Draft Regulation or Standard, both with a public comment period. From this a Final Regulation or Standard is completed. All these are published in the *Federal Register* with the citation in the form x FR y where x

271

indicates the volume and *y* the page. The current regulations and standards are published each year in the *Code of Federal Regulations (CFR)* with ancillary publications during the year. Citations are given in the form *a CFR b*, referring to the title and section where the regulation or standard is published. 10 *CFR* (Energy) is revised in January of every year. The current publication is from 2001.

Familiarity with the *Federal Register* and *Code of Federal Regulations* is needed by those working with radiation and radioactive materials because standards and regulations continually change. A number of services are available to assist individuals in keeping abreast of regulations which may affect them such as *Nuclear Licensing Reports* (3) or *NMSS Licensee Newsletter* (4). In addition the Health Physics Society keeps their members informed of the activities of federal agencies through a number of means including a monthly newsletter.

Historically, the United States Government established broad authority for radiation control with the Atomic Energy Act of 1946. This Act created the Atomic Energy Commission which had authority to control source and by-product materials and issue licenses (5). This Act had been amended frequently and then rewritten in 1954 (6). In 1959, the Federal Radiation Council (FRC) was established and guidance promulgated by this organization and signed by the President was the basis for federal regulation on radiation exposure (7).

In 1970, the Environmental Protection Agency (EPA) was established and all functions of the FRC and Atomic Energy Commission which regulated environmental limits for radiation were transferred to it. In 1974 the Atomic Energy Commission was abolished. In its place the Energy Research and Development Administration (ERDA) and the Nuclear Regulatory Commission (NRC) were formed. ERDA had the responsibility of studying and promoting nuclear science, and the NRC had the responsibility of regulation (8).

Today, the NRC regulates the production, distribution, use and disposal of source, by-product and special nuclear material. Individual states have authority over all other sources of radiation including naturally-occurring and accelerator-produced radioactive materials and the use of X-ray-producing devices. Some states have entered into agreement with the NRC to assume part of the NRC's authority over certain by-product material, source material and special nuclear material (in quantities less than a critical mass). These states are known as 'Agreement States'. The state must be willing to assume regulatory control and the Governor must certify that the state has a regulatory control program for radiation which is adequate to protect public health and safety. The NRC determines if the state program is compatible with the NRC's regulatory program and adequate to protect public health and safety (7). As of 1998, there are 30 Agreement States which have assumed responsibility from the NRC.

All regulations of the NRC can be found in 10 *CFR,* Title 10 of the *Code of Federal Regulations* (9). The primary references for NRC regulations are 10 *CFR* 19 Notices, Instructions and Reports to Workers: Inspection and Investigations, 10 *CFR* 20 Standards for Protection Against Radiation and 10 *CFR* 30 Rules of General Applicability to Domestic Licensing of By-product Material. Depending on the kind of license required, the following parts may be relevant: 10 *CFR* 31 General

Domestic Licenses for By-product Material, 10 *CFR* 33 Specific Domestic Licenses of Broad Scope for By-product Materials and 10 *CFR* 35 Medical Use of By-product Material. More information will be given in the following sections.

2 Licensing requirements

Any individual or organization wishing to possess or use radioactive materials will require a license from the NRC or Agreement State. Except for very small quantities or specially approved devices containing radioactive materials, it is illegal to receive, acquire, own, possess, use or transfer any radionuclide if you do not have the appropriate license.

Part 30 of 10 *CFR* lists types of licenses, general and specific, and requirements for licensing to obtain, keep and use radioactive materials. The general license does not require the filing of an application with the NRC or receipt of a license. Specific licenses for by-product material are issued to organizations or individuals upon submission and acceptance of an application filed with the NRC, and in accordance with Parts 32–36 and 39 of Chapter 10(9). This is more fully explained in Sections 4 and 5 below. It should be noted that by-product material is any radioactive material (except special nuclear material) made radioactive by exposure to radiation incident to the process of producing or utilizing special nuclear material. Usually, this refers to reactor-produced radionuclides, the radioactive material likely to be used by readers of this book. Special nuclear material is usually referred to as weapons-grade material and includes plutonium, uranium-233, uranium enriched in isotope 233 or 235, and other material designated by the NRC. Source material is usually referred to as material of reactor fuel grade and includes any uranium or thorium as well as ores which contain by weight, one-twentieth of 1% or more of uranium, thorium or any combination thereof.

3 General domestic license

General licenses for possession and use of by-product material contained in certain items and a general license for ownership of by-product material are given in 10 *CFR* 31.

Items included for possession under a general license, if the manufacturer has complied with appropriate NRC regulations, are:

- static elimination devices containing not more than 500 μCi (18.5 MBq) of polonium-210;
- ion-generating tubes containing not more than 500 μCi (18.5 MBq) of polonium-210 or 50 mCi (1.85 GBq) of tritium;
- aircraft luminous safety devices containing not more than 10 Ci (370 GBq) of tritium or 300 mCi (1.11 GBq) of promethium-147;
- calibration or reference sources of americium-241, not more than 5 μCi (185 kBq) in one location;

- ice detection devices containing not more than 50 μCi (1.85 MBq) of strontium-90;

- devices designed for detecting, measuring, gauging or controlling thickness, density, level, interface location, radiation, chemical composition, or for producing light or an ionized atmosphere. If these devices contain only krypton, they need not be tested for leakage of radioactive material. If these devices contain only tritium or <100 μCi (3.7 MBq) or any other beta- and/or gamma-emitting material or <10 μCi (370 kBq) of alpha-emitting material, held in storage in the original shipping container prior to initial installation, they need not be tested for any purpose. All other devices will require further action and record-keeping by the holder of the device (see 10 *CFR* 31.5);

- pre-packaged clinical and laboratory test units containing not more than 10 μCi (370 kBq) of iodine-125, iodine-131, carbon-14 or selenium-75, 50 μCi (1.85 MBq) of tritium, 20 μCi (740 kBq) of iron-59, or a reference or calibration source of 0.05 μCi (1.85 kBq) of iodine-129 or 0.005 μCi (0.185 kBq) of americium-241.

For clinical and laboratory use the recipient must file an NRC form, 'Registration Certificate—*In Vitro* Testing with By-product Material under a General License', and show adequate capability to monitor radiation in the laboratory. Further requirements can be found by consulting 10 *CFR* 31 in the publication for the current year.

4 Specific licenses of broad scope

The provisions of the NRC in 10 *CFR* 33 for granting an individual a specific license are:

(1) the application is for a purpose authorized by the Act;

(2) the applicant's proposed equipment and facilities are adequate to protect health and minimize danger to life or property;

(3) the applicant is qualified by training and experience to use the material for the purpose requested in a safe manner;

(4) any special requirements in Parts 31–35 and 39 are satisfied; and

(5) if the application involves conduct of any activity the NRC determines will significantly affect the quality of the environment, a special determination must be made.

Each license specifies the by-product material and quantity that the licensee may hold. Three types of license—A, B and C, are available, as specified in 10 *CFR* 33.11. All applicants must comply with the requirements of 10 *CFR* 30.3.

Type A licenses require the most comprehensive radiation safety program and are issued for quantities in the multicurie range. This license requires the establishment of a Radiation Safety Committee which reviews and approves safety evaluations of proposed use of radioactive materials, the appointment of a radiological safety officer who is qualified by training and experience in radiation protection, and extensive administrative controls to assure safe operation. Once this license is issued, the NRC grants broad latitude to the Radiation Safety Committee to make certain programmatic changes without further approvals.

Type B licenses are issued if only one radionuclide is possessed and it is less than the quantity listed in 10 *CFR* 33.100, Schedule A, Column 1. When two or more radionuclides are possessed, the sum of the ratio of each radionuclide to the limit in Schedule A must not exceed 1. Type B licenses also require the appointment of a radiological safety officer who reviews and approves the use of radioactive materials and administrative controls to assure safe operation. However, there is no requirement for a Radiation Safety Committee and the NRC is more restrictive in the types of changes which may be made to the program without prior approval from the NRC.

Type C licenses are issued for very small quantities and limited use of radioactive materials. Supervision of the use of radioactive material is made by an individual with training and experience in the safe handling of radioactive materials. Very little latitude is granted by the NRC outside the specifics which may have been given in the license application.

Additional types of specific license are issued by the NRC for other purposes. These include manufacture of devices containing by-product material (10 *CFR* 32), license and radiation safety of industrial radiography (10 *CFR* 34), license and radiation safety for irradiators (10 *CFR* 36) and license and radiation safety for well logging (10 *CFR* 39).

5 Medical uses

A special type of specific license is covered by 10 *CFR* 35, Medical Uses of By-product Material. The requirements for licensing under Part 35 are extensive and cover licenses for use in institutions and by individual physicians, dentists or podiatrists; use of sealed sources for certain types of medical uses; for teletherapy sources; and general licenses for certain quantities of by-product material. The applicant should consult appropriate portions of 10 *CFR* 35 for specific requirements.

The license for institutions will require: adequate facilities for clinical care and radiation safety, a Radiation Safety Officer, and a Radiation Safety Committee; also that the physician designated as an authorized user has appropriate experience working with radioisotopes, and that general technical requirements of 10 *CFR* 35 have been met (9).

There are extensive training and experience requirements listed in 10 *CFR* 35.900 for the Radiation Safety Officer and physicians, dentists or podiatrists who are identified as authorized users. The appropriate sections of 10 *CFR* 35 should be consulted if any use of radioisotopes for medical use is considered. The sub-parts listed under Part 35 are:

- Sub-part A—General Information
- Sub-part B—General Administrative Requirements

- Sub-part C—General Technical Requirements
- Sub-part D—Uptake, Dilution, and Excretion
- Sub-part E—Imaging and Localization
- Sub-part F—Radiopharmaceuticals for Therapy
- Sub-part G—Sources for Brachytherapy
- Sub-part H—Sealed Sources for Diagnosis
- Sub-part I—Teletherapy
- Sub-part J—Training and Experience Requirements
- Sub-part K—Enforcement

The NRC also requires immediate telephone notification of therapy misadministration and written notification of therapy and diagnostic misadministration and maintenance of records of misadministration (9).

6 Exemptions

There are several different exemptions from the regulations governing radioactive materials. For example, common and contract carriers, freight forwarders, warehousemen and the United States Postal Service in the case of normal transport are exempt. There are also some exemptions for use of by-product material under certain Department of Energy and Nuclear Regulatory Commission contracts (9)

Exempt concentrations of by-product material a person or organization may hold are listed in 10 *CFR* 30 Schedule A. These materials may not be introduced into any commodity designed for ingestion, inhalation by or application to a human being, nor may they be transferred to persons exempt from regulations. Exempt quantities of radioactive material a person may hold are listed in 10 *CFR* 30 Schedule B. These by-product materials may not be used in production of products intended for commercial distribution nor may they be transferred to persons exempt from regulations. These exempt quantities are in the microcurie range (9). Quantities of licensed material requiring labeling are listed in Appendix B to Part 30 and are also in the microcurie range (9).

Items such as self-luminous watch dials or hands, gas or smoke detectors that contain by-product materials have maximum quantities or levels of radiation as listed in 10 *CFR* 30 Parts 15, 16, 19 and 20.

7 Radiation protection—Nuclear Regulatory Commission

Radiation protection requirements are listed in 10 *CFR* 20 and include: permissible doses, levels and concentrations; precautionary procedures; waste disposal; records, reports and notification; exemptions and additional requirements; and enforcement. A summary of important points will be made here; however, the provisions in this Part are extensive and should be consulted for specific requirements (9).

The total effective dose equivalent for the whole body is the sum of the dose resulting from external exposure to ionizing radiation, namely the effective dose equivalent from external radiation, and the committed effective dose equivalent from that year's intake of radionuclides. The annual occupational dose limit is the total effective dose equivalent of 5 rem (0.05 Sv) or the sum of the deep-dose equivalent and the committed dose equivalent of 50 rem (0.5 Sv) to any individual organ or tissue other than the eye. The annual limit to the lens of the eye is 15 rem (0.15 Sv). The skin and any extremity also have a shallow dose equivalent of 50 rem (0.50 Sv).

An adult worker may receive doses in addition to the occupational doses listed above if special provisions and record-keeping requirements are met. These are known as 'planned special exposures' and are described in 10 *CFR* 20.1206.

The annual occupational dose limits for minors are one-tenth of the annual dose limits for adults, as specified above. The dose limit to an embryo/fetus during the entire pregnancy due to occupational exposure of a declared pregnant woman is 0.5 rem (5 mSv).

For individual members of the public, the total effective dose equivalent from licensed operations may not exceed 0.1 rem (1 mSv) in a year. This is exclusive of background radiation, medical administration of radiation or radiation from disposal of radioactive materials into the sanitary sewer in accordance with regulations.

Individual monitoring of external and internal occupational dose is required when it is likely the individual will receive 10% of the dose limit, enter a high or very high radiation area or receive an intake of 10% of the applicable Annual Limit on Intake (ALI), which are listed in Appendix B to 10 *CFR* 20.

Whole body or extremity personal dosimeters such as film badges or thermoluminescent dosimeters (TLDs) are most often used to determine external dose. Personal dosimeters that are used for official dose data must be processed and evaluated by a dosimetry processor which holds accreditation from the National Voluntary Laboratory Accreditation Program (NVLAP).

Internal dose is often determined by measuring concentrations of radioactive materials in the air, by direct measurement of radionuclides in the body or by measurement of radionuclides excreted from the body. In each case, the dose to the individual is calculated. These are generally sophisticated calculations requiring a health physicist knowledgeable in internal dosimetry.

Limits of exposure to airborne radioactive materials are presented as Derived Air Concentrations (DACs) and ALIs, given in 10 *CFR* 20 Appendix B, *Table 1*. This information may be used to determine dose from intake of radionuclides and may be summed with the external exposure to ensure compliance with the total occupational dose limit. If a person is required to be monitored, external and internal doses are required to be recorded and summed. The sum of external and internal doses must be below the limits.

The regulations also define Restricted and Unrestricted Areas. In a Restricted

Area, access is limited by the licensee for the purpose of protecting individuals against undue risks from exposure to radiation and radioactive material. A Restricted Area must not include residential quarters. In an Unrestricted Area, access is neither limited nor controlled by the licensee. If an individual were continuously present in an Unrestricted Area, the dose from external sources may not exceed 0.002 rem (0.02 mSv) in 1 h and 0.05 rem (0.5 mSv) in 1 year. In addition, the annual average concentration of radioactive materials released in gaseous and liquid effluents at the boundary of the Unrestricted Area may not exceed the values given in 10 *CFR* 20 Appendix B, *Table 2*.

Licensees are required to make appropriate surveys and to provide caution signs, labels and controls. Magenta and yellow radiation trefoil emblems must be posted with the warning 'Caution Radiation Area', 'Caution High Radiation Area', 'Caution Very High Radiation Area', 'Caution Airborne Radioactivity Area' or 'Caution Radioactive Materials', as appropriate for the hazard. Exceptions to these requirements are given in 10 *CFR* 20.1903 and 20.1904. A container of radioactive material in excess of quantities listed in Appendix C to 10 *CFR* 20.1001-2401, and not continually attended by an individual, must be labelled with the radioactive symbol and the words 'CAUTION, RADIOACTIVE MATERIAL'.

Special procedures are required for receiving and opening packages. Packages should be received or collected expeditiously. With exceptions noted in 10 *CFR* 20.1906, packages must be monitored within 3 h during the working day or ≤3 h from the beginning of the next working day if it is received after working hours. If removable radioactive contamination is in excess of 0.4 Bq (22 d.p.m.)/cm^2 with 300 cm^2 of the package surface wiped or dose rates in excess of 200 mrem (2 mSv)/h at the surface or 10 mrem (0.1 mSv)/h at a distance of 1 m from the surface of the package, the licensee must immediately notify the NRC and the final delivering carrier.

Waste disposal into sanitary sewerage systems is possible but cannot exceed the limits of 10 *CFR* 20 Appendix B, *Table 3*. Amounts released are related to sewage flow and volume. 10 *CFR* 20.2003 should be consulted for details. In any case, there are annual limits, not to exceed 1 Ci (37 GBq) of any radionuclide per year. If more than one radionuclide is released, special conditions must be met. In addition, up to 5 Ci (185 GBq) of tritium and 1 Ci (37 GBq) of carbon-14 may be released each year. Excreta from persons undergoing medical diagnosis or therapy are exempt from this section.

No licensed materials can be disposed of by incineration except material with ≤0.05 μCi (1.85 kBq) of tritium or carbon-14 per gram of scintillation counting fluid or per gram of animal tissue averaged over the weight of the entire animal. Animal tissue not exceeding these concentrations can be disposed of by any means except that it may not be used as human food or animal feed. Records of receipts, transfer and disposal must still be maintained and all regulations governing toxic or hazardous material must be followed.

Requirements for disposal in a licensed land disposal facility and the manifest tracking system for control can be found in 10 *CFR* 20.2006 which should be consulted for the detailed requirements. Sections 10 *CFR* 20.2101–2110 gives requirements on records of surveys, radiation monitoring and disposal and 10 *CFR* 20.2201-2214 on reports of theft or loss of licensed material. For the theft, loss or releases of radioactive material or exposures, the following reports to the NRC must be made. In each case, a written report must be made within 30 days after making the telephone report.

7.1 Immediate telephone report

Immediate telephone reports are required when one of the following conditions exist:

- when 1000 times the quantity specified in Appendix C to Part 20 is lost, missing or stolen;
- any event which causes, or threatens to cause, an individual to receive 25 rem (0.25 Sv), an eye dose of 75 rem (0.75 Sv) or a shallow dose to skin or extremities of 250 rads (2.5 Gy); or
- release of radioactive material so that an individual, present for 24 h, could have received an intake of five times the ALI;
- no later than 4 h from discovery, any event that prevents immediate protective actions necessary to avoid exposures to radiation or radioactive materials that could exceed any regulatory limit. Examples may be fires, explosions, toxic gas releases, etc.

7.2 Twenty-four hour telephone report

Twenty-four hour telephone reports are required when one of the following conditions exits:

- any event which causes or threatens to cause an individual to receive 5 rem (0.05 Sv), an eye dose of 15 rem (0.15 Sv) or a shallow dose to skin or extremities of 50 rem (0.5 Sv);
- release of radioactive material so that an individual, present for 24 h, could have received an intake of one occupational ALI;
- unplanned contamination that:
  (a) requires restriction of the area for >24 h;
  (b) involves a quantity of radioactive material greater than five times the ALI; and
  (c) the area is restricted for a reason other than to allow the radioisotope with a half-life of <24 h to decay prior to decontamination;
- an event that requires unplanned medical treatment at a medical facility with spreadable radioactive contamination;
- an unplanned fire or explosion damaging radioactive materials or any device, container or equipment with radioactive material when:
  (a) the quantity of material is greater than five times the ALI; and
  (b) the damage affects the integrity of the radioactive material or its container; or

- equipment which is disabled when:
  - (a) it is required to prevent releases or exposures exceeding regulatory limits;
  - (b) the equipment is required to be operable when it fails to function; and
  - (c) no redundant equipment is available.

7.3 Telephone report within 30 days

Telephone reports are required within 30 days when 10 times the quantity specified in Appendix C to Part 20 is lost, missing or stolen.

7.4 Thirty day written report

A written report must be made to the NRC within 30 days when any of the following occurs:

- any occupational dose to an adult, minor or embryo/fetus of a declared pregnant woman is exceeded;
- the limit for a member of the public is exceeded;
- the levels of radiation or concentration of radioactive materials in a restricted area are in excess of any applicable license limit;
- the levels of radiation or concentrations of radioactive materials in an unrestricted area are >10 times any regulatory limit; or
- releases to the environment in excess of license limits or EPA standards (constraint dose on air emission of 10 mrem).

Requirements for personnel monitoring reports are given in 10 *CFR* 20.2106. Notification and reports to individuals of radiation exposure data are given in 19 *CFR* 13.

8 Other sections of Title 10, *Code of Federal Regulations*

The regulations of the NRC are extensive. In addition to the parts listed and discussed above, other Parts will have to be consulted for specific requirements; for example 10 *CFR* 9 Public Records, 10 *CFR* 10 Clearances, 10 *CFR* 19 Notices, instructions, and reports to workers and inspections, 10 *CFR* 21 Reporting of defects and non-compliance, and 10 *CFR* 71 Packaging and transportation of radioactive material.

9 Radiation protection—Department of Labor

The Department of Labor, through its Occupational Safety and Health Administration (OSHA), also regulates and enforces radiation protection in the occupational setting under Title 29 of the *Code of Federal Regulations* in 29 *CFR* (10).

The occupational exposure limits enforced by OSHA are the same as those for the NRC in 10 *CFR* 20 as are the requirements for posting of signs and warnings and employee record-keeping. In addition, in the case of an overexposure which requires notification to the NRC, the Assistant Secretary of Labor must also be notified.

It should be noted that the requirements given in 29 *CFR* 1910.1096 were not officially updated when the NRC updated its regulations on Standards for Protection Against Radiation in the early 1990s.

10 Transportation

10.1 Department of Transportation

The transport of radioactive materials is jointly regulated at the federal level by the NRC and Department of Transportation (DOT) (11). The DOT, through 49 *CFR*, regulates carriers, certain packages (such as Type A packages), radiation training programs and training of shippers and other than NRC-licensed carriers, and issues Certificates of Competent Authority for International Shipments. The NRC, through 10 *CFR*, is responsible for regulating certain packages (such as Type B and Fissile packages), transportation safeguards, investigating accidents/incidents and serving as a technical advisor to the DOT. *The Code of Federal Regulations* are extensive and must be consulted. They cover requirements on labeling, packaging, manifesting, warning labels and all aspects of packaging and offering for transport. For the purpose of transportation, the definition of radioactive material is any material which has a specific activity of >0.002 μCi/g (70 Bq/g) uniformly distributed. Specific activities lower than this are not regulated in transport; however, other hazards may have to be considered. For transportation of radioactive materials, the most appropriate sections of the *Code of Federal Regulations* are:

- 49 *CFR* 100–177 and Part 178–199 on Hazardous Materials Regulations;
- 49 *CFR* 300-399 on Motor Carrier Safety Regulations, and regulations pertaining to driver qualification; and
- 10 *CFR* 71 on Packaging of Radioactive Materials for Transport and Transportation of Radioactive Materials Under Certain Conditions.

Other useful documents are the International Air Transport Association (IATA) Dangerous Goods Regulations (12) and International Civil Aviation Organization (ICAO) Technical Instructions for the Safe Transport of Dangerous Goods by Air (13). These are not published by a governmental regulatory agency, but are a simplification of the rules of many nations. The ICAO Technical Instructions are followed for transport by air and ground shipments to and from the airport as long as the criteria of 49 *CFR* 171.11 are followed.

10.2 United States Postal Service

The United States Postal Service regulates under Title 39 of the *Code of Federal Regulations*. In summary, mailable packages of radioactive material must meet the same requirements as limited quantity and excepted articles as specified in the DOT regulations 49 *CFR* 173 but only one-tenth of the activity is permitted in the package. The Postal Service Regulations in the *Domestic Mail Manual*, Part 124.37, state that any package of radioactive materials required to bear the DOT Radioactive White-I, Radioactive Yellow-II or Radioactive Yellow-III labels or which contain radioactive material in quantities in excess of those authorized in the *US Postal Publication 6* are non-mailable (14). The *Domestic Mail Manual* should be available at any Post Office in the USA; *US Postal Service Publication 6* (15) would have to be obtained from a library or from the United States Government Printing Office. Many countries prohibit the movement of radioactive materials in their postal service. It is wise to ensure that any international shipment through the mail is allowed in the receiving country.

11 Environment—Environmental Protection Agency

The EPA issues regulations in Title 40 of the *Code of Federal Regulation* (16). Of particular interest is Sub-part I (40 *CFR* 61) which regulates radionuclides as hazardous air pollutants. This rule applies to NRC-licensed facilities and facilities owned or operated by any federal agency other than the Department of Energy. It does not apply to NRC-licensees that possess and use only sealed sources of radionuclides, low-energy accelerators or facilities such as uranium mill tailings piles or disposal facilities.

These regulations require that emissions of radionuclides, excluding radon-222 and radon-220, may not exceed those amounts that would cause any member of the public to receive an effective dose equivalent of 10 mrem/year (0.1 mSv/year). If radioiodine is emitted, it may not exceed those amounts that would cause any member of the public to receive 3 mrem/year (0.03 mSv/yr). These values may seem at odds with NRC regulations which require that licensees not exceed 0.1 rem/year to a member of the public (10 *CFR* 20.1301). However, the NRC recently established a dose constraint of 10 mrem/year (0.1 mSv/yr) to meet EPA's Sub-part I requirements (17). If the constraint number is exceeded, then the licensee must report to the NRC within 30 days after learning of the dose in excess of the constraint.

Compliance with the emission standard may be determined through the use of either the EPA computer code COMPLY, alternative requirements given in Appendix E to Sub-part I, or through the use of computer models equivalent to COMPLY and approved by EPA. If a facility determines that emissions are <10% of the standard, they are exempt from reporting requirements; however, a facility must perform this determination annually.

12 Human uses—Department of Health and Human Services

The Food and Drug Administration (FDA), part of the Department of Health and Human Services, regulates medical use of radioactive materials in Title 21 of the

Code of Federal Regulations (16). Under Part 310—New Drugs, the FDA listed a number of radioactive drugs, for which safety and effectiveness could be demonstrated by new-drug application or by licensing by the Public Health Service. These drugs were required to have appropriate labeling by isotope, chemical form and use, and adequate evidence of safety and effectiveness had to be furnished. The NRC and the FDA then concluded that these isotopes should not be distributed under investigational-use labelling if they were actually intended for use in medical practice. Additional provisions for manufacture or distribution of radioactive drugs and requirements for a new-drug application or application for approval to market a new drug are also detailed in this regulation.

The FDA has regulated research use of radioactive drugs in humans in 21 *CFR* 361. In addition to being safe and effective, such drugs may not be intended for immediate therapeutic, diagnostic or similar purposes, nor to determine the safety and effectiveness of the drugs in humans (clinical trial). They may be used in a research project to obtain basic information on metabolism, physiology, pathophysiology or biochemistry.

Conditions under which radioactive drugs for research are considered safe and effective are:

- the project is approved by a Radioactive Drug Research Committee;
- the pharmacological dose is limited to cause no clinically detectable pharmacological effect;
- the limits on annual radiation dose in adults are,
  - (a) *for whole-body, active blood-forming organs, lens of eye, and gonads*: single dose, 3 rem (30 mSv); annual and total dose commitment, 5 rem (50 mSv);
  - (b) *for other organs*: single dose, 5 rem (50 mSv); annual and total dose commitment, 15 rem (150 mSv).

  (Persons under 18 years of age at their last birthday shall receive a radiation dose not exceeding 10% of that listed above) and
- any radiation doses from X-ray procedures that are part of the research study must be included when determining total doses and commitments.

Detailed requirements on composition of the Radioactive Drug Research Committee, reports to be made, approval of the Radioactive Drug Research Committee, qualifications of investigators, requirement for NRC licensing, selection of human subjects, research protocol, treatment of adverse reactions and institutional review board approval are also included in this section.

13 Historical perspective and future developments

In 1934 the International Commission on Radiological Protection (ICRP) published its first recommendations on exposure limits. They were originally recommended as daily, and then weekly, limits and on the basis that there was a tolerance dose for all the biological effects of radiation. What is now known as

Table 1 Change in stochastic limits from 1934 to 1990

Year	Stochastic limits	
1934	0.1 roentgen/day	300 mSv/year
1950	0.3 rem/week	150 mSv/year
1958	5 rem/year	50 mSv/year
1977		50 mSv/year
1990		20 mSv/year

the stochastic limit has changed over the years, as shown in *Table 1* (the earlier limits were set in the older units used at these times).

The concept of *no threshold* was introduced in 1950 and the concept of *as low as is readily achievable (ALARA)* in 1977. The ideas inherent in ALARA were encapsulated in 1977 in a system of dose limitation which included: all exposures shall be kept as low as is readily achievable, economic and social factors being taken into account; the dose equivalent to individuals shall not exceed the dose limits recommended for the particular circumstances.

In 1977 the ICRP concluded (19) that the risk of an exposure leading to a fatal cancer was 1×10^{-2} Sv^{-1} and that the risk of an exposure leading to severe hereditary ill effects in the next two generations was 4×10^{-3} Sv^{-1}. The combined risk was 1.4×10^{-2} Sv^{-1}. The ICRP decided that a risk of 10^{-4} per year would be acceptable because this was a risk of accidental death which seemed to be accepted in industries regarded as having a high standard of safety. The acceptable exposure limit would then have been derived by dividing the acceptable risk of death per year (10^{-4}) by the combined risk per sievert of fatal cancers plus severe hereditary effects. This risk was rounded to 1×10^{-2}; the result would have been 0.01 Sv/year (= 10 mSv/year). However, an annual limit of 50 mSv was chosen for reasons stated as follows:

(a) the only possible hazards of radiation at these levels are the development of cancers or hereditary damage; in a workplace where the annual risk of accidental death is 10^{-4}, accidents also cause serious injuries and disablements;

(b) the latency involved in cancer development leads to an average loss of life of 15 years for radiation-induced cancers; fatal accidents in industry lead to an average loss of life of 35 years;

(c) if exposure **limits** are set at 50 mSv, the average dose in practice is likely to be no more than 5 mSv.

In 1990 the ICRP recommended (20) a decrease to an annual limit of 20 mSv averaged over defined periods of 5 years, with the added provision that the effective dose should not exceed 50 mSv in any year. The decrease was recommended because the estimate of the cancer risk per sievert had increased and a different model was used for assessing the extra number of cancers in an exposed population. The risk of a fatal cancer is now estimated as 4×10^{-2} Sv^{-1} and that of causing a serious hereditary effect in *all succeeding generations* as 8×10^{-3} Sv^{-1}. (Note that the genetic risk was previously estimated for only the next *two* generations). This now gives a combined risk of 4.8×10^{-2} Sv^{-1} compared with 1.4×10^{-2} Sv^{-1} in 1977. The ICRP introduced a weighted risk factor for the detriment of non-fatal cancers; this factor is 8×10^{-3} Sv^{-1}. The total risk is therefore now taken to be 5.6×10^{-2} Sv^{-1}.

This latest annual limit of 20 mSv is said to represent the dose at which the risks are just tolerable. Above it they become unacceptable and below it just acceptable. The policy of doses being as low as is readily achievable still applies and exposures should be appropriately below the limit.

Another addition was the concept of *constraint*. This is related to the idea of *optimization*, which was also a feature of the 1977 recommendations. The term 'constraint' indicates a restriction to be applied to the individual doses resulting from a single source of exposure, such as a type of work activity, as part of the procedure of optimization. A constraint therefore differs from a *dose limit*, which refers to the total dose from all relevant sources. The application of constraints is intended to give extra control over workplaces where the dose limit would be unnecessarily high. It is possible that a constraint significantly lower than the 20 mSv dose limit will be assigned to laboratories, where good working practices can easily ensure annual doses very much lower than 20 mSv.

With the changes proposed by ICRP in 1977, the USA slowly changed its regulatory limits. In 1987, President Reagan—through the 'Memorandum for the President' (21)—approved the document *Federal Radiation Protection Guidance to Federal Agencies for Occupational Exposure* (22). In this document, the EPA Administrator had recommended the following:

(a) There should be no exposure without expectation of an overall benefit from the activity.

(b) ALARA (as low as reasonably achievable) practices are applicable, economic and social factors being taken into account.

(c) Radiation doses received as a result of occupational exposure should not exceed the limiting values for assessed doses to individual workers specified below:

(i) *for career and genetic effects*, the effective dose equivalent, H_E, received in any one year by an adult worker should not exceed 50 mSv; and

(ii) *for other health effects*, in addition to the limitation on effective dose equivalent, the dose equivalent, H_T, received in any year by an adult worker should not exceed 150 mSv to the lens of the eye, and 500 mSv to any other organ, tissue (including skin) or extremity of the body.

The effective dose equivalent is defined as:

$$H_E = \Sigma_T W_T H_T,$$

where Σ_T is the sum over T, W_T is a tissue weighting factor and H_T is the annual dose equivalent averaged over organ or tissue T. Values of W_T for different organs and tissues are shown in *Table 2*.

Table 2 Tissue weighting factors (W_T)

Tissue or organ	W_T
Gonads	0.25
Breasts	0.15
Red bone marrow	0.12
Thyroid	0.03
Bone surfaces	0.03
Remainder	0.30[a]

[a] The W_T value of 0.30 for 'remainder' applies to the five other organs receiving the highest doses in any situation. All other tissues or organs are neglected.

(d) In the case of committed doses from internal sources of radiation, intake of radionuclides by an adult worker will be controlled so that:

(i) the anticipated committed effective dose equivalent from the annual intake plus any annual effective dose equivalent from external sources will not exceed 50 mSv; and

(ii) the anticipated magnitude of the committed dose equivalent to any organ or tissue from such intake plus any annual dose equivalent from external exposure will not exceed 500 mSv; the committed effective dose equivalent from internal sources of radiation, $H_{E,50}$ is

$$H_{E,50} = \Sigma_T W_T H_{T,50}$$

where W_T is defined as in Recommendation (c) and $H_{T,50}$ is the sum of all dose equivalents to an organ or tissue T that accumulate in an individual's life-time (taken as 50 years).

(e) Occupational dose equivalents to individuals under 18 years of age are limited to one-tenth the values specified in Recommendations (c) and (d) for adult workers.

(f) The dose-equivalent to an unborn as a result of occupational exposure of a woman who has declared she is pregnant should be kept as low as reasonably achievable and in any case should not exceed 5 mSv; however, no discrimination in employment will be allowed.

(g) Individuals occupationally exposed to radiation should be instructed on basic health risk from radiation and basic radiation protection principles; the degree and type of instruction should depend on the potential exposures involved.

(h) There should be appropriate personnel and workplace monitoring and records kept to ensure conformity to these Recommendations.

(i) Radiation exposure control measures should be designed, selected, utilized and maintained to ensure anticipated and actual doses meet the objectives of the guidance.

(j) The numerical values recommended should not be exceeded lightly, only in emergencies or after review of reactions by the federal agency having jurisdiction (22). The Recommendations also permit use of the ICRP quality factors, dosimetric conversions, models for reference persons, metabolic models and dosimetric methods in determining conformance with the Recommendations (22).

The Federal Guidance has been incorporated in the regulations of the various federal agencies and the requirements listed in the *Code of Federal Regulations* have changed to reflect the new Guidance. After the 1987 Federal Guidance was issued, the ICRP subsequently recommended a decrease to the annual limit of 20 mSv averaged over a period of 5 years. In general, the United States regulations have not kept pace with ICRP.

14 Conclusion

The regulations governing radiation protection and the possession and use of radioactive materials are extensive and complex, involving many federal as well as state and local agencies. Selected regulations on radiation of the NRC, the Department of Health and Human Services and the EPA have been extracted here. This has not been a detailed explanation of the regulations and the reader is cautioned to consult the original references for details. Most federal and state agencies have public information services or professional information services to answer questions or provide copies of regulations. These may be contacted by mail, telephone or, more recently, over the Internet. Many agencies have their own websites.

Finally, it should be remembered that the *Code of Federal Regulations* is republished every year. As new regulations are developed they are incorporated in the next issue of the *Code of Federal Regulations*. A continual review of these regulations and corresponding state regulations is necessary to maintain a current knowledge of the requirements.

References

1. World Health Organization (1972). *Protection Against Ionizing Radiation*. WHO, Geneva.
2. Ballitine, J.A. (1969). *Ballentine's Law Dictionary*, 3rd edn. Bancroft Whitney Co., San Francisco, CA.
3. Brennan, R.F. (ed.) *Nuclear Licensing Reports*. Central Publishing Company, Rockville, MD.
4. US Nuclear Regulatory Commission. *NMSS Licensee Newsletter*.US Nuclear Regulatory Commission, Washington, DC.
5. United States Code Annotated (1978). *Title 42: The Public Health and Welfare*, Parts 1400–1890. West Publishing Co., St Paul, MN [see Parts 1801–1819.5].
6. United States Code Annotated (1973). *Title 42: The Public Health and Welfare*, Parts 2011–3100. West Publishing Co., St Paul, MN [see Parts 2011–2296].
7. Eisenhower, D. D. (1959) 2.1 E.O. 10831 (14 August 1959). *Establishment of the Federal Radiation Council. Federal Register* **24**, 6669.
8. Code of Federal Regulations (1971). *Title 3: The President, 1966–1970 Compilation*, pp. 1072–1075. US Government Printing Office, Washington, DC.

9. Code of Federal Regulations (1988). *Title 10: Energy*, Chapter I: Nuclear Regulatory Commission, Parts 0–50 and 51–199. US Government Printing Office, Washington, DC.

10. Code of Federal Regulations (1978). *Title 29: Labor*, Chapter XVII: Occupational Safety and Health Administration. Department of Labor, Parts 1900–1999. US Government Printing Office, Washington, DC.

11. Code of Federal Regulations (1987). *Title 49: Transportation, Subtitle B—Other Regulations Relating to Transportation*. Chapter I: Research and Special Programs Administration, Department of Transportation, Parts 100–199. US Government Printing Office, Washington DC.

12. International Air Transport Association (1995). *Dangerous Goods Regulations*, 36th edn. IATA, Montreal.

13. International Civil Aviation Organization (1995). *Technical Instructions for the Safe Transport of Dangerous Goods by Air, 1994–1995*. DOC 9284-AN/905. ICAO, Montreal.

14. US Postal Service (September 1983). *US Postal Service Publication 6—Radioactive Materials*. US Government Printing Office, Washington, DC.

15. US Postal Service (3 April 1988) *Domestic Mail Manual*, issue 26. US Government Printing Office, Washington, DC.

16. Code of Federal Regulations (1 July 1987). *Title 40: Protection of the Environment*, Chapter I: Environmental Protection Agency, Parts 61–80 and 190–399. US Government Printing Office, Washington, DC.

17. US Nuclear Regulatory Commission Regulatory (December 1996). Guide 4.20: *Constraint on Releases of Airborne Radioactive Materials to the Environment for Licensees Other Than Power Reactors*. US Nuclear Regulatory Commission, Washington, DC.

18. Code of Federal Regulations (1 April 1988). *Title 21: Food and Drug*, Chapter I—Food and Drug Administration, Department of Health and Human Services, Parts 300–499. US Government Printing Office, Washington, DC.

19. International Commission on Radiological Protection (1977). *Recommendations of the International Commission on Radiological Protection*. Publication 26. *Annals of the ICRP*, Vol. 1, no. 3. Pergamon Press, New York.

20. International Commission on Radiological Protection (1991). *1990 Recommendations of the International Commission on Radiological Protection*. Publication 60. *Annals of the ICRP*, Vol. 21, nos 1–3. Pergamon Press, New York.

21. Reagan, R. (1987). Title 3: The President Recommendations Approved by the President Radiation Protection Guidance to Federal Agencies for Occupational Exposure. *Federal Register* **52**(17), 2822.

22. Thomas, L.M. (1987) Memorandum for the President, Federal Radiation Protection Guidance for Occupational Exposure. *Federal Register* **52**(17), 2822–2834.

Appendices

A1 Properties and radiation protection data of isotopes commonly used in the biological sciences

Radioisotopes vary considerably with respect to their properties and potential hazard. Table 1 below summarizes some generally useful information about isotopes used in biological experiments.

Table 1

Property/radiation protection criteria	^{3}H	^{14}C	^{35}S	^{32}P	^{33}P	^{125}I	^{131}I
$T_{1/2}$	12.3 years	5730 years	87.4 days	14.3 days	25.4 days	59.6 days	8.04 days
Mode of decay	β	β	β	β	β	X (EC)	γ and β
Max β energy (MeV)	0.019	0.156	0.167	1.709	0.249	Auger electrons 0.035	0.806
Monitor	Swabs counted by LSC	β-counter	β-counter	β-counter	β-counter	γ-probe	β-counter or γ-probe
Biological monitoring	Urine	Urine, breath	Urine	Urine	Urine	Thyroid	Thyroid
ALI^{a}	480^{b}	34	15	6.3	14	1.3^{c}	0.9^{c}
Critical organ	Whole body	Whole body, fat	Whole body, testis	Bone	Bone	Thyroid	Thyroid
Maximum range in air	6 mm	24 cm	26 cm	790 cm	49 cm	>10 m	>10 cm
Shielding required	None	1 cm Perspex	1 cm Perspex	1 cm Perspex	1 cm Perspex	Lead 0.25 m or lead impregnated Perspex	Lead 13 mm
γ dose rate (μSv\h from1 GBq at 1 m)	–	–	–	(β dose rate 210 mSv at surface of 1MBq/ml)	–	41	51
Special considerations	Monitoring difficulties lead to potential internal hazard. DNA precursors more toxic than tritiated water	Avoid generation of CO_2	Avoid generation of SO_2	Potential high source of external radiation. Lead shielding and finger dosimeters for quantities greater than 300 and 30 MBq respectively		Iodine sublimes, work in fumehood. Spills should be treated with sodium thiosulfate prior to decontamination. Iodine compounds may penetrate rubber gloves, wear two pairs.	

[a] Based on a dose limit of 20mSv using the most restrictive dose coefficients for inhalation or ingestion, 30, see Chapter 2; some figures differ in Germany

[b] organically bound H-3

[c] Based on dose equivalent limit of 500 mSv to thyroid

A2 Types of monitor

1. Personal dosimeters

1.1 Film badges and thermoluminescent dosimeters (TLDs)

Film badges consist of an X-ray film, resembling that used by a dentist, held in a plastic badge-holder pinned to the lab coat at waist or chest height. The holder incorporates a number of different thicknesses of shielding to facilitate assessment of the dose once the film is developed. Films give an estimate of dose up to 1 Gy.

TLDs contain a phosphor (such as LiF) that becomes and remains excited once exposed to radiation. Heat treatment results in light emission whose intensity is proportional to the radiation absorbed. TLDs can record a dose up to 20 Gy.

Both types of dosimeter are processed by an approved laboratory (e.g. NRPB in the UK) on a regular basis such as monthly or quarterly depending on the relative likelihood of exposure. TLDs are becoming increasingly popular but film badges remain superior in some respects as they provide more information about the type and energy of radiation absorbed. This is likely to be of little importance to a biology laboratory, however, as the worker may only be exposed to one or two relevant isotopes such as ^{32}P or ^{125}I. Neither type of dosimeter is appropriate for ^{3}H, and they are of little or no value for work with ^{14}C or ^{35}S, as the energy of radiation is so low.

It must be remembered that personal dosimeters only give information retrospectively. Their use means that a record is kept of doses received and is transferred with change in employment. If a dosimeter is lost, an appropriate proportion of the maximum permitted dose is assigned to that period.

1.2 Finger badges

These are small TLDs held in plastic holders worn under disposable gloves. They are processed by an approved laboratory as for above. They are particularly useful for work with ^{32}P where finger doses can be potentially high.

1.3 Instantaneous dosimeters

These are pocket ionization chambers or quartz fibre electroscopes. They generally have a metal case and are therefore only appropriate for gamma emitters. They are useful for measuring doses over short periods if radiation exposure is likely to be relatively high.

2. Laboratory monitoring

Most laboratory monitors are one of three basic types:

Geiger–Müller counters,

gas proportional counters, or

scintillation counters.

The first two types of instrument are based on gas ionization as the mode of detection. Briefly, they consist of a small gas chamber containing two electrodes, a voltage supply and a scaler. When ionizing radiation enters the chamber through the thin end window of the chamber, gas is ionized and a current pulse is recorded (c.p.m). Proportional counters are so called because the degree of ionization in the chamber is proportional to the applied voltage. The relationship differs between alpha and beta particles as the former has greater ionizing power. Geiger–Müller counters operate at a higher voltage, but the signal is independent of voltage within the operating range and there is no discrimination between alpha and beta radiation. They operate at a higher gas pressure than proportional counters and thus have lower sensitivity because the end window is thicker. However, they are usually lighter and cheaper than proportional counters.

Both types of gas ionization counter are suitable for detecting ^{32}P and will pick up ^{14}C and ^{35}S if an instrument with a thin end window is used. Neither counter will detect ^{3}H. Monitoring of this low energy emitter is carried out by taking solvent or aqueous wipe tests (or 'swabs') followed by liquid scintillation counting of the effluent. Alternatively, a gas flow proportional counter can be used, but they are bulky and not particularly reliable or sensitive for ^{3}H monitoring.

Most of the so-called contamination monitors are proportional counters with a large end window on the lower face of the instrument. They give a reading in Bq cm^{-2}, counts sec^{-1}, or Sv h^{-1}. All radioisotope laboratories should have at least one of these but the cheaper Geiger counters should also be available. Their lower cost means that more can be made available and they generally have a probe on a flexible cable for convenience.

Laboratories using gamma-emitters require a hand-held scintillation counter. These have a probe that is either integral or connected by a flexible cable. Depending on the instrument, readings are given in counts sec^{-1} or Sv h^{-1}.

Contamination monitors must be calibrated annually by an approved laboratory (e.g. NRPB in the UK).

A3 Curie–becquerel conversion chart

μCi/mCi/Ci	kBq/MBq/GBq	μCi/mCi/Ci	MBq/GBq/TBq
0.1	3.7	30	1.11
0.2	7.4	40	1.48
0.25	9.25	50	1.85
0.3	11.1	60	2.22
0.4	14.8	70	2.59
0.5	18.5	80	2.96
1.0	37	90	3.33
2.0	74	100	3.7
2.5	92.5	125	4.625
3.0	111	150	5.55
4.0	148	300	7.4
5.0	185	250	9.25
6.0	222	300	11.1
7.0	259	400	14.8
8.0	296	500	18.5
9.0	333	600	22.2
10.0	370	700	25.9
12.0	444	750	27.75
15.0	555	800	29.6
20.0	740	900	33.3
25.0	925	1000	37.0
		3000	111.0
		5000	185

Examples	1 μCi = 37 kBq
	50 μCi = 1.85 MBq
	10 mCi = 370 MBq
	3000Ci = 111 TBq

A4 Calculation of amount of radioactivity remaining (R_R) after a given half-life (H_L) has passed

H_L	R_R	H_L	R_R
0.00	1.00	1.55	0.342
0.02	0.986	1.60	0.330
0.04	0.973	1.65	0.319
0.06	0.959	1.70	0.308
0.08	0.946	1.75	0.297
0.10	0.933	1.80	0.287
0.12	0.920	1.85	0.277
0.14	0.903	1.90	0.268
0.16	0.895	1.95	0.259
0.18	0.883	2.00	0.250
0.20	0.871	2.10	0.233
0.25	0.851	2.20	0.218
0.30	0.812	2.30	0.203
0.35	0.785	2.40	0.189
0.40	0.758	2.50	0.177
0.45	0.732	2.60	0.165
0.50	0.707	2.70	0.154
0.55	0.683	2.80	0.144
0.60	0.660	2.90	0.134
0.65	0.638	3.00	0.125
0.70	0.616	3.10	0.117
0.75	0.595	3.20	0.109
0.80	0.574	3.30	0.102
0.85	0.555	3.40	0.095
0.90	0.535	3.50	0.088
0.95	0.518	3.60	0.083
1.00	0.500	3.70	0.077
1.05	0.483	3.80	0.072
1.10	0.467	3.90	0.067
1.15	0.451	4.00	0.063
1.20	0.435	4.10	0.058
1.25	0.421	4.20	0.054
1.30	0.406	4.30	0.051
1.35	0.393	4.40	0.047
1.40	0.379	4.50	0.044
1.45	0.367	4.60	0.041
1.50	0.354	4.70	0.039
		4.80	0.036
		4.90	0.034
		5.00	0.031

A5 Units commonly used to describe radioactivity

Unit	Abbreviation	Definition
Counts per minute or second	c.p.m. c.p.s.	The recorded rate of decay
Disintegration per minute or second	d.p.m. d.p.s.	The actual rate of decay
Curie	Ci	The number of d.p.m. equivalent to 1 g of radium (3.7×10^{10} d.p.s.)
Millicurie	mCi	$Ci \times 10^{-3}$ or 2.22×10^9 d.p.m.
Microcurie	μCi	$Ci \times 10^{-6}$ or 2.22×10^6 d.p.m.
Becquerel (SI unit)	Bq	1 d.p.s.
Gigabecquerel (SI unit)	GBq	10^9 Bq or 27.027 mCi
Megabecquerel (SI unit)	MBq	10^6 Bq or 27.027 μCi
Electron volt	eV	The energy attained by an electron accelerated through a potential difference of 1 volt. Equivalent to 1.6×10^{-19} joules.
Roentgen	R	The amount of radiation which produces 1.61×10^{15} ion pairs per kg of air (2.58×10^{-4} coulombs kg^{-1})
Rad	rad	That dose which gives an energy absorption of 0.01 joule kg^{-1} (J kg^{-1})
Gray (SI unit)	Gy	That dose which gives an energy absorption of 1 joule per kilogram. Thus, 1 Gy = 100 rad.
Rem	rem	That amount of radiation which gives a dose in man equivalent to 1 rad of X-rays.
Sievert (SI unit)	Sv	That amount of radiation which gives a dose in man equivalent to 1 gray of X-rays. Thus, 1 Sv = 100 rem.

A6 Licensing and advisory authorities

The Environment Agency (UK)
Rio House
Waterside Drive
Aztec West
Almondsbury
Bristol
BS32 4UD

The Environmental Protection Agency (USA)
ANR 46041 M. St. SW
Washington DC 2046
USA

International Commission on Radiological Protection
SE-171 16 Stockholm
Sweden

National Radiological Protection Board
Chilton
Didcot
Oxon
OX11 0RQ, UK

A7 List of suppliers

Aldrich, 1001 West Saint Paul Avenue, Milwaukee, WI 53233, USA

Ambion Inc, 2130 Woodward, Austin, TX 78744-1832, USA

American Radiolabeled Chemicals, (*see* Tocris Cookson)

Amersham Pharmacia Biotech UK Ltd, Amersham Place, Little Chalfont, Buckinghamshire HP7 9NA, UK (*see also* Nycomed Amersham Imaging UK; Pharmacia).
Tel: +44 (0)800 515313
Fax: +44 (0)800 616927
URL: http//www.apbiotech.com/

Anderman and Co. Ltd, 145 London Road, Kingston-upon-Thames, Surrey KT2 6NH, UK.
Tel: +44 (0)181 5410035
Fax: +44 (0)181 5410623

Beckman Coulter
Beckman Coulter (UK) Ltd, Oakley Court, Kingsmead Business Park, London Road, High Wycombe, Buckinghamshire HP11 1JU, UK.
Tel: +44 (0)1494 441181
Fax: +44 (0)1494 447558
URL: http://www.beckman.com/
Beckman Coulter Inc., 4300 N. Harbor Boulevard, PO Box 3100, Fullerton, CA 92834-3100, USA.
Tel: +1 714 8714848
Fax: +1 714 7738283
URL: http://www.beckman.com/

Becton Dickinson
Becton Dickinson and Co., 21 Between Towns Road, Cowley, Oxford OX4 3LY, UK.
Tel: +44 (0)1865 748844
Fax: +44 (0)1865 781627
URL: http://www.bd.com/
Becton Dickinson and Co., 1 Becton Drive, Franklin Lakes, NJ 07417-1883, USA
Tel: +1 201 8476800
URL: http://www.bd.com/

Bio 101
Bio 101 Inc., c/o Anachem Ltd, Anachem House, 20 Charles Street, Luton, Bedfordshire LU2 0EB, UK.
Tel: +44 (0)1582 456666
Fax: +44 (0)1582 391768; URL: http://www.anachem.co.uk/
Bio 101 Inc., PO Box 2284, La Jolla, CA 92038-2284, USA.
Tel: +1 760 5987299   Fax: +1 760 5980116
URL: http://www.bio101.com/

Bio-Rad
Bio-Rad Laboratories Ltd, Bio-Rad House, Maylands Avenue, Hemel Hempstead, Hertfordshire HP2 7TD, UK.
Tel: +44 (0)181 3282000
Fax: +44 (0)181 3282550
URL: http://www.bio-rad.com/
Bio-Rad Laboratories Ltd, Division Headquarters, 1000 Alfred Noble Drive, Hercules, CA 94547, USA.
Tel: +1 510 7247000   Fax: +1 510 7415817
URL: http://www.bio-rad.com/

Boehringer-Mannheim (*see* Roche)

Brunner Scientific, Unit 4C, Hunmanby
Industrial Estate, Bridlington Road,
Hunmanby, nr Filey, Yorkshire YO14 0PH,
UK.

Cambridge Research Biochemicals,
Gadbrook Park, Northwich, Cheshire CW9
7RA, UK.

Christison Scientific Equipment Ltd, Albany
Road, East Gateshead Industrial Estate,
Gateshead, Tyne and Wear NE8 3AT, UK.

CP Instrument Co. Ltd, PO Box 22, Bishop
Stortford, Hertfordshire CM23 3DX, UK.
Tel: +44 (0)1279 757711
Fax: +44 (0)1279 755785
URL: http//:www.cpinstrument.co.uk/

Dupont
Dupont (UK) Ltd, Industrial Products Division,
Wedgwood Way, Stevenage, Hertfordshire
SG1 4QN, UK.
Tel: +44 (0)1438 734000
Fax: +44 (0)1438 734382
URL: http://www.dupont.com/
Dupont Co., Biotechnology Systems Division,
PO Box 80024, Wilmington, DE 19880-002,
USA.
Tel: +1 302 7741000   Fax: +1 302 7747321
URL: http://www.dupont.com/

Eastman Chemical Co., 100 North Eastman
Road, PO Box 511, Kingsport, TN
37662–5075, USA.
Tel: +1 423 2292000
URL: http//:www.eastman.com/

Fisher Scientific
Fisher Scientific UK Ltd, Bishop Meadow Road,
Loughborough, Leicestershire LE11 5RG, UK.
Tel: 01509 231166   Fax: 01509 231893
URL: http://www.fisher.co.uk/

Fisher Scientific, Fisher Research, 2761 Walnut
Avenue, Tustin, CA 92780, USA.
Tel: +1 714 6694600
Fax: +1 714 6691613
URL: http://www.fishersci.com/

Fluka
Fluka, PO Box 2060, Milwaukee, WI 53201,
USA.
Tel: +1 414 2735013
Fax: +1 414 2734979
URL: http://www.sigma-aldrich.com/
Fluka Chemical Co. Ltd, PO Box 260, CH-9471,
Buchs, Switzerland.
Tel: 0041 81 7452828
Fax: 0041 81 7565449
URL: http://www.sigma-aldrich.com/

Gibco-BRL (see Life Technologies)

Hybaid
Hybaid Ltd, Action Court, Ashford Road,
Ashford, Middlesex TW15 1XB, UK.
Tel: 01784 425000
Fax: 01784 248085
URL: http://www.hybaid.com/
Hybaid US, 8 East Forge Parkway, Franklin,
MA 02038, USA.
Tel: +1 508 5416918
Fax: +1 508 5413041
URL: http://www.hybaid.com/

HyClone Laboratories,
1725 South HyClone Road, Logan,
UT 84321, USA.
Tel: +1 435 7534584
Fax: +1 435 7534589
URL: http//:www.hyclone.com/

ICN Phamaceuticals Ltd, Biomedical
Research Products, 1 Elmwood, Chineham
Business Park, Crockford Lane, Basingstoke,
Hampshire RG2 8WG, UK.
URL: http://www.icnbiomed.com

Invitrogen

Invitrogen Corp., 1600 Faraday Avenue, Carlsbad, CA 92008, USA.
Tel: +1 760 6037200
Fax: +1 760 6037201
URL: http://www.invitrogen.com/
Invitrogen BV, PO Box 2312, 9704 CH Groningen, The Netherlands.
Tel: 00800 53455345
Fax: 00800 78907890
URL: http://www.invitrogen.com/

Jencons PLS, Cherrycourt Way Industrial Estate, Stanbridge Road, Leighton Buzzard LU7 8UA, UK.

Kodak Ltd, Kodak House, Station Road, Hemel Hempstead, Hertfordshire HP1 1JO, UK.

Life Technologies

Life Technologies Ltd, PO Box 35, 3 Free Fountain Drive, Inchinnan Business Park, Paisley PA4 9RF, UK.
Tel: 0800 269210   Fax: 0800 243485
URL: http://www.lifetech.com/
Life Technologies Inc., 9800 Medical Center Drive, Rockville, MD 20850, USA.
Tel: +1 301 6108000
URL: http://www.lifetech.com/

LJL Biosystems Ltd, Dorset House, Regent Park, Kingston Road, Leatherhead, Surrey KT22 7PL, UK.

Merck Eurolab Ltd, Laboratory supplies, Merck House, Poole, Dorset BN15 1TO, UK.

Merck Sharp & Dohme Research Laboratories, Neuroscience Research Centre, Terlings Park, Harlow, Essex CM20 2QR, UK.
URL: http://www.msd-nrc.co.uk/

Merck Sharp & Dohme GmbH, Lindenplatz 1, D-85540, Haar, Germany.
URL: http://www.msd-deutschland.com/

Millipore

Millipore (UK) Ltd, The Boulevard, Blackmoor Lane, Watford, Hertfordshire WD1 8YW, UK.
Tel: 01923 816375
Fax: 01923 818297
URL: http://www.millipore.com/local/UKhtm/
Millipore Corp., 80 Ashby Road, Bedford, MA 01730, USA.
Tel: +1 800 6455476   Fax: +1 800 6455439
URL: http://www.millipore.com/

Nalge Co., 75 Panorama Creek Drive, PO Box 20365, Rochester, NY14602-0365, USA.

National Diagnostics, 305 Patten Drive, Atlanta, Georgia 30336, USA
URL: http://www.nationaldiagnostics.com

NEN

NEN Life Science Products, BRU/BRU/40349, PO Box 66, Hounslow TW5 9RT, UK.
NEN Life Science Products, 549–3 Albany Street, Boston, MA 02118, USA.
URL:http://www.nen.com

Neosystem Laboratoire

SNPE England, Suffolk House, George Street, Croydon, Surrey CR0 1PE.
SNPE North America, 103 Carnegie Center, Route One, Princeton, NJ 08540, USA.

New England Biolabs, 32 Tozer Road, Beverley, MA 01915–5510, USA.
Tel: +1 978 9275054.

Nikon

Nikon Inc., 1300 Walt Whitman Road, Melville, NY 11747–3064, USA.
Tel: +1 516 5474200
Fax: +1 516 5470299
URL: http://www.nikonusa.com/
Nikon Corp., Fuji Building, 2–3, 3-chome, Marunouchi, Chiyoda-ku, Tokyo 100, Japan.
Tel: 00813 32145311
Fax: 00813 32015856
URL: http://www.nikon.co.jp/main/index_e.htm/

Nycomed Amersham

Nycomed Amersham Imaging, Amersham Labs, White Lion Rd, Amersham, Buckinghamshire HP7 9LL, UK.
Tel: 0800 558822 (or 01494 544000)
Fax: 0800 669933 (or 01494 542266)
URL: http//:www.amersham.co.uk/
Nycomed Amersham, 101 Carnegie Center, Princeton, NJ 08540, USA.
Tel: +1 609 5146000
URL: http://www.amersham.co.uk/

Packard

Canberra Packard Ltd, Brook House, 14 Station Road, Pangbourne, Berkshire RG8 7AN, UK.
Packard Instrument Company, 800 Research Parkway, Meriden, CT 06450, USA.
URL: http://www.packardinst.com

Perkin Elmer Ltd, Post Office Lane, Beaconsfield, Buckinghamshire HP9 1QA, UK.
Tel: 01494 676161
URL: http//:www.perkin-elmer.com/

Pharmacia, Davy Avenue, Knowlhill, Milton Keynes, Buckinghamshire MK5 8PH, UK (also see Amersham Pharmacia Biotech).
Tel: 01908 661101
Fax: 01908 690091
URL: http//www.eu.pnu.com/

Pierce Chemical Co., PO Box 117, Rockford, IL 61105, USA.

Promega

Promega UK Ltd, Delta House, Chilworth Research Centre, Southampton SO16 7NS, UK.
Tel: 0800 378994
Fax: 0800 181037
URL: http://www.promega.com/
Promega Corp., 2800 Woods Hollow Road, Madison, WI 53711–5399, USA.
Tel: +1 608 2744330
Fax: +1 608 2772516
URL: http://www.promega.com/

Qiagen

Qiagen UK Ltd, Boundary Court, Gatwick Road, Crawley, West Sussex RH10 2AX, UK.
Tel: 01293 422911   Fax: 01293 422922
URL: http://www.qiagen.com/
Qiagen Inc., 28159 Avenue Stanford, Valencia, CA 91355, USA.
Tel: +1 800 4268157
Fax: +1 800 7182056
URL: http://www.qiagen.com/

Roche Diagnostics

Roche Diagnostics Ltd, Bell Lane, Lewes, East Sussex BN7 1LG, UK.
Tel: 0808 1009998 (or 01273 480044)
Fax: 0808 1001920 (01273 480266)
URL: http://www.roche.com/
Roche Diagnostics Corp., 9115 Hague Road, PO Box 50457, Indianapolis, IN 46256, USA.
Tel: +1 317 8452358
Fax: +1 317 5762126
URL: http://www.roche.com/
Roche Diagnostics GmbH, Sandhoferstrasse 116, 68305 Mannheim, Germany.
Tel: +49 621 7594747
Fax: +49 621 7594002
URL: http://www.roche.com/

Schleicher and Schuell Inc., Keene, NH 03431A, USA.
Tel: +1 603 3572398.

Semat Technical (UK) Ltd, One Executive Park, Hatfield Road, St Albans, Hertfordshire AL1 4TA, UK.

Shandon Scientific Ltd, 93–96 Chadwick Road, Astmoor, Runcorn, Cheshire WA7 1PR, UK.
Tel: +44 (0)1928 566611
URL: http//www.shandon.com/

Shirks Laboratories, Buechstrasse 10, CH-8645, Jona, Switzerland.

Sigma

Sigma–Aldrich Co. Ltd, The Old Brickyard,
New Road, Gillingham, Dorset SP8 4XT, UK.
Tel: +44 (0)800 717181 (or (0)1747 822211)
Fax: +44 (0)800 378538 (or (0)1747 823779)
URL: http://www.sigma-aldrich.com/
Sigma Chemical Co., PO Box 14508, St Louis,
MO 63178, USA.
Tel: +1 314 7715765
Fax: +1 314 7715757
URL: http://www.sigma-aldrich.com/

Stratagene

Stratagene Inc., 11011 North Torrey Pines
Road, La Jolla, CA 92037, USA.
Tel: +1 858 5355400
URL: http://www.stratagene.com/
Stratagene Europe, Gebouw California,
Hogehilweg 15, 1101 CB Amsterdam
Zuidoost, The Netherlands.
Tel: +31 800 91009100
URL: http://www.stratagene.com/

Titertec, 330 Wynn Drive, Huntsville,
Alabama 35805, USA.

Tocris Cookson Ltd, Northpoint, Fourth Way,
Avonmouth, Bristol BS11 8TA, UK.
URL: http://www.tocris.com

Tocris Cookson Inc., 114 Holloway Road,
Suite 200, Ballwin, MO 63001, USA.

United States Biochemical (USB), PO Box
22400, Cleveland, OH 44122, USA.
Tel: +1 216 4649277.

Vector Laboratories Inc., 30 Ingold Road,
Burlingame, CA 94010, USA.

Wallac *Wallac and Berthold*, EG&G
Instruments Ltd, 20 Vincent Avenue,
Crownhill Business Centre, Crownhill,
Milton Keynes MK8 0AB, UK.
Wallac Inc., 9238 Gaither Road, Gaithersburg,
MD 20877, USA.

Index

Index

303